WATER RESOURCES MANAGEMENT IN ETHIOPIA

WATER RESOURCES MANAGEMENT IN ETHIOPIA

Implications for the Nile Basin

EDITED BY
Helmut Kloos and
Worku Legesse

CAMBRIA
PRESS

Amherst, New York

Requests for permission should be directed to:
permissions@cambriapress.com, or mailed to:
Cambria Press
20 Northpointe Parkway, Suite 188
Amherst, NY 14228

Library of Congress Cataloging-in-Publication Data

Water resources management in Ethiopia : implications for the Nile Basin / edited by Helmut Kloos and Worku Legesse.
p. cm.
Includes bibliographical references and index.
ISBN 978-1-60497-665-6 (alk. paper)
1. Water resources development—Ethiopia. I. Kloos, Helmut. II. Legesse, Worku.

HD1696.E8W38 2010
333.9100963—dc22

2009046084

"Let there be work, bread, water and salt for all."
—Nelson Madela

"The availability of water is a concern for some countries. But the scarcity at the heart of the global water crisis is rooted in power, poverty, and inequality, not in physical availability."
—United Nations Human Development Report 2006

Table of Contents

List of Figures

List of Tables

Preface

The impetus for this book originated during the 2006 Annual Conference of the Ethiopian Public Health Association in Harar, Ethiopia, which focused on the role of the changing Ethiopian environment in health, and also during side visits that revealed the aftermath of the devastating 2006 flood in nearby Dire Dawa Town and the dry bed of Lake Haramaya, which finally disappeared after years of water withdrawal for irrigation and urban needs. But our interest in the development and management of the country's water resources goes back several decades to field research on issues of the relationship between water development and water-borne diseases, dam construction and ecological changes, population movements and resettlement, community initiatives and government approaches to health development (Helmut Kloos), and the biological integrity of streams and rivers subject to pollution and hydropower development (Worku Legesse). These studies, involving extensive travel in the highland and lowland regions, revealed to us water-related problems in irrigated areas, man-made reservoirs, highland and lowland ecosystems, and in domestic water supplies and sanitation in a country emerging from medieval feudalism and trying to find its way into the modern world after the collapse of Haile Selassie's imperial regime in 1974. Worku Legesse's upbringing and early life in a village in the former Wello Administrative Region in the northeastern drought belt of Ethiopia provided him with valuable insights into the realities of that area's degraded environment and rural people's coping behavior during seasonal water scarcities and periodic droughts in the absence of effective and sustainable preventive government interventions. While many studies have been carried out by academics, government agencies, and aid organizations on water resources development and management in Ethiopia, most of them have been narrowly focused on disciplinary and sector-specific issues. Their scopes have been limited to examining the role of the natural environment, socioeconomic conditions, and political factors in the country's water problems across all water sectors (agriculture, hydropower, domestic water supply, health, and the environment), a shortcoming largely stemming from the scarcity of interdisciplinary studies in Ethiopia. Moreover, a broader transboundary perspective which recognizes that most of Ethiopia's water resources are located in the wider Nile Basin and must, therefore, be studied as an integral part of this larger hydrological and economic area is lacking.

In this book, we focus on managerial aspects in recognition of the overriding importance of sound management practices in achieving water security and, consequently, food and energy security, the top priorities of the Ethiopian government and many other governments in Africa, and the relatively neglected area of ecosystem water needs. Although Ethiopia and the other Nile Basin countries are keen on developing their water resources' potential to meet urgent societal needs, there is an increasing emphasis on proper management as the most promising path to achieving water security rather than on the quantity of water that can be harnessed. This puts greater importance on government policy, social, economic, and cultural practices, needs, and preferences. A major theme of this book is the role the Ethiopian government and communities play in water resources management.

While this is primarily a review of the literature, conceptualization of the research problem and the chapter presentations are guided by both the experiences of the editors and the other contributors and the integrated water resources management approach, described in chapter 1. Of the 16 contributors, 15 have extensive research experience in Ethiopia and one (Stuart McFeeters) prepared most of the maps using GIS tools. All but one of the chapters are prepared by multiple authors, including water specialists, engineers, foresters, and social scientists, fostering a multisectoral perspective which integrated water resources planning and management demands.

The 13 main chapters of the book sequentially examine water management issues at the world, African, Nile Basin, and Ethiopian levels in an effort to introduce major issues and problems common to all developing countries, particularly those in Sub-Saharan Africa and the Nile Basin, before examining in depth the situation in Ethiopia. Chapter 1 provides background information on pertinent global and Sub-Saharan water management issues and outlines the objectives, approaches, and methodology used in this book. Using a transboundary perspective of water management, chapter 2 examines the water resources and management practices in the 10 Nile Basin countries, with particular attention to the distribution, management, and use of water resources, development activities, and conservation efforts. Chapter 3 takes a brief inventory of water resources in Ethiopia in relation to environmental and socioeconomic conditions and discusses prospects and problems associated with large-scale hydroelectric and irrigation development. Chapter 4 summarizes Ethiopia's experience with both traditional and modern small-scale irrigation systems, water management institutions, and microdams since the 1974 revolution, with a focus on productivity, environmental impacts, and sustainability. Chapter 5 reviews progress with rainwater harvesting, particularly surface runoff-fed ponds, roof water harvesting, and sand dams, three technologies receiving increasing attention from the Ethiopian government and development organizations. Using a sector management perspective rather than a technical approach, chapter 6 provides an insider view of the coverage and use of water supply and sanitation services in the context of development plans, performance, and program deficiencies. Chapter 7 examines the history, causes, impacts, and possible remediation of deforestation and woodland degradation on water resources in Ethiopia's highland region. Chapter 8 evaluates biophysical, socioeconomic, and political determinants of land degradation and water scarcity in the highlands, with particular emphasis on the ecologically fragile mountain environment, different land use systems and practices, indigenous and government soil and water conservation strategies, resettlement, villagization, and food relief programs. Chapter 9 summarizes the history of river basin development, particularly the construction of dams and irrigation systems, on pastoralists and the environment in the Awash Valley and makes predictions of the possible impact of these activities in the lower Omo Valley and around Lake Turkana, where the huge Gibe III Dam is currently under construction. Chapter 10 briefly outlines the recent history of flooding in Ethiopia and, from a hydrological perspective, analyzes the distribution, main causes, and consequences of the 2006 floods in three Ethiopian watersheds (Omo, Lake Tana, and Dire Dawa) and makes recommendations for flood mitigation. Chapter 11 presents a case study of the 2006 flood in Dire Dawa Town, the most severely affected community during that year, and addresses the problems of disproportionately high impacts on the poor and difficulties with using an integrated flood control approach at the river basin level. Chapter 12 reviews the growing literature on water pollution by chemicals and heavy metals in Ethiopia within a development and policy context and considers some treatment and mitigation options. Chapter 13 describes major negative health impacts of water resources development in Africa, examines the role of different water development programs and activities in Ethiopia in the transmission of vector-borne parasitic and geochemical diseases and evaluates some preventive and mitigating strategies. The concluding chapter 14 highlights major challenges to achieving equitable and sustainable water security in Ethiopia and offers cautious prospects for the future.

We have tried to minimize repetition in this multiauthored volume by asking all contributors to emphasize water resource management and by editing each chapter. We believe that the information presented here will be most helpful to policy makers, managers, and researchers interested in a broad perspective of urgent water issues in Ethiopia and the Nile Basin and hope that the book will contribute to some real, if modest, progress toward the beneficial management of the waters of this region.

Acknowledgments

We are indebted to many individuals for their help and encouragements. As editors, we want to thank above all the contributors, who enthusiastically responded to our invitation to contribute chapters in their fields of expertise.

During the conceptualization of the water resources management theme in the study area, discussions with Zerihun Woldu and Getachew Kassa of Addis Ababa University, Abebe Beyene of Vrije Universiteit Brussel, Wakgari Furi of Oromo Water Works Construction Enterprise, and James McCann of the African Studies Center of Boston University proved most useful. Mulugeta Haileselassie of the Ministry of Water Resources in Addis Ababa and Ephraim Alamerew of the Ethiopian Rainwater Harvesting Association offered unreserved advice, facilitated the selection of some of the contributors, and provided encouragement.

Nigussie Haregeweyn of Mekele University and Mohamed Ali of the Ethiopian Environmental Protection Agency generously provided unpublished information. Richard Carter, Mathew McCartney, and Leif Ohlsson kindly sent us helpful publications, and Linda Vida of the

Water Resources Center Archives, University of California, Berkeley, staunchly assisted with literature searches.

Yirga Mulat generously provided technical assistance and warmly hosted Worku Legesse during his stay in the United States. Worku Legesse also would like to thank his wife, Haimanot Temesgen, for her steadfast support during his stay in the United States.

Lastly, we are grateful to Lester Bradford for critically reading the manuscript and the staff of Cambria Press for their creativity and resourcefulness during the editorial process.

Abbreviations

Birka	underground tank
BOD, BOD_5	biological oxygen demand
COD	chemical oxygen demand
Derg	"The Committee," the common name by which the socialist government that ruled Ethiopia from 1974 to 1991 was known
EEPCO	Ethiopia Electric Power Corporation
EPDRF	Ethiopian People's Democratic Revolutionary Front
FFEWS	Flood Forecasting and Early Warning System
GDP	gross domestic product
GWh	gigawatt hours
GIS	geographical information system
HDI	human development index
IWRM	integrated water resources management
Kebele	smallest administrative unit in Ethiopia
MCL	maximum contaminant level
MDG	Millennium Development Goals

MSF	Multi-Stakeholder Forum
MW	megawatts
MWR	Ministry of Water Resources
NBI	Nile Basin Initiative
NGO	nongovernmental organization
PASDEP	Plan for Accelerated and Sustained Development to End Poverty
PSNP	Productive Safety Net Program
RWH	rainwater harvesting
SNNPR	Southern Nations, Nationalities, and People's Region
SWSI	social water stress index
SSA	Sub-Saharan Africa
UAP	Universal Access Plan
UNICEF	United Children's Fund
WSDP	Water Sector Sevelopment Program
Woreda	District
WASH	water supply, sanitation and hygiene

WATER RESOURCES MANAGEMENT IN ETHIOPIA

Chapter 1

Water Resources Development and Management in Sub-Saharan Africa

Overview and Global Context

Jonathan Lautze, Mark Giordano, and Helmut Kloos

Introduction

With 9% of the world's water resources and 11% of its population (United Nations Educational, Scientific, and Cultural Organization, 2003), Sub-Saharan Africa (SSA) cannot, on the whole, be considered water poor by global standards. Nonetheless, freshwater withdrawal, development of hydropower, and the area of irrigated land are lower on a per capita basis in SSA than any other major region of the world. At the same time, SSA is the world's poorest region and the only region in which per capita food production and income have actually declined in recent decades (United Nations Environmental Program, 2000). In light of these realities, it would seem that additional control of SSA's still abundant water resources would be a major focus in achieving African economic and social development goals, for example, as embodied in national poverty reduction strategies (Federal Democratic Republic of Ethiopia, 2006) and the Millennium Development Goals (MDGs) (United Nations Development Program, 2007).

In SSA, large-scale hydropower and irrigation development typically affects transboundary water management. Transboundary watersheds occupy more than 60% of the region's land mass and contain 93% of the continent's population and 77% of the continent's water resources (Lautze & Giordano, 2005; United Nations Economic Commission for Africa, 2000; Turton, Earle, Malzbender, & Ashton, 2006). As a result of the dependence on transboundary waters, a dependence arguably greater than for any other region in the world, transboundary water governance in Africa forms a core part of almost any regional water strategy (Lautze & Giordano, 2007).

Water management at all smaller geographic scales, down to the community level where management decisions must be made usually by government and nongovernment institutions (NGOs) and communities themselves, poses increasing challenges in SSA to meet local needs in the face of increasing water scarcity and limited capacities (Lenton & Muller, 2009; Pender, Place, & Ehui, 2006). Somewhat surprisingly, despite a continuing need for water resources development (e.g., for domestic use, on-farm storage, and large-scale dam construction for irrigation and hydropower), the rate of development in SSA has gradually declined in recent decades as the orientation of SSA's water institutions has swayed with global policy shifts (Food and Agriculture Organization, 2001; World Commission on Dams, 2000). This development has contributed to leaving the world's poorest countries hostage to hydrology (Grey & Sadoff, 2007).

Ethiopia, the major contributor to the flow of the Nile, has risen to the forefront in water resources development in Africa. In pursuing its objective of economic development and poverty alleviation it is trying to retain a larger share of its crossboundary water, which constitutes more than 90% of its renewable freshwater resources (Okubagzhi, 2008), and has launched accelerated water storage, irrigation, and domestic water supply programs (Awulachew et al., 2005; Legesse, Haile Mariam, & Kloos, 2006). Although several large dams, thousands of microdams, and irrigation schemes of different sizes have been constructed and feasibility studies are underway in most administrative regions, there is a large discrepancy between Ethiopia's abundant water resources and the widespread water scarcity and drought conditions this country is known for. Increasing encroachment of competing agencies and activities on dwindling water resources and destructive pollution of freshwater resources and rapid population increase are prominent features of the landscape of most of the Nile Basin (Howell & Allan, 1994; Okubagzhi, 2008; see also Chapters 2 and 3), indicating that past transboundary and national water management strategies have failed. Highly regionalized distribution and accessibility of water resources in the Nile Basin and accelerating demand for water may result in a water crisis if the Nile Basin Initiative, a plan to promote development in the ten nations of the Nile Basin fails to materialize, as pointed out in chapter 2. But even if all riparian nations approve this plan, its emphasis on the construction of numerous large dams, an approach the World Commission on Dams (World Commission on Dams, 2000) questions in Africa because of the poor record of dams in that region, and the overall weak capacity to implement water development strategies makes it unlikely that the water-related problems besetting this region will be solved soon.

According to Abate (1994), who wrote the only other book on Ethiopia's water resources development and management, the two most promising approaches to achieving water security for human populations and the environment are water conservation facilitated by policy reform and by managing Ethiopia's water resources within an institutional framework for national and transboundary uses. He noted that these measures are prerequisites for Ethiopia's agricultural, hydroelectric, and industrial progress. Some of these changes have been made in the meantime, but their impacts on water, energy, and environmental insecurity are not known. Moreover, Abate focused only on the three main water sectors (irrigation, hydropower, and domestic water supplies), without giving due consideration to such secondary water issues as floods, pollution, and health and did not inculcate the need for the development of rainwater harvesting systems and alternative energy sources that can reduce the need for hydropower. Even more recent studies are limited to irrigation (Awulachew et al., 2005; Awulachew, McCartney, Ibrahim, & Shiferaw, 2008), hydropower (Block, Strzepek, & Rajagopalan, 2007), environmental impacts of large dams in the highlands (Shiferaw, McCartney, Sileshi, & Woldu, 2008; Tefera & Sterk, 2006; Abebe, McCartney, Douven, & Lentvaar, 2008), or flood management through dam construction (Achamyelew, 2003). Only one study has been conducted on environmental and socioeconomic assessments of large dams on the lowland flood plains occupied by pastoral nomads (Kloos, 1982) and none has been conducted on flood monitoring and forecasting at the watershed level (Moges, 2007), hydrological impacts of deforestation and land degradation, chemical water pollution, or negative health aspects of water resources development at the national level. Information is also needed on technologies and approaches recently introduced in Ethiopia and the other Nile riparian countries, including community approaches to water management, rainwater water harvesting and storage, climate change, trade in virtual water, biofuel production, renewable energy technologies other than hydropower, and antipollution measures, described in chapters 2, 3, and 12.

This book argues for a pluralist and integrated approach to water development and management, emphasizing the need for greater civil society participation in the planning, management, and governance of water projects in Ethiopia, where the government has traditionally been the predominant actor. Strategies are needed that substantively integrate local communities, socioeconomic and health parameters, as well as ecosystems into intersectoral development plans to meet water needs at the local, national, and river basin levels. The integrated water resources management (IWRM) approach, which emerged in the 1990s as a response to the worldwide water crisis, can meet these goals. Its key objectives are social equity, economic efficiency, and environmental sustainability (Butterworth, 2006; Lenton & Muller, 2009). The IWRM approach does not provide a common blueprint of how to meet the goals because the water resources, development priorities, levels of infrastructure, institutions, and social and economic challenges are country and locality specific and require strategies best suited to the each country's particular context. However, numerous case studies worldwide show that effective strategies and good practices for better water management employing the IWRM approach include one or more of the following four elements: 1) adequate investments in infrastructure to store, abstract, control, conserve, and protect water resources, 2) a strong enabling environment for setting goals for these water management parameters and strengthening legislative and financial support, 3) clear, robust, and comprehensive institutional roles to build institutional capacity, develop human resources, and establish transparent processes for decision making and for informed stakeholder participation, and 4) effective use of available management and instruments for planning, allocation, water demand management, and conflict resolution. These strategies go far beyond earlier hydrologic approaches focusing on the expansion of hydraulic infrastructure, calling for attention to both water and development challenges. As a result, IWRM has significantly contributed to reducing poverty, promoting economic growth, and improving livelihoods in different countries, watersheds, and communities worldwide (Lenton & Muller, 2009).

The Global Context

Water resources development usually refers to the capture of rainfall, stream flow, or groundwater to meet human needs. Common forms of development include diversions to abstract river water for domestic, industrial, or irrigation supply and dams to store water for hydropower generation, irrigation, flood control, and river navigation. The development of water resources has often been integral to economic development, for example, by providing reliable water supplies to increase agricultural production, promoting the growth of large urban centers and industries, improving hygiene and controlling disease, mitigating floods, and providing power for both domestic and industrial needs. Irrigated agriculture is typically the dominant user of water in arid and semiarid areas (Food and Agriculture Organization, 2005).

Trends in Dam Construction

Over the past three decades, the focus on large-scale water development worldwide has declined substantially. The construction of large dams, a common method for water control and development, highlights this point. As shown in figure 1.1, large dam construction exploded after World War II. By the 1960s, the rate of growth began to slow and by the 1980s the number of new dams constructed each year actually began to fall, further

FIGURE 1.1. Large dam construction by decade, from before 1900 to the 1990s.

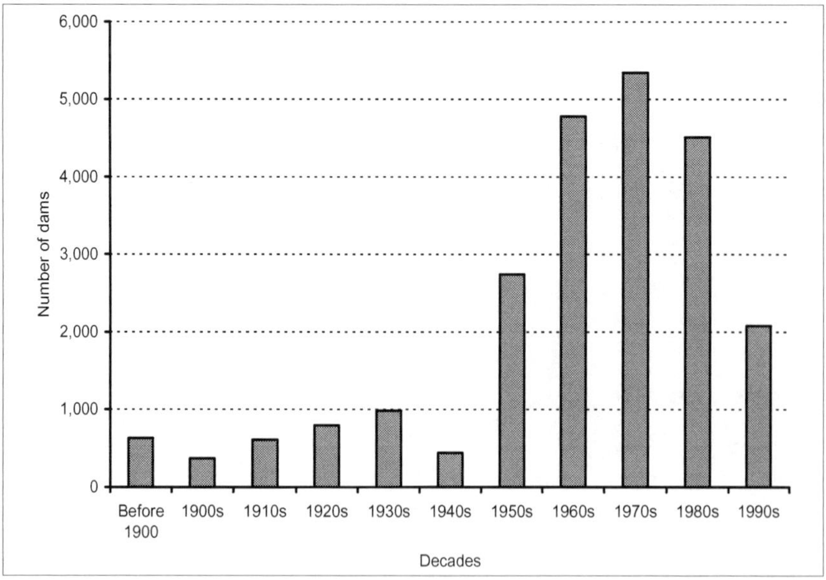

Source. World Commission on Dams (2000).

decelerating in the 1990s. The spike in dam construction, spurred by the desire to use water to increase economic development and improve human wellbeing and supported by increasing technical and financing abilities (Allan, 1999), was increasingly questioned for a number of reasons.

The decline in dam construction after the 1970s can be attributed to the exhaustion of the most suitable dam sites, increasing recognition and criticism of the social and environmental externalities attached to dams, and the growing value attached to the environment (World Commission on Dams, 2000; Gleick, 2004; Conca, 2006). By the end of the 20th century, many countries had witnessed a transformation in the way that water decisions were made. This change has accelerated in recent years because of some highly publicized water project controversies in which decisions were made by centralized government that adversely affected local populations. In addition to the well known debate over the Three Gorges Dam, which displaced between 1.3 and 1.9 million people in China, public protests took place over the construction of dams in India, Southern Africa, and elsewhere in the developing world (Gleick, 2006), and negative impacts have been reported for various dams in Africa (see chapters 9 and 13). In the U.S. and other developed countries, discussions of dam decommissioning are now common.

While the world is now much more aware of the negative impacts of large dams and irrigation schemes, their construction and use has had clear positive impacts on key aspects of economic development and global food security. For example, dam construction helped per capita water withdrawals keep pace with a rapidly rising global population (United Nations Educational, Scientific, and Cultural Organization, 2003). Further, dam construction helped provide water to expand irrigated areas, important not only for increasing food production with existing technologies but also because it allowed the widespread adoption of the high yield variety seeds of the 1960s Green Revolution. The result is that average food supply is above the per capita malnutrition threshold for the world as a whole, perhaps for the first time in history, and is at that threshold for the developing world (Food and Agriculture Organization, 2003).

From Development to Management of Water Resources

Corresponding with the shift away from water resources development on the ground, water laws and policies have experienced a shift that has seen the development agenda replaced by increased emphasis on allocation of water resources, on ensuring water quality, and on the joint management of shared water resources, at both national and international levels (Giordano, 2003; Mostert, 2003; Lautze & Giordano, 2007). Thus, while the number of transboundary agreements dealing with water management doubled between the 1950s and 1990s, the number of agreements focused on water development decreased more than five fold (Lautze & Giordano, 2007). Although water development and water management are not exclusive categories, these changes are significant and may reflect an evolving set of hydrologic, economic, and environmental realities. In the specific field of water resources, there is now a growing body of literature documenting the policy shift from development to utilization and then to allocation, management, and environmental conservation as new supplies become increasingly scarce and priorities change (Shah, Makin, & Sakthivadivel, 2005; Mostert, 2003; Molden, Sakthivadivel, Samad, & Burton, 2005).

New approaches in water management reflecting this shifting paradigm include changing policy priorities to protect watersheds, floodplains, and wetlands and establish flow requirements for the environment; providing universal access to safe drinking water and adequate sanitation by supporting community-based initiatives; developing strong regulatory frameworks that ensure equitable access to services; striving for conservation; enabling access to affordable, small-plot irrigation in conjunction with community-based watershed restoration and rainwater harvesting to improve food security and reduce poverty; increasing water productivity through greater efficiency in irrigation; and good water governance that is guided by the principles of stewardship, sharing, sustainability, and accountability (Postel, 2002; Rockström, et al., 2009). Increasing agricultural productivity in the degraded African lands is increasingly considered as a more water-efficient approach to achieving many of these objectives than the development of new areas for agriculture (InterAcademy Council, 2009).

Specific issues that need to be addressed in increasing the viability and efficiency of irrigation farming in Africa include greater participation of local communities in management decisions, reducing water waste due to overwatering, institutional strengthening, and the control of water-related diseases (Rached, Rathgeber, & Brooks, 1996). Crop irrigation uses more than 90% of all renewable water resources in Egypt, 56% in Sudan and 5% or less in the more humid Sub-Saharan African countries, including all upper Nile riparians (Food and Agriculture Organization, 2009). Irrigation is characterized by low water use efficiency (in regard to crop transpiration), with about 70% of all water lost to evaporation, surface runoff and deep percolation (Falkenmark & Rockström, 2006). Reuse of irrigation drainage and treated wastewater, widely practiced in Egypt and the Middle East, and greater water use efficiency are two particularly promising areas in irrigation water conservation (Bahri, 2008; Shahin, 2007). In the more humid, cooler highlands of the Nile Basin, where rainfed agriculture produces nearly all food crops, and forests and woodlands the bulk of the energy, additional challenges in water productivity and conservation must be dealt with. They include the integration of soil and water management focused on soil fertility, soil tillage for improved rainfall infiltration to reduce soil erosion and excess water runoff on steep slopes, and supplementary irrigation through water harvesting to meet dry season water needs (Rockström, et al., 2009; Pender, et al., 2006). Comprehensive examination of water needs and water management at the national level also needs to address the growing problem of water contamination from industrial, urban, and agricultural sources, and all water plans need to consider adaptive responses to climate change, which is increasingly challenging existing water resources management practices and expectations of transboundary water transfer (Draper & Kundell, 2007). The urgency to maintain the integrity of ecosystems cannot be overemphasized, considering their crucial role in ameliorating the impact of climate change and the fact that 60% of the world's ecosystems have been impacted by manmade activities to the point where they no longer provide some or all of their life enhancing services (Toepfer, 2005). "Water security," defined by Grey and Sadoff (2007, p. 545) as "the availability of an acceptable quantity and quality of water for health, livelihoods, ecosystems and production, coupled with an acceptable level of water-related risks to people, environments and economies" may be achieved in developing countries by pursuing the various approaches and activities mentioned above (Grey & Sadoff, 2007; Lenton & Muller, 2009). This includes access to water guaranteeing fundamental individual and national sovereignty rights over this resource (Wouter, 2005).

This tall order of measures may be implemented after diagnosing both physical and socioeconomic forms of water scarcity. Physical water scarcity, usually attributed to limited access to water resources due to either climatic conditions or unsustainable management of resources, is most often addressed by storage reservoir construction in Africa. Socioeconomic water scarcity, caused either by economic constraints to develop additional water resources or the lack of a society's capacity to adapt to the conditions related to physical scarcity, requires responses by policy makers and society at large that may involve the employment of innovative technology, policies, and management (Assimacopoulos, 2008). Appelgren and Klohn's (1999) study, which includes an examination of factors and conditions driving socioeconomic water scarcity, may lead to a better understanding of water governance and how science may inform policy makers and water resource managers.

WATER MANAGEMENT IN SUB-SAHARAN AFRICA

Underutilization of Water

Sub-Saharan Africa's levels of water resources development and poverty are far different from the corresponding global conditions. These differences become immediately apparent when per capita water withdrawal is examined. Water withdrawal per capita in SSA ranks the lowest of any region in the world and is only one fourth of the global average (figure 1.2). The region with the second lowest water withdrawal per capita, Southeast Asia and the Pacific, uses approximately twice the water per person while wealthier regions such as North America, Europe, and even the water-scarce Arab states withdraw more than five times as much.

FIGURE 1.2. Freshwater withdrawals per capita per year, by region, 1987–1995.

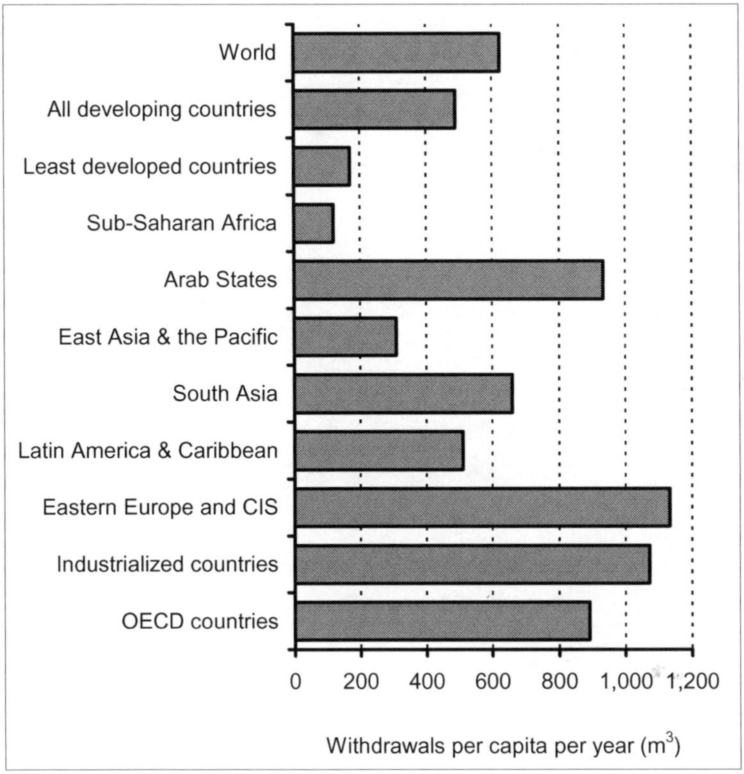

Source. World Commission on Dams (2000).

Water Availability and Storage

While it could be argued that the size of SSA's renewable water resources limits absolute levels of potential water withdrawals, the percent of SSA's renewable water resources that are stored is nevertheless small compared to other regions (table 1.1). Generally speaking, the largest proportion of withdrawn water comes from water storage, and low levels of water storage will usually lead to low levels of water withdrawal. SSA's low level of storage, therefore, imposes severe limitations on the

TABLE 1.1. Global water resources and reservoir storage, by region.

Region	Storage (km^3/year)	Renewable water resources (km^3/year)	Storage/ renewable water resources (%)
Sub-Saharan Africa	388	5,447	7
North Africa/Middle East	392	654	60
South America	891	17,195	5
Oceana	107	1,694	6
Asia	1,262	14,508	9
Central America/ Caribbean	148	1,251	12
Europe	1,083	7,771	14
North America	1,845	5,953	31

Source. White (2005).

quantity and reliability of water which can be withdrawn, and SSA's low water withdrawal is in all likelihood more a function of the region's paucity of storage than limitations related to the size of its natural water endowment.

Compounding SSA's water challenges is the region's high level of rainfall variability (Brown & Lall, 2006). Annual and seasonal variability in rainfall subjects the region to more droughts and floods than other regions. The consequences of these extreme weather events reveal the region's low storage capacities, which could be potentially increased to reduce vulnerability to droughts and floods. High seasonal variability renders annual data unreliable in the context of low storage. Theoretically abundant water supplies, when measured on an annual basis, may be available only during a few months of high rainfall. As a result, water scarcity may be theoretically absent on an annual basis yet practically acute on a seasonal basis.

It should be further noted that water resources development in SSA is distributed highly unevenly across regions. Of the 980 large dams in SSA contained in the World Commission on Dams database (World Commission on Dams, 2000), for example, 539 (or 55%) are located in South Africa. By comparison, Tanzania, a country only slightly smaller in area and population than South Africa, possesses only two large dams. More generally, while withdrawal in southern African countries and Egypt is quite high, eastern, western, and central Africa withdrawal tends to be low (World Resources Institute, 2007). Thus, for most countries in SSA, even the low average numbers presented in table 1.1 vastly overstate conditions on the ground.

The logical corollary to SSA's relatively low levels of total water withdrawal and storage is its comparatively low level of irrigation and hydropower development. Per capita irrigated area is lower in SSA than any other region (Food and Agriculture Organization, 2003), limiting the quantity of food which can be produced and the ability of the continent to take the same advantage of the Green Revolution as did Asia. This fact partially contributes to the region's low levels of food consumption and high rates of malnutrition (Food and Agriculture Organization, 2003). Similarly, hydropower generation in SSA lags behind other regions of the world. Indeed, 75% of Europe's hydropower potential has been developed, 69% of North America's, 33% of South America's, and 22% of Asia's. In Africa (both SSA and North Africa), however, only 7% of the continent's hydropower potential has been harnessed (International Hydropower Association, 2003).

Socioeconomic Constraints

Corresponding with SSA's globally low levels of irrigation and hydropower development is its high level of poverty. Similarly, the poorest and most hunger-prone countries worldwide, all of which are agriculture-based economies, face the most serious freshwater problems (Falkenmark & Rockström, 2006). SSA stands out as the world's poorest region by almost any socioeconomic indicator, for example, per capita food consumption as just mentioned, the Poverty Index, and the Human Development Index

(HDI) that includes life expectancy, education parameters, and GDP per capita, all of which are only about half as high as in high-income countries (United Nations Development Program, 2007; Food and Agriculture Organization, 2003). Given its low level of current water development, still high potential for future development, and its extraordinarily high level of poverty, one might expect the laws and policies governing SSA's water resources to follow a different path from the global trends outlined above. Indications are, however, that this is not the case. At a transboundary level, for example, the recent decline in the water resources development agenda embodied in SSA's agreements parallels that in the rest of the world (Lautze & Giordano, 2007).

Involvement of the North

From the 1950s through the 1980s, the predominant paradigm in global water governance, at both national and international levels, was development (Conca, 2006), and this was reflected in Northern water investments in SSA. For example, the U.S. provided loans essential for the construction of the Akosombo Dam between 1963 and 1966 (Barry, Pluquet, & Andreini, 2004), and the former Soviet Union provided funds for the Aswan Dam between 1960 and 1971 (Said, 1993). More recently, France financed the Diama Dam on the Senegal-Mauritania border, and Germany financed the Manantali Dam in Mali in the early 1980s (King, 1996). In most of these cases, international development financing was consistent with the objectives stated in national or regional water policies and laws of the time. For example, an agreement was signed to develop the Zambezi's hydropower potential in 1953, Ghanaian president Nkrumah was clear about his intentions to further his country's development aims through Volta Basin development (Barry et al., 2004), and the Senegal River Development Organization (OMVS) was created to *mise en valeur* (develop) the basin's waters in 1972.

Over the past two decades, outside involvement in SSA's transboundary and national waters has continued. However, the emphasis of that involvement has shifted with changing global paradigms away from development and toward conservation and management. One example of this is the 1987 Agreement on the Action Plan for Environmentally Sound Management of the Common Zambezi River System. The agreement focuses on coordinating environmental assessments, management plans, and legislation, and makes repeated mention of external actors such as the United Nations Environmental Program and World Bank. Similarly, two agreements signed in 1994 applying to Lake Victoria are heavily environmentally oriented in their outlooks, and Okaru-Bisant (1998) has outlined outside pressure as a factor in their creation and implementation. More recently, several agreements were signed in 2004 and 2005 among Volta Basin riparians to "create conditions necessary for the setting-up of an organization for the management of the water resources, ecosystems, and other related resources in the Volta basin with financial support from the European Union (Direction de l'Inventaire des Resources Hydrauliques, 2005).

In the area of water use, the North significantly influences water use in SSA through trade, as mentioned above, and the export of technology. International trade involves the production of large quantities of agricultural products in SSA countries for export to industrialized countries and food-deficient countries in the Near East, and the rapid expansion of disaster relief programs since the 1970s has resulted in the shipment of food to the South. While prolonged relief programs have become a disincentive for African farmers to increase food production and to carry out soil and water conservation activities (see chapter 8), the importation of foods has resulted in substantial water savings through "virtual water," defined as "the volume of water required to produce a commodity or service" (Chapagain & Hoekstra, 2004, p. 12). Virtual water comprises about 16% of the global water use that is not used for domestic consumption but for export, mainly of agricultural products, and provides an alternative for river basin management (Chapagain & Hoekstra, 2004).

Many African countries have begun to develop alternative energy sources using technologies imported from the North in their effort to harness reliable and sustainable energy sources, reduce the use of biomass (fuel wood, charcoal, and animal waste), and restore watersheds.

The abundance of solar, geothermal, and wind sources and their environmentally friendly development and use make them appropriate alternatives to hydropower (Karekezi & Kithyoma, 2003). The production of biofuel, on the other hand, competes with both water and food production. Even though the amount of water withdrawn for biofuel production globally was still modest in 2008, impacts may be significant in localities where bioethanol crops are irrigated (Hoogeveen, Faurès, & Van de Giessen, 2009; see also chapters 2, 3, and 8).

Balancing the Local and Global in Managing Sub-Saharan Africa's Water Resources

To be clear, water resources development is not a panacea that will solve all of SSA's problems. Neither is consideration of Integrated Water Resources Management (IWRM) concepts, joint management, concern for environmental impacts, and value for the natural state of rivers in some way wrong. On the contrary, such concerns are particularly valid in SSA given the continued dependence of many livelihoods on direct uses of natural resources. Manifestations of neglect to environmental requirements range from siltation of dams in degraded watersheds, increased risk of flooding as a result of intensive cultivation methods and deforestation, to the pollution of rivers, lakes, and groundwater with domestic, agricultural, and industrial wastes and chemicals (Showers, 2002; World Health Organization & United Nations Environmental Programme, 2008). The economic, environmental, and social value of "in situ" water left in the environment has received little attention in Africa, where the adoption and implementation of valuation of water for ecosystems and livelihoods are still in their infancy (Ampomah, 2004).

These activities may be carried out as part of integrated development programs which aim to value, coordinate, and integrate different uses of water in order to strike a desirable balance. While all water requirements may be theoretically factored into integrated development programs, in practice, national requirements may be prioritized over more local water requirements (International Water Management Institute, 2007).

Statutory and Customary Laws in Water Resources Management

Central governments, using formal statutory laws modeled after Western legal systems but hampered by limited capacity, are nowadays the dominant player in decision making and the management of water resources, particularly for mega hydropower and irrigation projects and to meet basic needs of urban populations in most African countries. Since independence, governments have increasingly made inroads into customary laws and decision making, which were developed locally, managed on a day-to-day basis and still prevail in many rural areas (Sokile & Van Koppen, 2004; Maganga, Kiwasila, Juma, & Butterworth, 2004). The diminution of their importance, observed throughout the developing world, has been associated with the transformation of informal indigenous water economies from rural to urban and from household self supply to modern, centralized water supply. Yet in SSA, more than 95% of the region's household water economy continues to rely on self supply and community institutions (International Water Management Institute, 2007) and must be considered in water policies, especially for rural areas (Nkonya, 2008). Integrated water resources management (IWRM) promotes the integration of local laws and institutions into management systems (Butterworth, 2006). Another important feature of IWRM is the formation of partnerships, including public-private partnerships (the predominant model in services provision) and water users associations (United Nations Educational, Scientific, and Cultural Organization, 2009). This trend has not, however, abolished the dominating top-down, authoritarian approach continuing to prevail in many SSA countries, posing a major impediment to the evolution of equitable, sustainable, and environmentally sound management of water resources, as described in most chapters.

Leapfrogging of African Water Development Policies

It might be argued that SSA can "leapfrog" in the development of its water policies by learning from the knowledge and lessons acquired elsewhere in the world (Shah et al., 2005). If one takes a teleological view of water resources development, one could even contend that applying policies now from regions with high levels of water resource and economic development will allow SSA to anticipate future conditions and circumvent the problems of following a more organic policy evolution process. It can be equally asserted, however, that prematurely employing policies more suitable for projected future water resources conditions may impede economic growth and poverty alleviation—the overriding goals of most SSA countries. Concern for demand management and a focus on conservation, in particular, may have steered SSA's policies away from water resources development. As a result, while the direct environmental externalities associated with water development are certainly reduced, the region as a whole remains as food insecure and poor as ever, with the increased long-term pressure for environmental destruction that is associated with such conditions.

As described by Shah et al. (2005), industrialized Europe and North America underwent a 200-year process of water resources development and natural resources degradation that corresponded with their substantial increases in per capita income and food production. Only in the latter part of the 20th century did the importance of natural resources management and conservation begin to outweigh development and its environmental consequences. Most of SSA launched its water resources development efforts much later, largely in the latter half of the 20th century. However, the region has not been afforded the same time to achieve the income and security levels of more industrialized regions. Instead, SSA's period of water resources development was "telescoped" and an agenda of development was integrated with demand management and conservation within just a few decades.

The corollary is that the focus of SSA's water policies has increasingly diverged from addressing conditions in SSA to embodying global "best practices" in transboundary water management. While insights contained in these best practices should clearly be appreciated, the most appropriate approaches in SSA are likely those that are more closely aligned with local and regional conditions rather than the "state of the art" elsewhere. To be clear, however, this is not to imply that the policy guidance from developed countries is meant maliciously, nor is it to imply that the environment is not important in its own right or as a producer of economic value. Nonetheless, global paradigms and values may be applied too uniformly as the severe poverty and low levels of water resources development found in much of SSA would appear to call for more emphasis, and investment, on supply-side interventions than management projects.

Following the decline in new large dam construction in recent decades, large-scale irrigation projects were increasingly de-emphasized in the 1980s and 1990s as part of changing rural development strategies that increasingly focused on poverty. As donors supported economic reform, integrated rural development projects, public sector agricultural credit, and large irrigation projects were phased out. While the performance of irrigation schemes in SSA has generally been mixed, it was particularly poor for large-scale projects (Inocencio et al., 2007). Small-scale irrigation, including simple traditional systems of water harvesting, has been increasingly promoted as potentially more effective to achieve food security. Farmer-managed irrigation using simple, low-cost methods of water management has been enthusiastically adopted by farmers, nongovernmental organizations, and international agencies, and increasingly by African governments (Carter, 1989; Tafesse, 2003).

Recent Developments and Moving Forward

Interestingly, the policies of important international lenders such as the World Bank appear to be making a shift back toward water development. In the late 1980s and 1990s, several World Bank Operational Directives (OD) focused on environmental concerns and effectively reduced possibilities for water resources development lending. However, the recent Water Resources Sector Strategy of the World Bank (2004) indicates a greater openness to development. Further, former president Wolfowitz, at the World Economic Forum (2006) in Davos, acknowledged a call by African leaders for more infrastructure lending, a sentiment echoed by Maria Mutagamba (World Water Forum, 2006), recent chair of the African Ministers Council for Water, who declared that Africa is currently held hostage by its hydrology. Although it may be impossible to definitively ascertain the extent to which this shift reflects a new alignment between the interests of global lenders and African countries, there is growing evidence that the shift is translating to increased water resources development on the ground in SSA. Further, governments as well as international and national development agencies have belatedly begun to recognize the value of accumulated knowledge and participatory approaches in decision making for socially just and sustainable natural resources management (Knox & Meinzen-Dick, 2001).

Given the recent rate of prefeasibility and feasibility assessments in SSA countries, combined with the level of on-the-ground construction, one could even go so far as to say that the region has embarked on a new era of dam building. An immediate impetus for such recent developments is mitigation of the crippling power shortages faced due to growing demand, which disrupt economic activities and are virtually unavoidable in African capitals such as Accra, Kinshasa, Nairobi, and Addis Ababa. Tangible developments are now taking place in the Volta, Niger, Nile, Congo, and Omo/Lake Turkana basins and speculation is rife that development is forthcoming in the Zambezi Basin.

Renewed dam construction and irrigation development requires that water, environmental, and health planners, administrators, and community workers must play a greater role in creating an environment in SSA that is conducive to the evolution of sustainable water resources management by contributing to reducing negative impacts of widespread water-related diseases, pollution, and flood hazards and increasing human capital (King, 1996; World Commission on Dams, 2000). One or more of these issues are addressed in most chapters and discussed in greater detail in chapters 6 and 10 to 14.

References

Abate, Z. (1994). *Water resources development in Ethiopia: An evaluation of present experience and future planning*. Reading, U.K.: Ithaca Press.

Abebe, W. A., McCartney, M., Douven, W. J. A. M., & Lentvaar, J. (2008). Environmental impact assessment follow-up in the Koga Irrigation Project, Ethiopia. In R. S. Bayout & E. Humphreys (Eds.), *Proceedings of the CGIAR Challenge Program on Water and Food. Second international forum on water and food: Vol. 3. Fighting poverty through sustainable water use* (pp. 52–55). Addis Ababa: International Livestock Research Institute and International Water Management Institute.

Achamyelew, K. (2003). *Ethiopia: Integrated flood management*. Report prepared for the World Meteorological Organization, Associated Program on Flood Management, Geneva.

Allan, J. A. (1999). *The Nile Basin: Evolving approaches to Nile waters management.* (Occasional Paper 20, SOAS Water Issues Group). London: University of London.

Ampomah, B. Y. (2004). *A 'new economy' of water for food and ecosystems: Making well-informed, transparent and fair choices based on appropriate methodologies for assessing multi-dimensional water values*. Paper presented at the African Preconference on Water for Food and Ecosystems, Addis Ababa.

Appelgren, B., & Klohn, W. (1999). Management of water scarcity: A focus of social capacities and options. *Physics and Chemistry of the Earth, 24*, 361–373.

Assimacopoulos, D. (2008). Water scarcity: Research opportunities for competing and conflicting demands. Retrieved from http://environ.chemeng.ntua.gr/en/UserFiles/2008/D.%20Assimacopoulos%20-20 Water%20Scarcity%20EU%20Water%20Research%20Day.pdf

Awulachew, S. B., McCartney, M., Ibrahim, Y., & Shiferaw, Y. S. (2008). Evaluation of water availability and allocation in the Blue Nile Basin. In *Proceedings of the CGIAR Challenge Program on Water and Food, 2nd International Forum on Water and Food* (Vol. 3, pp. 6–10). Addis Ababa: Consultative Group of International Agricultural Research.

Awulachew, S. B., Merrey, D. J., Kamara, A. B., Van Koppen, B., de Vries, F. P., & Boelee, E. (2005). *Experiences and opportunities for promoting small-scale/micro irrigation and rainwater harvesting for food security in Ethiopia.* Colombo, Sri Lanka: International Water Management Institute.

Bahri, A. (2008). No single magic bullet can solve the developing world's water problems. *Nature, 456*, 39. doi:10.1038/twaso8.39a

Barry, B., Pluquet, P., & Andreini, A. (2004). *Water development projects in the Volta Basin.* Accra: International Water Management Institute.

Block, P. J., Strzepek, K., & Rajagopalan, B. (2008). *Integrated management of the Blue Nile Basin in Ethiopia: Hydropower and irrigation modeling.* Washington, DC: International Food Research Institute.

Brown, C., & Lall, U. (2006). Water and development: The role of interannual variability and a framework for resistance. *Natural Resources Forum, 30*(4), 306. doi: 10.1111/j.1477-8947.200600118.x

Butterworth, J. (2006, October). People-centered integrated water resources management (IWRM). *Well factsheet.* Retrieved from http://www.lboro.ac.uk/well/resources/fact-sheets/fact-sheets-htm/IWRM.htm

Carter, R. C. (1989). The development of small-scale irrigation in sub-Saharan Africa. *Public Administration and Development, 9*, 543–555.

Chapagain, A. K., & Hoekstra, A. Y. (2004). *Water footprint of nations* (Research Report Series No. 16, Vol. 1). Delft, The Netherlands: UNESCO-IHE.

Comprehensive Assessment of Water Management in Agriculture. (2007). *Water for food, water for life: A comprehensive assessment of water management in agriculture*. London: Earthscan, and Colombo: International Water Management Institute.

Conca, K. (2006). *Governing water*. Cambridge: MIT Press.

Direction de l'Inventaire des Resources Hydrauliques (2005). *Memorandum of understanding to establish a Volta Basin Authority*. Ouagadougou, December 6, 2005. Provided by Francis Bougaire, Director of Direction de l'Inventaire des Resources Hydrauliques.

Draper, S., & Kundell, J. E. (2007). Impact of climate change on transboundary water sharing. *Journal of Water Resources Planning and Management, 133,* 405–415.

Falkenmark, M., & Rockström, J. (2006). The new blue and green water paradigm. *Journal of Water Resources Planning and Management, 132,* 129–132.

Federal Democratic Republic of Ethiopia, Ministry of Finance and Economic Development. (2006). *Ethiopia: Building on progress, a plan for accelerated and sustained development to end poverty (PASDEP) (2005/06-2009/10)*. Addis Ababa: Author.

Food and Agriculture Organization of the United Nations. (2001). *Atlas of water resources and irrigation in Africa.* Land and Water Digital Media Series (No. 13). Retrieved from http://www.fao.org/nr/water/aquastat/main/index.stm

Food and Agriculture Organization of the United Nations. (2003). *Agriculture, food and water*. Retrieved from www.ftp://ftp.fao.org/agl/aglw/docs/agricfoodwater.pdf

Food and Agriculture Organization of the United Nations. (2005) *Irrigation in Africa in figures: Aquastat survey-2005*. Retrieved from http//:www.fao.org/nr/water/aquastat/regions/África/index.stm

Food and Agricultural Organization of the United Nations, Aquastat. (2009). *Review of global agricultural water use per country*. Retrieved from http://www.fao.org/nr/water/aquastat/water_use/print5.stm

Giordano, M. A. (2003). Managing the quality of international rivers: Global principles and basin practice. *Natural Resources Journal, 43*, 111–146.

Gleick, P. (2004). *The world's water: 2002–2003. The biennial report on freshwater resources.* Washington, DC: Island Press.

Gleick, P. (2006). *The world's water: 2006–2007. The biennial report on freshwater resources.* Washington, DC: Island Press.

Grey, D., & Sadoff, C. W. (2007). Sink or swim? Water security for growth and development. *Water Policy, 9*, 545–571.

Hoogeveen, J., Faurès, J.-M., & Van de Giessen, N. (2009). Increased biofuel production in the coming decade: To what extent will it affect global freshwater resources? *Irrigation and Drainage, 58*, S148–S160.

Howell, P. P., & Allan, J. A. (Eds.). (1994). *The Nile: Sharing a scarce resource*. Cambridge: Cambridge University Press.

Inocencio, A., Kikuchi, M., Tonosaki, M., Maruyama, A., Merrey, D., Sally, H., & de Jong, I. (2007). *Costs and performance of irrigation projects: A comparison of Sub-Saharan Africa and other developing regions.* Colombo, Sri Lanka: International Water Management Institute.

InterAcademy Council. (2009). *Realizing the promise and potential of African agriculture*. Amsterdam: Author.

International Hydropower Association. (2003). *The role of hydropower in sustainable development.* Sutton, Surry, U.K.: Author.

Karekezi, S., & Kithyoma, W. (2003). *Renewable energy development in Africa: Prospects and limits.* Paper presented at the Workshop for African Energy Experts on Operationalizing the NEPAD Energy Initiative, Dakar, Senegal.

King, C. (1996). *The incorporation of health concerns into African river basin planning*. (Unpublished doctoral dissertation). Massachusetts Institute of Technology, Cambridge, MA.

Kloos, H. (1982). Development, drought and famine in the Awash Valley of Ethiopia. *African Studies Review, 25*, 21–48.

Knox, A., & Meinzen-Dick, R. (2001). *Collective action, property rights, and devolution of natural resources management: Exchange of knowledge and implications for policy*. International Food Policy Research Institute. Retrieved from http://capri.cgiar.org/pdf/CAPRIWP11.pdf

Lautze, J., & Giordano, M. (2005). Transboundary water law in Africa: Development, nature, and geography. *Natural Resources Journal, 45*, 1053–1087.

Lautze, J., & Giordano, M. (2007). Demanding supply management and supplying demand management: Transboundary waters in Africa. *Journal of Environment and Development, 16*, 290–306.

Legesse, W., Haile Mariam, D., & Kloos, H. (2006). Water supply and sanitation. In Y. Berhane, D. Haile Mariam, & H. Kloos (Eds.), *The epidemiology and ecology of health and disease in Ethiopia* (pp. 129–151). Addis Ababa: Shama Books.

Lenton, R., & Muller, M. (Eds.) (2009). *Integrated water resources management in practice: Better water management for development.* London: Earthscan.

Maganga, F. P., Kiwasila, H. L., Juma, I. H., & Butterworth, J. A. (2004). Implications of customary laws in implementing IWRM: Findings from Pangani and Rufii basins, Tanzania. *Physics and Chemistry of the Earth, 29*, 1335–1342.

Moges, S. A. (2007, May). *Flood Forecasting and Early Warning System (FFEWS): An alternative technology for flood management systems and damage reduction in Ethiopia: A concept note.* Paper presented at the 2007 Lake Abaya Research Symposium, Arba Minch University, Arba Minch, Ethiopia.

Molden, D., Sakthivadivel, R., Samad, M., & Burton, M. (2005). Phases of river basin development: The need for adaptive institutions. In M. Svendsen (Ed.), *Irrigation and river basin management: Options for governance and institutions* (pp. 19–29). Wallingford, U.K.: CABI and Colombo, Sri Lanka: International Water Management Institute.

Mostert, E. (2003). Conflict and cooperation in international freshwater management: A global review. *International Journal of River Basin Management, 1*, 1–12.

Nkonya, L. K. (2008). *Rural water management in Africa: The impact of customary institutions in Tanzania.* Amherst, NY: Cambria Press.

Okaru-Bisant, V. (1998). Institutional and legal frameworks for preventing and resolving disputes concerning the development and management of Africa's shared river basins. *Colorado Journal of International Environmental Law and Policy, 9*, 331–369.

Okubagzhi, Y. (2008). *Water resources and inter-riparian relations in the Nile Basin*. New York: SUNY Press.

Pender, J., Place, F., & Ehui, S. (Eds.). (2006). *Strategies for sustainable land management in the East African Highlands*. Washington, DC: International Food Policy Research Institute.

Postel, S. (2002). *From Rio to Johannesburg: Securing water for people, crops and ecosystems*. Worldwatch Institute, 2002. Retrieved from http://www.worldwatch.org/node/1732

Pretty, J. N., & Shaw, P. (1998). Making soil and water conservation sustainable: From coercion and control to partnership and participation. *Land Degradation and Development, 8*, 39–58.

Rached, E., Rathgeber, E., & Brooks, D. B. (Eds.) (1996). *Water management in Africa and the Middle East: Challenges and opportunities.* Ottawa: International Development Research Center.

Rockström, J., Falkenmark, M., Karlberg, L., Hoff, H., Rost, S., & Gerten, D. (2009). Future water availability for global food production: The potential of green water for increasing resilience to global change. *Water Resources Research*, *45*, W00A12, doi:10.1029/2007WR006767

Said, R. (1993). *The river Nile: Geology, hydrology, and utilization.* Oxford: Pergamon.

Shah T., Makin, I., & Sakthivadivel, R. (2005). Limits to leapfrogging: Issues in transposing successful river basin management institutions in the developing world. In M. Svendsen (Ed.). *Irrigation and river basin management* (pp. 31–50). Wallingford, U.K. and Cambridge, MA: CAB International.

Shahin, M. (2007). *Water resources and hydrometeorology of the Arab region*. Dortrecht: Springer.

Shiferaw, A., McCartney, M., Sileshi, Y., & Woldu, Z. (2008). Impacts of the Chara Chara Weir and their environmental implications at the source of the Blue Nile, Ethiopia. *Proceedings of the CGIAR Challenge Program on Water and Food, 2nd International Forum on Water and Food* (Vol. 3, pp. 75–78). Addis Ababa: Consultative Group of International Agricultural Research.

Showers, K. B. (2002). Water scarcity and urban Africa: An overview of urban-rural water linkages. *World Development, 30*, 621–648.

Sokile, C. S., & van Koppen, B. (2004). Local water rights and local water user entities: The unsung heroines of water resources management in Tanzania. *Physics and Chemistry of the Earth*, *29*, 1349–1356.

Tafesse, M. (2003). *Small-scale irrigation for food security in sub-Saharan Africa* (CTA Working Document No. 8031). Addis Ababa: CTA and Mekele: Tigray Water Resources Development Bureau.

Tefera, B., & Sterk, G. (2006). Environmental impact of a hydropower dam in Fincha's watershed, Ethiopia: Land use changes, erosion problems, and soil and water conservation adoption. In *Proceedings of the second international conference on sustainable sloping lands and watershed management* (pp. 89–101). Luang Prabang, Lao PDR: National Agriculture and Forestry Research Institute.

Toepfer, K. (2005). Foreword. In F. Dodds & T. Pippard (Eds.), *Human and environmental security: An agenda for change.* London: Earthscan.

Tolcha, T. (1989). Results of soil conservation experiments and the scope for implementation: The case of Chercher Highlands. *Ethiopian Journal of Development Research, 20*, 25–63.

Turton, A. R., Earle, A., Malzbender, D., & Ashton, P. J. (2006). Hydropolitical vulnerability and resilience along Africa's international waters. In A. T. Wolf (Ed.), *Hydropolitical resilience and vulnerability along international waters* (Report No. UNEP/DEWA 2006, chapter 6). Nairobi: United Nations Environmental Program.

United Nations Development Program. (2007). *Human development report 2007/2008*. New York: Author.

United Nations Economic Commission for Africa. (2000). *Transboundary river/lake basin water development in Africa: Prospects, problems and achievements*. Addis Ababa: Author.

United Nations Educational, Scientific, and Cultural Organization. (2003). *World water development report.* Paris: Author.

United Nations Educational, Scientific, and Cultural Organization. (2009). *The 3rd United Nations world development report: Water in a changing world.* New York: Author.

United Nations Environmental Program. (2000). *Global environment outlook.* New York: Author. Retrieved from http://www.unep.org/GEO2000/index.htm.

White, W. R. (2005). *World water storage in man-made reservoirs*. Marlow, U.K.: Foundation for Water Research.

World Bank. (2001). *Ethiopia: Woreda studies*. Draft report. Addis Ababa: World Bank Country Office in Ethiopia. Retrieved from http//info.worldbankk.org/etools/docs/library/5687/Ethiopia%20Woreda%20Studies,%20WB.htm

World Bank. (2004). *Water resources sectors strategy: Strategic directions for World Bank engagement.* Washington, DC: Author.

World Commission on Dams. (2000). *Dams and development: A new framework for decision-making.* London: Earthscan.

World Economic Forum. (2006). *Next steps for Africa.* Retrieved from http://www.weforum.org/site/knowledgenavigator.nsf/Content/_S16175?open&event_id=1462&year_id=2006.

World Health Organization & United Nations Environmental Programme. (2008). *First inter-ministerial conference in health in Africa: Health security through healthy environments.* Retrieved from www.unep.org/health-env/pdfs/Media_FactFigures.pdf

World Resources Institute. (2007). *Earth trends, the environmental information portal: Water resources and freshwater ecosystems.* Retrieved from http://earthtrends.wri.org/features/index.php?theme=2

World Water Forum. (2006). *World Water Forum addresses water for growth and development.* Retrieved from http://www.iisd.ca/ymb/worldwater4/mar17.html

Wouter, P. (2005). Water security: What role for international law? In F. Doods & T. Pippard (Eds.), *Human and environmental security: An agenda for change.* London: Earthscan.

Chapter 2

Water Resources Management in the Nile Basin

Helmut Kloos, Worku Legesse, and Aynalem Adugna

Introduction

The Nile Basin is a resource rich but climatically harsh and economically poor region experiencing rapid population growth, great disparities in water resources at the national and local levels, and frequent droughts. Despite the Nile Basin's large size (3.1 million km^2) and containing the world's longest river, runoff is less than half that of the Zambezi and Niger rivers, even though they drain smaller catchments (Andjelic, 1999). Great spatial and seasonal variations in rainfall characterize most of the Nile Basin. The population in the Nile Basin, including nearly the total population of Egypt, Sudan, and Uganda, and variable proportions of those of Tanzania, Kenya, Burundi, Rwanda, The Democratic Republic of the Congo, Ethiopia, and Eritrea (figure 2.1), was about 202 million in

FIGURE 2.1. The Nile Basin.

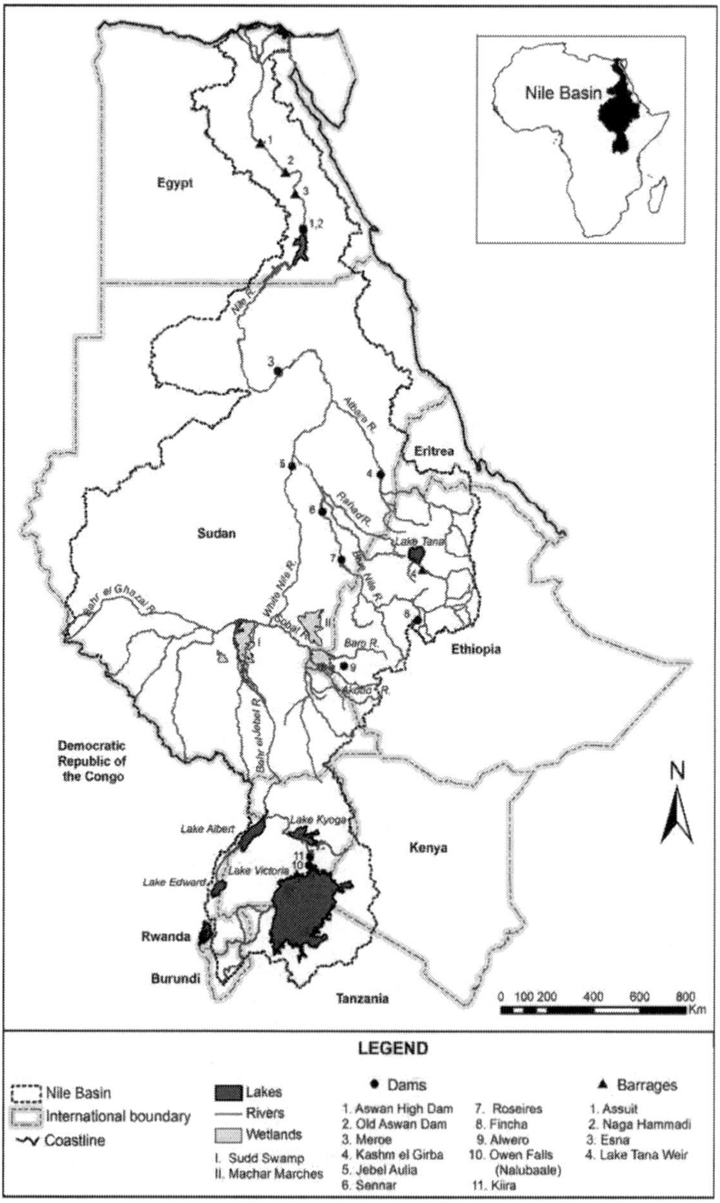

2005 and is expected to reach 336 million by 2030 (table 2.1), suggesting that water scarcity may reach crisis proportions if water needs cannot be met in a timely manner.

Assessment of water stress and water scarcity is inherently associated with difficulties because of the interplay of hydrological and social determinants. The classification by Falkenmark and Widstrand (1992), based on estimated annual renewable water resources and average rainfall within a country, according to which a country is considered to be chronically water scarce if it has between 500–1,000 m^3 renewable water per capita per year and water stressed at 1,000–1,700 m^3, must be used with caution in the Nile Basin. These indicators, based solely on traditional hydrological indices, are inadequate in gauging the capacity to access and utilize available water resources. The social water stress index (SWSI) incorporating both hydrological indices and the United Nations Development Program Human Development Index (HDI), provides a measure of social adaptive capacity, a crucial requirement for accessing and managing available water resources (Ohlsson, 1998). The HDI contains three important determinants of successful water resources management, namely life expectancy (an indicator of general level of development), education attainment (an indicator of institutional capacity), and real GDP per capita. Thus, while Ethiopia is relatively water sufficient when using hydrological stress criteria, it is classified as "water scarce" according to the social water stress index (SWSI), and Egypt's position improves from water scarce under the former to water stressed under the latter index. Rwanda and Burundi, with some of the smallest renewable water resources and lowest HDI scores, had the lowest SWSI scores (table 2.1). Disparities in food consumption levels between Egypt and all other Nile Basin countries largely reflect the SWSI scores. Per capita calorie consumption in 1995–1996 was about one-third higher in Egypt than in the countries upstream on the Nile, although national food demand is projected to rise twice as fast in the upper riparian countries than in Egypt between 1989 and 2010 due to relatively faster population growth (Appelgren, et al., 2000). Although the upper riparians, particularly Sudan, Ethiopia, Uganda, Kenya and

TABLE 2.1. Renewable water resources (RWR), human development index (HDI), social water stress index (SWSI), land and water use, and population in the Nile Basin countries.

Country	Total RWR*	RWR per capita (in m^3 per year) in 2000	HDI 2005	SWSI 1995	Irrigated land in Nile Basin (1,000 ha) in 2000#	Land used for agriculture nationally (1,000 ha) in 2000	Population not using an improved water source (%) in 2005	Population in the Nile Basin (millions) 2005##
Egypt	58.3	851	0.716	17	3,078 (4,420)	3,300	2	72.6 (2.0)
Sudan	149.0	4,966	0.526	5	1,935 (60,500)	21,000	30	32.4 (2.7)
Kenya	30.2	1,003	0.521	19	6 (180)	25,820	39	13.4 (2.2)
Uganda	66.0	3,027	0.505	9	9 (202)	8,610	40	28.5 (3.1)
Tanzania	89.0	2,656	0.467	9	10 (30)	39,650	38	7.9 (2.3)
Rwanda	6.3	818	0.452	44	9++ (150)	1,615	26	7.7 (2.3)
Eritrea	8.8	2,256	0.442	ND**	15 (150)	7,467	40	1.7 (2.6)
Burundi	3.6	537	0.413	68	21++ (80)	2,200	21	4.6 (2.8)
D.R. Congo+	1,019.0	19,709	0.411	ND	ND	22,900	54	1.9 (3.2)
Ethiopia	123.2	1,969	0.406	21	76 (2,220)	30,600	78	31.0 (2.5)

Sources. Amer et al. (2005), Food and Agriculture Organization (2005); Kundell (2008a, 2008b); Population Reference Bureau (2005); United Nations Development Program (2007); Boukerrou (2008, 2009); and Ohlsson (1998).
Notes. *in km^3; +Democratic Republic of the Congo; **no data; ++these data are for the whole country; #potentially irrigable land in Nile Basin in parentheses; ##annual national growth rate (percent) for 2005 in parentheses.

Tanzania, have large areas of irrigable and/or rain-fed arable land, they all share low HDI scores, impeding efforts to achieve water and food security (table 2.1; Appelgren, Klohn, & Alam, 2000).

Agriculture is the economic base of all 10 Nile riparian countries and the major consumer of water in all countries except Uganda. But whereas Egypt irrigated 3.08 million ha and Sudan 1.94 million ha around 2000, Ethiopia irrigated only about 76,000 ha within the Nile Basin, and the remaining riparians irrigated between 6,000 ha and 15,000 ha each (table 2.1). Irrigation agriculture uses between 79% and 97% of all blue water consumed in Egypt, Sudan, Ethiopia, Kenya, and Tanzania, indicating the significant water savings that may be possible with increased conservation. The huge amounts of water allocated for irrigation and the rapidly increasing energy demands and hydropower development have major implications for all countries in the Nile Basin, not only for those depending largely on the Nile (Sudan and Egypt), but also for the increasingly water-scarce upper riparians themselves, who are determined to develop the Nile's water to promote socioeconomic development. Water scarcity threatens to curtail future irrigation development and is most contentious in the international arena. This requires that demographic, socioeconomic, political, and ecosystems issues be addressed for the success and sustainability of projects (Amer et al., 2005). Renewable water resources available per person in the Nile Basin are estimated to decline between 28% and 51% in different riparian countries between 2000–2025 due to population growth and increasing demand for water with increases in socioeconomic level and dietary changes toward more water-demanding food (United Nations Educational, Scientific, and Cultural Organization, 2009; table 2.1). In Ethiopia, with the lowest HDI, a larger proportion of the population (78%) did not use potable water than in any other Nile country. In Egypt, only 2% of the population was without access to potable water. Population growth rates were also lowest in Egypt (table 2.1; Food and Agriculture Organization, 2005). These data point out the need for strengthening the social adaptive capacities of most Nile riparians that will allow them to achieve acceptable levels of water and food security in spite of increasing scarcity. The development of the large irrigable areas in Sudan and Ethiopia (table 2.1) has given rise to many speculations about potential food production, but their utilization is not ensured due to increasing water scarcity, lack of capacity, and a number of other constraints discussed in this chapter.

THE NILE BASIN INITIATIVE

The Nile is the life line for Egypt, which obtains practically all of its freshwater from this river. Egypt has always tried to protect its water supply and control the Nile's source since ancient times, prompting it to repeatedly invade Ethiopia, especially in the 19th century (Tved, 2004). Renewed tensions, but also attempts to reach an acceptable agreement over the use of the Nile's waters, emerged in the riparian nations after the 1960s. These developments resulted from a combination of rapidly increasing demands for water to irrigate, generate hydropower, and for domestic and industrial uses and the determination of the newly independent countries of East Africa to challenge the 1959 water agreement between Egypt and Sudan. This agreement, which allocates 55.5 km^3 to Egypt and 18.5 km^3 to Sudan without giving water rights to the other eight riparians, was based on the 1929 colonial Nile Water Agreement between Great Britain and Egypt (Howell, 1994; Kagwanja, 2007). The 1929 agreement was one of 10 bilateral agreements made between 1891 and 1953 between Great Britain and the Nile riparians that prevented them from interfering with the flow of the Nile (Hefny & Amer, 2005). After independence, the countries of the Lake Victoria Basin and the Western Rift Valley (Uganda, Kenya, Tanzania, Rwanda, Burundi, and Democratic Republic of the Congo) began to collaborate among themselves over their resources. These efforts became stronger after the revival of the East African Community in 1999, the year the Nile Basin Initiative was launched (Kagwanja, 2007).

The Nile Basin Initiative (NBI) seeks to develop the river in a cooperative manner, share substantial socioeconomic benefits, and promote regional peace and security (Nile Basin Initiative, 2009). This includes the equitable and sustainable development of the Nile Basin's water resources, efficient water management and use, cooperation and joint action among the riparians, and the control of poverty and economic integration. All countries recognize that interstate collaboration is essential for effective watershed management, soil erosion and flood control, and to facilitate economic development at the regional level (Arsano & Tamrat, 2005; Hefny & Amer, 2005; Kagwanja, 2007). The first economic model of optimal Nile water utilization estimated the annual gross economic benefits accruing from irrigation, flood control, and hydroelectric power production at US$7–11 billion (Whittington, Wu, & Sadoff, 2007).

The three main programs of the Nile Basin Initiative are the Shared Vision Program (SVP), the Eastern Nile Subsidiary Action Program (ENSAP), and the Nile Equatorial Lakes Subsidiary Action Program (NELSAP). The SVP is focused on building the institutions, sharing information, providing training, and identifying possibilities for dialogue and collaboration necessary to develop water resources in a sustainable manner. ENSAP aims to develop the water resources of the Eastern Nile Basin (Ethiopia, Eritrea, Sudan, and Egypt) in an equitable and sustainable manner by establishing environmentally sound management of watersheds, reducing poverty, and connecting the power grids of Ethiopia and Sudan to expand cross-border energy trade, develop regional irrigation and drainage projects, and improve flood forecasting and preparedness. This program also includes activities in the areas of alternative livelihood creation, water conservation at the watershed level, and ecosystem maintenance. By 2008, four fast-track projects had been started: 1) the Watershed Management Project in the watersheds of the Blue Nile, Baro/Akobo/Sobat, and Atbara basins with the objective of promoting their sustainable management; 2) the Ethiopia Power Export Project to connect the Ethiopian and Sudanese power grids to permit cross-border energy trade and to ameliorate the severe electricity shortage in both countries; 3) the Regional Irrigation and Drainage Project, including water conservation and irrigation rehabilitation projects in Egypt's western Nile Delta, two irrigation and drainage projects on Lake Tana, and an irrigation project on the Atbara River in Sudan;

and 4) the Flood Preparedness and Early Warning Project aimed at disseminating flood warning information to at-risk communities and strengthening institutions and preparedness and response plans (World Bank, 2009a). NELSAP focuses on reducing poverty, promoting growth, and controlling environmental degradation through integrated water resources management, fisheries, lake protection, and transboundary hydropower development in Uganda, Kenya, Tanzania, Rwanda, Burundi, and the Democratic Republic of the Congo (World Bank, 2009b). It must be pointed out that while the Nile Basin Initiative emphasizes the need for blue water (surface and groundwater) sharing and conservation it fails to address possibilities for green water (rainwater directly used in the soil and evaporated by nonirrigated land use systems) and renewable energy development other than hydropower. This neglect and the push for large hydraulic structures may diminish the viability of the initiative as a comprehensive and sustainable water management strategy in the long term.

The Nile Basin Initiative had not been signed by late 2009 due to the insistence by Egypt that its current water uses and rights be preserved that prohibit upper riparians from implementing projects that would reduce the volume of water reaching Egypt (Musinguzi, 2009; see also Howell, 1994). Nevertheless, the regional structures in place help reduce conflict and the potential for water wars. This achievement has been described as "a triumph of regionalism over stateism, and multilateralism over unilateralism" (Kagwanja, 2007, p. 333). However, continuing arguments of water rights and Egypt's and Sudan's reluctance to sign the protocol threatens to disrupt momentum toward a final agreement (Musoni, 2009). In the absence of a signed protocol, states continue to make unilateral decisions about management of the Nile's water within their territories which may jeopardize the agreement (Eguavoen, 2009). Because the most contentious issue facing the Nile Basin Initiative is achieving agreement on a principle for the allocation of water and water benefits, a discussion of the spatial and temporal distribution of water resources in the Nile Basin and recent changes due to new construction of hydraulic projects follows.

Water Resources and Major Uses

The Nile Basin, extending from the highlands in tropical East Africa and Ethiopia to the subtropical lowland deserts in Egypt, is characterized by a wide range of climates. Heavy bimodal and monomodal rainfall with annual totals of 1,100–1,300 mm prevail in the Lake Victoria area and other parts of the East African highlands, with peaks between March and December and a short dry season. Rainfall in the Ethiopian highlands, by contrast, although producing similar annual totals, is concentrated during a shorter monomodal period between June and September. Rainfall progressively declines northward through Ethiopia, Eritrea, and Sudan and becomes rare in Egypt's desert. Stream density decreases with decreasing rainfall, causing most smaller streams in the lowlands to either dry up during the long dry season or to disappear (Sutcliffe & Parks, 1999).

Groundwater use for irrigation is relatively insignificant in most of the Nile Basin and, until now, has been limited mostly to domestic use of springs in the high-rainfall highlands and wells along some rivers in the lowlands, where groundwater levels are high and stream and river waters are unreliable or unpalatable. The extensive but nonrenewable Nubian Sandstone Aquifer, estimated to contain 150,000 km^3 of fossil water that is to be managed and shared by Egypt, Sudan, Chad, and Libya, has so far been tapped mainly by the latter, to a lesser extent by Egypt, and only slightly by Sudan (Nubian Sandstone Aquifer System, 2009; International Atomic Energy Agency, 2009). Renewable groundwater in Sudan is estimated at 7 km^3 per year, including some shallow aquifers around Khartoum that have become polluted (Hamad & El-Battahani, 2005; Kundell, 2008a). Groundwater assumes considerable importance in Egypt, where it supplied 14.3% of all available water in 2000, although water degradation by agro-chemicals and salts is posing increasing health and agronomic problems (Kundell, 2008b). Ethiopia's groundwater resources are described in chapter 3.

Mean annual and seasonal river flows in most of the Nile Basin have been measured since the beginning of the 20th century, but the hydrographic record in Egypt goes back to Pharaonic times. Sutcliffe and Parks (1999) analyzed the published data up to the late 1990s using the water balance method, which measures rainfall and river flow inputs and evaporation and river flow outputs to examine the characteristics of river section, lake, and tributary. Their results showed that the Nile's hydrology is strongly associated with the region's climate, geology, water storage, and evaporation and that river regulation must be considered in basin-wide water management.

The Upper Riparians Around the Equatorial Lakes

The seasonally fairly even flow of the two main tributaries of the upper White Nile, the Kagera River originating in Rwanda and Burundi, and the Semliki River in the Lake Edwards Basin, as well as those in southern Sudan reflect the humid equatorial climate regime. The flow of the upper White Nile is attenuated by a series of large lakes and wetlands, particularly lakes Victoria, Kyoga, and Albert, the Sudd Swamp, and the Machar Marches. The completion of Owen Falls Dam (renamed Nalubaale Dam by President Museveni) at the outflow of Lake Victoria at Jinja Town in the early 1960s as part of an agreement between Uganda and Egypt increased the lake's storage capacity, assuring adequate water supplies in Egypt during droughts.

The sharp rise in Lake Victoria's level between 1961 and 1964, which corresponded with increased levels of other East African lakes and higher rainfall and runoff across the Sahel and in Ethiopia, and its subsequent decline caused a hydrological debate over the causes of this anomaly. Whereas the rise was attributed to global atmospheric and oceanic properties and processes (Young, 2004; Grijsen, 2007), the lake's decline was due to a combination of drought, increasing releases of lake water from the new Kiira Dam to meet Uganda's hydroelectric needs (Kull, 2006; Grijsen, 2007), and apparent abstraction for irrigation in Tanzania (The National Association of Professional Environmentalists in Uganda, 2009; Sutcliffe & Parker, 1999). Releases between 2004 and 2006 were nearly double those specified by the Agreed Curve, a measure based on Lake Victoria levels the Nile countries agreed on (Kull, 2006).

Sutcliffe and Petersen (2007) computed the impact of the overabstraction and pointed out the need to adjust releases accordingly to reestablish base level flow.

Lake Victoria's changing hydrology illustrates the need for proper management of water resources impacted by both drought and increasing anthropogenic uses, a scenario increasingly reported from other parts of the Nile Basin. Plans and ambitions of all three countries bordering the lake (Kenya, Uganda, and Tanzania) point to greater use of Lake Victoria for irrigation and other uses in the future (Okubagzhi, 2008). Droughts were common throughout East Africa from 1997 to 1999 and from 2005 to 2006, resulting in up to 50% declines in hydropower production in Uganda, Tanzania, and Kenya. Uganda, Kenya, and Tanzania are planning to construct at least eight additional dams in the Lake Victoria Basin, including one dam on the Kagera River at Rusumo Town in Tanzania and the Bujagali Dam downstream of the Owen Falls and Kiira dams (International Water Power and Dam Construction, 2008; Grijsen, 2007).

Not only are quantity parameters of Lake Victoria at stake but also the quality of its waters. Accelerating nutrient loading from agricultural runoff, deforestation, industries, mines, and sewage from the rapidly increasing population around the lake have resulted in massive fish kills and pose a threat to its biological integrity (Kolding, Van Zwiete, Mkumbo, Silsbe, & Hecky, in press; Zhang, 2007). Deterioration of water quality due to these and other anthropogenic activities in other, smaller lakes and in rivers in the Nile Basin is contributing to water scarcity, a situation we examine for Ethiopia in chapter 12.

Sudan, Ethiopia, and Eritrea

Water resources and uses in these three countries vary greatly and in Sudan and Ethiopia promise to become a key to water resources management in the entire Nile Basin, particularly as it affects Egypt. Three large wetlands, the Sudd and Bahr el Ghazal swamps and the Machar Marches, are causing the decline by about half of the flow of the Bahr el Ghazal, Bahr el Jebel, and Sobat rivers (figure 2.1). Various proposals, starting in 1904, to construct a drainage canal in the Sudd and, thus, save what has been described as 'lost' Nile water, failed to materialize (Collins, 2002). An expanded drainage plan is currently being considered within the context of the Nile Basin Initiative. The potential environmental impact of draining these wetlands is not fully known due to lack of information on their size and evaporation capacity during different times of the year. A recent modeling effort using remote sensing techniques estimated the potential gain in the flow of the White Nile at 36 km^3 (Mohamed, Savenije, Bastiaanssen, & Van den Hurk, 2006). This study also predicted a significant impact on the local climate, with temperatures rising by 4–6° C and the relative humidity decreasing 30–40% during the dry season. All plans for draining these wetlands have been opposed by local pastoralists and by environmentalists, who argue that draining them would result in significant ecological and economic damage (Collins, 2002; Dixon & Wood, 2003). The seasonal floodplains *(toiches)* provide essential pastures for nomadic pastoralists and nutrient-rich nursery grounds for many fish and bird species. The basket of economic services provided by wetlands in the Nile Basin has been estimated at US$15,000 per hectare (Appelgren, et al., 2000). It remains to be seen if the negative attitude of the international donor community in the 1990s to developing these wetlands (Mallat, 1994) will continue to block financial and technical assistance in the future if the current water use in agriculture in Sudan and Egypt cannot be reduced and if industrial and other demands continue to increase.

The Blue Nile, associated since ancient times with the annual inundation of the Nile's flood plains in Egypt, together with the three other major rivers draining the western highlands of Ethiopia—the Baro, Akobo, and Atbara—supplies about four fifths of the main Nile's water at Aswan. The average annual flow of the Blue Nile at the Sudan border for the period 1921–1989 was 52.0 km^3, but Ethiopia's consumptive use in recent years was only around 0.3 km^3 (Awulachew, McCartney, Ibrahim, & Shiferaw, 2008). The average annual contribution of the Blue Nile to the main Nile's flow decreased from about 65% to 55% during the 20th century, the result of increased releases from Lake Victoria after

1961 and the recent decline in Blue Nile discharge (Sutcliffe & Parks, 1999). River flow data indicate that while the mean annual volume of the Blue Nile at Khartoum increased during the first six decades of the 20th century, it decreased by 13% between 1961 and 1995, largely due to abstractions for irrigation in the Gezira cotton scheme and other large irrigation schemes in Sudan. Lake Tana, the official source of the Blue Nile, supplies only 7% of the river's runoff, most of its water coming from tributaries in Ethiopia. Between 80% and 85% of the Blue Nile's discharge occurs between July and October. There are only two hydraulic structures on the Blue Nile and its tributaries between Lake Tana and Roseires Dam in Sudan, a weir constructed at Lake Tana's outlet for hydroelectric production at the two Tis Abbay hydropower plants on the Blue Nile and Fincha Dam on a tributary. They have, unlike the dams on Lake Victoria, little mediating influence on river flow (Awulachew et al., 2008; Sutcliffe & Parks, 1999). But abstraction of Lake Tana water for the Tana-Beles hydropower plant, to be completed in 2009, is expected to decrease the lake's contribution to the flow of the Blue Nile (Eguavoen, 2009). Absence of dams on the Blue Nile and its many tributaries in Ethiopia has allowed for continuing inundation of the riverine plains in Sudan during the river's summer spike and caused reservoir siltation. The heavy silt load of the Blue Nile caused a 60% reduction in the storage capacity of Sennar Dam and a 35% capacity reduction of Roseires Dam on the Blue Nile in Sudan and interrupted turbine operation and hydropower production (United Nations Environmental Programme, 2007). Similarly, silt from the Atbara River reduced the capacity of Kashm-El Girba Reservoir by 42% after only 12 years (Conway, 2000).

The Ethiopian government was nearing completion of Koga Dam in the Lake Tana Sub-Basin in 2009 and plans to construct one dam each on the Dabus and Didessa tributaries and in the long term four mega multipurpose dams (Karadobi, Mabil, Mendaia, and Border) on the Blue Nile between Lake Tana and the Sudan border. These four dams, described further in chapter 3, are expected to serve as major storage facilities for downstream irrigation and flood control purposes, as well as Ethiopia's major source of electricity for export (Arsano & Tamrat, 2005; Amer et al.,

2005). In 2010, prior to the construction of these dams, Ethiopia is planning to abstract an estimated 2.2–3.8 km^3 for the development of about 220,000 ha for irrigation (Awulachew et al., 2008), only 4.4% of the area irrigated in Sudan and Egypt in 2000 (table 2.1). Ethiopia's irrigable area in the Nile Basin compared to those of Sudan and Egypt is even smaller (1.2%) (table 2.1; see also table 3.4 in chapter 3).

The four planned Ethiopian dams may have an even greater impact on water management in Sudan by adding an additional 9.3 km^3 above its current use of 14.6 km^3, and 5.4 km^3 above its allotment under the 1959 water agreement with Egypt (Cascão, 2009). The Sudanese government, earlier attempting to become the "bread basket of the Arab world," is determined again to develop its huge land and water resources to the greatest extent possible. This renewed interest in large-scale water projects rests on the prospect of a seasonally more reliable and less silty water supply from the proposed Ethiopian dams, as well as Sudan's increased oil revenues generated after the peace agreement with Southern Sudan in 2005, favorable hydro construction contracts by Chinese firms, and financial assistance from Arab countries. Meroe Dam was completed in 2008, the old Sennar Dam is being heightened, and current plans call for the construction of two additional dams on the Nile (Kajbar and Shereiq), one dam on the Atbara River (Rumala), one barrage each on the Rahad and Dinder rivers, and the extension of the irrigated areas. These unilateral developments run counter to the 1959 water agreement with Egypt and signify greater independence of Sudan in water planning (Cascão, 2009).

The seasonal flow pattern of the Rahad, Atbara, and other rivers originating in the northern Ethiopian and Eritrean highlands are similar to that of the Blue Nile but with an even more pronounced seasonal peak. Completion in 2009 of the Tekeze Dam in Ethiopia will significantly reduce its seasonal peak flood. Start of the dam's operation was delayed by one year due to landslides of the reservoir walls and the drought in 2008. Mean annual runoff of the Atbara at its confluence with the Nile decreased by 30%, from 12.3 km^3 during the period 1911–1960 to 8.6 km^3 between 1961 and 1994, less than half of the decline from abstractions for irrigation in Ethiopia and Sudan and the rest from a natural decrease in river runoff (International Water Power and Dam Construction, 2009; Sutcliffe & Parks, 1999).

Eritrea does not produce hydropower at this time (all energy comes from biomass fuels and imported petroleum), but at least three potential dam sites were identified on the Tekeze-Setit and Anseba rivers (National Investment Brief, 2008). Additional potential utilization of the country's major rivers is expected for irrigation development to more effectively address the worsening food insecurity. In recent years, Eritrea depended on food imports and food aid to meet 40–50% of its food needs (Okubagzhi, 2008).

Egypt

No perennial rivers enter the main Nile downstream of the Atbara confluence, making the Nile and Lake Nasser practically the only sources of all surface freshwater in Egypt. The mean annual Nile volume at Wadi Halfa above Aswan Dam declined from 97.3 km^3 during the period 1890–1910 (prior to the construction of any dams on the Nile) to 86.1 km^3 between 1911 and 1960, and to 73.1 km^3 during the 1961–1995 period. These decreases have been attributed to increasing abstraction for irrigation and evaporation from channels and reservoirs (Sutcliffe & Parks, 1999). Most of these losses were compensated for by the construction of Aswan High Dam in the early 1960s, which increased the annual flow of the Nile by 22 km^3 (after deducting 10 km^3 for estimated evaporation from Lake Nasser) by storing the seasonal floods, which used to flow to the Mediterranean. The storage not only eliminated the old flood problem in Egypt but also permitted the rapid extension of the perennially irrigated area and the production of hydropower for domestic and industrial purposes (Amer et al., 2005; Sutcliffe & Parks, 1999) and significantly increased Egypt's water security overall (Strzepek, Yohe, Tol, & Rosegrant, 2006). This effect was clearly revealed during the 1984 famine in Ethiopia, when annual Nile runoff declined from a mean of 84 km^3 or billion m^3 to 42 km^3, but Egypt continued to use its 55.5 km^3 allotment under the 1959 agreement due to between-year storage, made possible by Aswan Dam

(Chesworth, 1994). No other high dam has been constructed in Egypt since Aswan High Dam, but a new barrage was completed in 2008 at Naga Hammadi to provide for year-round irrigation for more than two million people and to drive a 64 MW power plant (Stanford, 2008).

In addition to the 55.5 km^3 guaranteed annually under the 1959 Nile Water Agreement, Egypt receives additional inflow due to incomplete withdrawal upstream, a surplus which will eventually be eliminated with continued water abstractions, especially in Ethiopia and Sudan. Egypt also utilizes 2.8 km^3 from renewable groundwater resources in the Nile Valley and Delta, the Nubian Sandstone Aquifer, renewable surface water resources within Egypt, and desalinized seawater, providing an annual total of more than 60 km^3 before recycling. Because of extensive recycling mandated by Egypt's government, an additional 12.8 km^3 became available for use in 2000. Of the water withdrawn in 2000, 86% was used for agriculture, 8% for domestic needs, and 4% for industry. Of the irrigation water, 83% was surface water, 11% groundwater, and 6% was from mixed sources, including surface water, water harvesting, and fossil water from the Nubian Sandstone Aquifer (Kundell, 2008b).

Development of the Nubian Sandstone Aquifer, the world's largest fossil water aquifer, and desalinized seawater can meet some of Egypt's future water needs. The Nubian Sandstone Aquifer has been identified as the largest source of water to meet growing demands and development goals in Egypt, Sudan, Libya, and Chad. Libya has most aggressively pursued plans to exploit the aquifer for the irrigation of 160,000 ha of coastal land in its quest for self-sufficiency in food production. However, this system is poorly understood and although Libyan officials estimate that this groundwater source will last 50 years under projected withdrawal, overabstraction has already been reported, potentially making its equitable management across international boundaries difficult (International Atomic Energy Agency, 2003–2006; Killgore, 2001). Desalinized sea water is increasingly being tapped along the Mediterranean and Red Sea coasts for use in resorts, hotels, and sea-side towns in Egypt, Sudan, and other North African and Near Eastern countries. Although total annual production in Egypt was only about 100,000 m^3

in 2002 (Kundell, 2008b), the recent development of solar-powered desalinization systems that are expected to become more efficient and more cost effective than fossil fuel powered plants favor the adoption of sea water desalinization technologies (Lamei, Van der Zaag, & Munch, 2008; Fath et al., 2008).

The establishment of robust contingency planning and early warning systems allowed Egypt to assess its water uses and put in place plans to respond to the low Nile flows in the 1980s and 1990s. This included the pursuit of opportunistic policies to expand irrigation, particularly through new land reclamation, although this put the country at greater risk of climate change and reduced Nile flow (Conway, 2005). As part of a new policy of moving people out of the old lands of the Nile Valley toward new reclaimed lands in the desert, Egypt started to reclaim lands for irrigation in 1952 and is currently developing nearly 1 million *feddans* (420,000 ha) in the Sinai and an even larger area in the southwestern desert, the New Valley/Toshka Development Project. The immense Toshka Project, dubbed "Mubarak's Pyramid" after President Mubarek, who implemented it in 1997, was initially supplied with excess water from an unusually large Nile runoff stored in Lake Nasser and delivered via huge pumping stations and a spillway (Abdelsalam, Youseff, Arafat, & Alfarhan, 2008; United Nationals Environmental Programme, 2008). This project is increasingly using groundwater from the Nubian Sandstone Aquifer and plans to withdraw 5 km^3 per year from Lake Nasser to eventually irrigate 1.5 million acres. Several riparians, in particular Ethiopia, heavily criticized this and the other unilateral reclamation projects as hegemonic ambitions for a Nile water control project (Cascão, 2009). Other critics consider it as a drain on scarce water and other resources away from the Nile Valley (Nubian Sandstone Aquifer System, 2009; United Nations Development Program, 2007).

Water Conservation in the Nile Basin

Although all Nile riparian countries are increasingly formulating policies that include water conservation strategies, a trend seen most clearly in Egypt, whose dependence on the upper riparians for nearly all of its water needs makes water conservation particularly urgent, a brief review of current practices indicates that much more needs to be done to increase water use efficiency in the increasingly water-scarce Nile Basin.

Egypt's Leading Role

Egypt has enacted water policies designed to increase water use efficiency employing an integrated approach that seeks to meet agricultural, domestic, industrial, navigation, and ecosystem needs. With limited prospects for substantially increasing supplies from the Nile, management policies have been focused on both the supply and demand sides to optimize use of available surface and groundwater resources, minimize water losses, increase irrigation efficiency, improve cost sharing, change cropping patterns, and reuse agricultural drainage water and treated wastewater (Ministry of Water Resources and Irrigation, 2005). Recycling and reusing drainage and wastewater added 10 m^3 to the renewable water resource base of 58.3 m^3 in 2000 (Kundell, 2008b). Nevertheless, the great water consumption and conveyance losses in irrigation systems (estimated to have been around 76% of all losses in the 1990s), have not been reduced significantly. Water management initiatives aimed at sharply decreasing these losses will further reduce the Nile's discharge to the Mediterranean Sea, with significant environmental and ecological repercussions in the Nile Delta and the sea (Appelgren et al., 2000).

The major water conservation measures that have been implemented include the installation of an extensive underground drainage system in Lower Egypt, which was started in the 1970s, expanded throughout Egypt, and covered more than 65% of the total cultivated area by 2003, together with lining canals in sandy areas and other conservation measures. These measures have slowed the rise in the water table and salinization levels and provided additional water for irrigation. Changes in the management of the subsurface drainage system are estimated to increase water use efficiency by 15–20% without crop reduction (Wahba, Christen, & Amer, 2005). The government recently banned surface irrigation in all newly reclaimed areas with their predominantly sandy soils characterized by low water holding capacity, allowing only the sprinkler and drip methods. Increased irrigation efficiency of around 30% has been reported for sprinkler irrigation and by about 50% for drip irrigation from various countries (Brown, 2008; Arafa, Wasif, & Mehawed, 2008). Egypt uses the traditional surface irrigation method on 88.5% of the total irrigated area and localized irrigation (mostly drip irrigation) and sprinkler irrigation on 11.5% of the cultivated area (Kundell, 2008b), indicating that water use can be further reduced.

Numerous studies have shown that additional water savings are possible with more efficient water application methods, including deficit irrigation and surge flow rather than continuous flow, that reduce the amounts of water applied to crops without reducing yields (Ouda et al., 2008; Ismail, Depeweg, & Schults, 2004), improved subsoil irrigation methods (Barth, 1999), cultivation of more water-efficient crops, water user fees, participatory irrigation management, revised government crop subsidies, water pricing structures aimed at conservation, and water awareness campaigns (Batt & Merkley, 2009; Kundell, 2008b). The new West Delta Project incorporates these and other approaches necessary to reform the irrigation sector and is generating interest for possible use in other African nations, including Ethiopia (Baetti & Abdel-Dayem, 2008). But it will take time to overcome deeply entrenched practices caused by subsidized water and the cultural attitude that water should be free, which encouraged the cultivation of high-water demanding crops, overirrigation, and low cost recovery that resulted in the neglect of irrigation works and low irrigation efficiency (Malashkhia, 2003).

Imports of food products have contributed to Egypt's water and food security through virtual water since the 1960s. In 1994, the water equivalent of all Egyptian imports was 18.2 km^3, the largest of any country in the Middle East (Allan, 1999; Yang & Zehnder, 2008). Imports of wheat, the main staple in Egypt, vary between 6.0–7.5 million tons or an equivalent 6–7.5 km^3 of virtual water per annum (U.S. Department of Agriculture, 2006; Allan, 1999). Egypt is also a major importer of sugar (U.S. Department of Commerce, n.d.). Imported industrial goods water accounts for only about 5% of Egypt's virtual water savings (Hoekstra &

Chapagain, 2008). Egypt's net balance in virtual water trade was somewhat reduced by the export of around 1.4 million tons of rice annually in recent years, but a government ban on rice export in response to high prices on the world market in 2008 sharply reduced exports (Reuters, 2009). Although net water savings are substantial, resorting to virtual water (and massive food subsidies) primarily to stave off the food crisis and social unrest delayed major reforms needed to address underlying structural problems such as transforming the economy toward industrialization and more cost-effective use of water. The 20–70 times higher economic returns possible in industry than agriculture from one unit of water would argue for a shift from an emphasis on food self-sufficiency to food security based on the importation of food Egypt no longer has the resources to grow (Allen, 1999; Ohlsson, 2000). The total amount of virtual water per capita saved by Egypt by importing agricultural goods between 1997 and 2001 (197 m^3 per year) was nearly twice as high as in the five other largest Nile riparian countries combined (Chapagain & Hoekstra, 2004). Trade in virtual water as a policy option and as an alternative water source requires further research on socioeconomic and environmental ramifications (Allan, 1999; National Water Research Council, 2007).

Sudan's Conservation Challenges

The strengthening of the irrigation program will permit the revival of about 1 million km^3 of the total area of 1.86 million ha equipped for irrigation in 2000. In that year, only 801,000 ha (43%) were actually irrigated, and 500,000 ha were salinized by irrigation. In the Gezira, the oldest and largest irrigation scheme, both productivity and farm incomes sharply fell in the late 1990s due to water mismanagement, siltation, water hyacinth infestation, inadequate application of water, and wastage of water (Kundell, 2008a). The proposed construction of dams on the Blue Nile in Ethiopia and an erosion control program in Ethiopia and Sudan within the framework of the Nile Basin Initiative (Hamad & El-Battahani, 2005) may contribute to reducing the problem of water scarcity caused by siltation and aquatic weeds. However, the numerous technical, institutional, managerial, and political problems interfering with conservative water use and effective management in Sudan will have to be addressed to improve water use efficiency. Alternative approaches to achieving water and food security, particularly the improvement of rain fed farming in the south and full and supplemental irrigation, have not received much attention by planners (Hamad & El-Battahani, 2005) in the aftermath of the civil war.

The Upper Nile Countries: Going the Full Cycle

Water conservation in the upper riparian countries is becoming increasingly urgent to meet both their own rapidly rising demands and those of Sudan and Egypt. Although water availability, use, and management in the eight upper riparian countries vary considerably due to major environmental and socioeconomic differences, they share a number of characteristics relevant to the theme of this book. Most importantly, the populations of these countries are not, unlike those of Egypt and much of Sudan, settled along fertile, irrigated flood plains of the Nile and its tributaries but are dispersed in mountainous areas with limited access to hydraulic infrastructure.

Many of them developed over the centuries agricultural systems that included effective soil and water conservation methods that were disregarded by external development policies and practices during the last century. Most of these coercive programs were resisted by local populations and, thus, failed. New initiatives recognize that farmers are the solution rather than the problem in land degradation and water scarcity and that indigenous systems can be made productive and sustainable in the long term if farmers are viewed as the solution rather than the problem (Critchley & Graham, 1991; Pretty & Shah, 1998). Methods that not only conserve moisture and soil but also maintain soil fertility, are socially acceptable, and economically viable are readily accepted by farmers, a truism that spelled doom for the conservation programs of the colonial powers in East Africa (Ellis-Jones & Tengberg, 2000; Pretty & Shah, 1998) and the top-down, coercive approach employed by soil and water conservation and other programs of the socialist government in Ethiopia, as we will elaborate in this book. There is now an urgent need for government to work closely with farmers in a support rather than control capacity to improve and develop soil and water conservation techniques to ameliorate impending water, food, and environmental crises (Ellis-Jones & Tengberg, 2000).

All upper riparian countries have developed or are in the process of developing water policies and programs aimed at increasing water conservation and use efficiency. Recent efforts include the decentralization of water development and service provisions, the transfer of water management from government agencies to water user associations and other community-based organizations and private firms, expansion of rainwater harvesting, particularly in Tanzania and Kenya, and increased use of sprinkler irrigation (Matlock, 2008a, 2008b; Kundell, 2008c). Land reform, on the other hand, has not moved together with conservation programs, slowing down the adoption and sustainability of improved land management technologies and water conservation programs (Foundation for Environmental Security and Sustainability, 2006; Kimaru & Jama, 2006). One exception is Machakos District in Kenya, an area with secure tenure, where tree cover and soil conservation increased even though the population increased five fold (Tiffen, Mortimore, & Gichuki, 1994). Plans for electric power development in the Great Lakes Region are increasingly considering the development of alternative energy sources in an effort to reduce dependence on hydropower, conserve biofuels and forests, and thus ensure environmental security. This includes geothermal and wind energy (Kenya), natural gas (Lake Kivu area), and coal (Tanzania). These changes, although contributing to maintaining current flow patterns of the Nile and the integrity of the basin's aquatic system, have been adopted at a slow pace (World Bank, 2005), largely the result of lack of political will and research conducive to their development (Mariita, 2007; Otieno & Awange, 2006). Full participation of local populations as active partners in all water sector programs and initiatives promises to ensure water, energy, food, and environmental security in the long term.

References

Abdelsalam, M. G., Youseff, A. M., Arafat, A. M., & Alfarhan, M. (2008). Rise and demise of the New Lakes of the Sahara. *Geosphere, 4*, 375–386.

Allan, J. A. (1999). A convenient solution. *The UNESCO Courier*, February 1999, 29–32.

Amer, S. E.-D., Arsano, Y., El-Battahani, A., Hamad, O. E.-T., Hefny, M. A. E.-M., Tamrat, I., & Mason, S. (2005). Sustainable development and international cooperation in the Eastern Nile Basin. *Aquatic Sciences, 67*, 3–14.

Andjelic, M. M. (1999). *Water resources and flood monitoring in the Nile River Basin.* Retrieved from http://www.fao.org/docrep/004/ac146e/AC146E04.htm#chIII.7

Appelgren, B., Klohn, W., & Alam, U. (2000). *Water and agriculture in the Nile Basin.* Background Paper No. AGL/MISC/29/2000. Rome: Food and Agriculture Organization of the United Nations, Land and Water Development Division.

Arafa, Y. E., Wasif, E. A., & Mehawed, H. E. (2008). Maximizing water use efficiency in wheat yields based on drip irrigation systems. *Australian Journal of Basic and Applied Sciences, 3*, 790–796.

Arsano, Y., & Tamrat, I. (2005). Ethiopia and the Eastern Nile Basin. *Aquatic Sciences, 67*, 15–27.

Awulachew, S. B., McCartney, M., Ibrahim, Y., & Shiferaw, Y. S. (2008). Evaluation of water availability and allocation in the Blue Nile Basin. In R. S. Bayout & E. Humphreys (Eds.), *Proceedings of the CGIAR Challenge Program on Water and Food, 2nd International Forum on Water and Food* (pp. 6–10). Addis Ababa: CGIAR.

Baetti, A., & Abdel-Dayem, S. (2008, May). A demand-driven design for irrigation in Egypt: Minimizing risks for both farmers and private investors. *Gridlines,* Note No. 38. Retrieved from www.ppiaf.org/documents/gridlines/38demand-driven-design-irrigation-Egypt.pdf

Barth, H. K. (1999). Sustainable and effective irrigation through new sub-soil irrigation system. *Agricultural Water Management, 40,* 282–290.

Batt, H. A., & Merkley, G. P. (2009). Water management and user association analysis for irrigation improvement in Egypt. *Irrigation and Drainage.* Advance online publication. doi: 10.1002/ird.465

Boukerrou, L. (2008). Burundi. In *The encyclopedia of Earth.* Washington, DC: Encyclopedia of Earth. Retrieved from http://www.eoearth.org/article/Burundi

Boukerrou, L. (2009). Rwanda. In *The encyclopedia of Earth.* Washington, DC: Encyclopedia of Earth. Retrieved from http://www.eoearth.org/article/Rwanda

Brown, L. R. (2008). *Plan B 3.0: Mobilizing to save civilization.* New York: W. W. Norton.

Cascão, A. E. (2009). Changing power relations in the Nile River Basin: Unilateralism vs. cooperation? *Water Alternatives, 2,* 245–268.

Chesworth, P. M. (1994). The history of water use in the Sudan and Egypt. In P. P. Howell & J. A. Allan (Eds.), *The Nile: Sharing a scarce resource* (pp. 65–79). Cambridge: Cambridge University Press.

Collins, R. O. (2002). *The Nile.* New Haven, CT: Yale University Press.

Conway, D. (2000). The climate and hydrology of the upper Blue Nile River. *Geographical Journal, 166,* 46–62.

Conway, D. (2005). From headwater tributaries to international river: Observing and adapting to climate variability and change in the Nile basin. *Global and Environmental Change, Part A, 15,* 99–114.

Critchley, W., & Graham, O. (1991). *Looking after our land: Soil and water conservation in dryland Africa.* Oxford: Oxfam.

Dixon, A. B., & Wood, A. P. (2003). Wetland cultivation and hydrological management in eastern Africa: Matching community and hydrological needs through sustainable use. *Natural Resources Forum, 27*, 117–129.

Eguavoen, I. (2009). *The acquisition of water storage facilities in the Abay River Basin, Ethiopia* (ZEF Working Paper No. 38). Bonn, Germany: Bonn University, Center for Development Research.

Ellis-Jones, J., & Tengberg, A. (2000). The impact of indigenous soil and water conservation practices on soil productivity: Examples from Kenya, Tanzania and Uganda. *Land Degradation and Development, 11*, 19–36.

Falkenmark, M., & Widstrand, C. (1992). Population and water resources: A delicate balance. *Population Bulletin, 47*, 1–36.

Fath, H. E. S., Elsherbiny, S. M., Hassan, A. A., Rommel, M., Wieghaus, M., Koschikowski, J., & Vatansever, M. (2008). PV and thermally driven small-scale, stand-alone solar desalinization systems with very low maintenance needs. *Environmental Sciences and Pollution Management, 225*, 58–69.

Food and Agriculture Organization. (2005). *Population prospects in the Nile Basin. Information products for Nile Basin water resources management.* Retrieved from www.faonile.org/PopulationProspects.pdf

Foundation for Environmental Security and Sustainability. (2006). *Uganda's fading luster: Environmental security in the pearl of Africa.* Retrieved from http://www.fess_global.org/files/uganda_esaf_exec_summary.pdf

Grijsen, J. (2007, March 1). Potential impacts of hydrologic uncertainty and climate change on regional power options in the Lake Victoria Basin. PowerPoint presentation in *Water Week*, World Bank. Retrieved from www.worldbank.org/INTWRD/Resources/Johan_Grijsen_Hydrologic_Uncertainty.pdf

Hamad, O. E.-T., & El-Battahani, A. (2005). Sudan and the Nile Basin. *Aquatic Sciences, 67*, 28–41.

Hefny, M., & Amer, S.E.-D. (2005). Egypt and the Nile Basin. *Aquatic Sciences, 67*, 42–50.

Hoekstra, A. Y., & Chapagain, A. K. (2008). *Globalization of water: Sharing the planet's freshwater resources*. Oxford: Blackwell.

Howell, P. P. (1994). East Africa's water requirements: The Equatorial Nile Project and the Nile Waters Agreement of 1929. A brief historical review. In P. P. Howell & J. A. Allan (Eds.), *The Nile: Sharing a scarce resource* (pp. 81–107). Cambridge: Cambridge University Press.

International Atomic Energy Agency. (n.d.). *More people, more development: Pressures in Sudan*. Retrieved from http://www.-naweb.iaea.org/napc/ih/Nubian/IHS_nubian_more_people_sudan.html

International Atomic Energy Agency. (2003–2006). *Nubian Aquifer Project: About the Nubian Project.* Retrieved from http://www.iaea.org/nape/ih/Nubian/IHS_nubian.html

International Water Power and Dam Construction. (2008, October 3). Will integration spur dam construction? *Water Power Magazine*. Retrieved from http://www.waterpowermagazine.com/story.asp?storyCode=2051106

International Water Power and Dam Construction. (2009, May 11). Hydro power in Ethiopia—the staged construction of Tekeze Arch Dam. *Water Power Magazine*. Retrieved from http://www.waterpowermagazine.com/story.asp?sectioncode=46&storyCode=2052913

Ismail, B. M., Depeweg, H., & Schultz, B. (2004). Surge flow irrigation under short field conditions in Egypt. *Irrigation and Drainage, 53*, 461–475.

Kagwanja, P. (2007). Calming the waters: The East African community and conflict over the Nile resources. *Journal of Eastern African Studies, 1*, 321–337.

Killgore, A. I. (2001, March). Libya's great man-made river project. *Washington report on Middle East affairs*. Retrieved from http://www.wrmea.com/archives/march01/0103056.html

Kimaru, G., & Jama, B. (2006). *Improving land management in eastern and southern Africa: A review of practices and policies.* ICRAF Working Paper No. 18. Nairobi: World Agroforestry Centre.

Kolding, J., van Zwiete, P., Mkumbo, O., Silsbe, G., & Hecky, R. (in press). Are the Lake Victoria fisheries threatened by exploitation or eutrophication? Towards an ecosystem based approach to management. In G. Bianchi & H. R. Skjoldal (Eds.), *The ecosystem approach to fisheries*. Nairobi: CABI.

Kull, D. (2006). Connections between recent water level drops in Lake Victoria, dam operations and drought (Unpublished manuscript). Retrieved from internationalrivers.org/files/060208vic.pdf

Kundell, J. (2008a). Water profile of Sudan. In *The encyclopedia of Earth*. Washington, DC: *Encyclopedia of Earth*. Retrieved from http://www.eoearth.org/article/Water_profile_of_Sudan

Kundell, J. (2008b). Water profile of Egypt. In *The encyclopedia of Earth*. Washington, DC: Encyclopedia of Earth. Retrieved from http://www.eoearth.org/article/Water_profile_of_Egypt

Kundell, J. (2008c). Water profile of Kenya. In *The encyclopedia of Earth*. Washington, DC: Encyclopedia of Earth. Retrieved from http://www.eoearth.org/article/Water_profile_of_Kenya

Lamei, A., van der Zaag, P., & Munch, E. (2008). Impact of solar energy cost on water production cost of seawater desalinization plants in Egypt. *Environmental Sciences and Pollution Management, 36*, 1748–1756.

Malashkhia, N. (2003). Social and environmental constraints to the irrigation water conservation measures in Fayoum, Egypt (Unpublished master's thesis). Lund University, Lund, Sweden.

Mallat, C. (1994). Law and the Nile River: Emerging international rules and the Shari'a. In P. P. Howell & J. A. Allan (Eds.), *The Nile: Sharing a scarce resource* (pp. 365–384). Cambridge: Cambridge University Press.

Mariita, N. (2007). Hurdles to financing geothermal development in ARGeo countries with special focus on Kenya. *Transactions-Geothermal Resources Council, 31*, 57–62.

Matlock, M. (2008a). Water profile of Tanzania. In *The encyclopedia of Earth*. Washington, DC: Encyclopedia of Earth. Retrieved from http://www.eoearth.org/article/Water_profile_of_Tanzania

Matlock, M. (2008b). Water profile of Uganda. In *The encyclopedia of Earth*. Washington, DC: Encyclopedia of Earth. Retrieved from http//www.eoearth.org/article/Water_profile of Uganda

Ministry of Water Resources and Irrigation. (2005). *National water resources plan for Egypt—2017*. Cairo: Author.

Mohamed, Y. A., Savenije, H. H. G., Bastiaanssen, W. G. M., & Van den Hurk, J. J. M. (2006). New lessons on the Sudd hydrology learned from remote sensing and climate modeling. *Hydrological and Earth System Sciences, 10*, 507–518.

Musinguzi, J. (2009, June 2). Egypt differs with Nile basin colleagues. *East African Business Week*. Retrieved from http://www.busiweek.com/index.php?option=com_content&task=view&id=1604&Itemid=2

Musoni, E. (2009, February 23). Africa: Rift widens as Egypt, Sudan delay signing Nile Basin pact. *The New Times*. Retrieved from http//:allafrica.com/stories/200902230029.html

The National Association of Professional Environmentalists in Uganda. (2009, January 29). Who is to blame for falling Lake Victoria water levels? *News*. Retrieved from htpp://www.nape.or.ug/

National Investment Brief: Eritrea. (2008, December). Paper presented at the Ministerial Conference on Water for Agriculture and Energy in Africa: The Challenges of Climate Change, Sirte, Libya. Retrieved from www.sirtewaterandenergy.org/does/reports/Eritrea-Draft2.pdf

National Water Research Council. (2007). *Virtual water in the Arab world. CSD WAND web-based case study database*. Retrieved from http://www.csdwand.net/data/sheet.asp?cn=Egypt&fn=LA1505

Nile Basin Initiative. (2009). *NBI background.* Retrieved from http://www.nilebasin.org/index.php?option=com_content&task=view&id=13&ltemid=42

Nubian Sandstone Aquifer System. (2009). In *Wikipedia, the free encyclopedia.* Retrieved from http:en.wikipedia.org/wiki/Nubian_Sandstone_Aquifer_System

Ohlsson, L. (1998). *Water conflicts and social resource scarcity* (Unpublished manuscript). An issue paper commissioned by FAO AGLW, Department of Peace and Development Research, University of Göteborg.

Ohlsson, L. (2000). Water conflicts and social resource scarcity. *Physics and Chemistry of the Earth, 25*, 213–220.

Okubagzhi, Y. (2008). *Water resources and inter-riparian relations in the Nile Basin: The search for an integrated discourse*. Albany: State University of New York Press.

Otieno, H. O., & Awange, J. L. (2006). *Energy resources in East Africa: Opportunities and challenges.* New York and Heidelberg: Springer.

Ouda, S. A. E.-F., Khalil, F. A., Elenin, A., Shreif, M. K., Beni, B., & Qadir, M. (2008). Using yield-stress model in irrigation management for wheat grown in Egypt. *Journal of Applied Biological Sciences, 2*, 57–65.

Population Reference Bureau. (2005). *2005 world population data sheet.* Washington, DC: Author.

Pretty, J. N., & Shah, P. (1998). Making soil and water conservation sustainable: From coercion and control to partnerships and participation. *Land Degradation and Development, 8*, 39–58.

Reuters. (2009, February 2). Egypt extends ban on rice exports beyond April. *FlexNews*. Retrieved from http://www.flex-news-food.com/pages21871/Egypt/rice/egypt-extends-ban-rice-exports-beyond-april.html

Stanford, D. (2008, May 23). Nile barrage sets standard for community relations. *Daily News Egypt.* Retrieved from http://www.dailystaregypt.com/article.aspx?ArticleID=13923

Strzepek, K. M., Yohe, G. W., Tol, R. S. J., & Rosegrant, M. (2006). The value of the high Aswan Dam to the Egyptian economy. *Ecological Economics*, *66*, 117–126.

Sutcliffe, J. V., & Parks, Y. P. (1999). *The hydrology of the Nile*. Wallingford, U.K.: International Association of Hydrological Sciences.

Sutcliffe, J. V., & Petersen, G. (2007). Lake Victoria: Derivation of a corrected natural water level series/Lac Victoria: Derivation d'une série naturelle corrigée des niveaux d'eau. *Hydrological Sciences Journal, 52*, 1316–1321.

Tiffen, M., Mortimore, M., & Gichuki, F. (1994). *More people, less erosion: Environmental recovery in Kenya.* Nairobi: Acts Press and New York: Wiley.

Tved, T. (2004). *The Nile: An annotated bibliography*. 2nd ed. London: I. B. Tauris.

United Nations Development Program. (2007). *Human Development Report 2007/2008.* New York: Author.

United Nations Educational and Scientific Organization. (2009). The 3rd United Nations world water development report: Water in a changing world (WWDR-3). Retrieved from http://www.unesco.org/water/wwap/wwdr/wwdr3/tableofcontents.shtml.

United Nations Environmental Programme. (2007). *Sudan post-conflict environmental assessment.* Nairobi: Author.

United Nations Environmental Programme. (2008). *Africa's lakes: Atlas of a changing environment*. Nairobi: Author.

U.S. Department of Agriculture. (2006). *Egypt grain and feed.* Annual 2006 Report No. EG6007. Retrieved from www.fas.usda.gov/gainfiles/200603/146187048.pdf

U.S. Department of Commerce. (n.d.). *U.S. trade: Refined sugar imports by country of destination.* Retrieved from www.fas.usda.gov/htp2/sugar/1997/97-11refexp97.pdf

Wahba, M. A. S., Christen, E. W., & Amer, M. H. (2005). Irrigation water saving by management of existing subsurface drainage in Egypt. *Irrigation and Drainage, 54*, 205–215.

Whittington, D., Wu, X., & Sadoff, C. (2007). Water resources management in the Nile basin: The economic value of cooperation. *Water Policy, 7*, 227–252.

World Bank. (2005). *Nile Basin Initiative, Nile Equatorial Lakes Subsidiary Program: Strategic/sectoral, social and environmental assessment of power development options in the Nile equatorial lakes region. Stage II. Synopsis and final report.* Washington, DC: Author.

World Bank. (2009a). *ENSAP investment projects*. Retrieved from http://qo.worldbank.org/69)LX5AB00.

World Bank. (2009b). *Nile Equatorial Lakes subsidiary action program.* Retrieved from http://qo.worldbank.org/EIZSZONSNO.

Yang, H., & Zehnder, A. J. B. (2008). *Globalization of water resources through virtual water trade (*Unpublished manuscript). Swiss Federal Institute for Aquatic Science and Technology, Duebendorf.

Young, S. (2004). Using climate to predict Nile flow. Unpublished manuscript. Retrieved August 5, 2009, from web.mit.edu/eltahir/www2/Makana/papers/research_climate.pdf

Zhang, W. (2007). *Kenya's hydrological and environmental challenges in Lake Victoria Basin* (Unpublished manuscript). Cornell University.

FURTHER READING

Dumind, H. J. (Ed.). (2009). *The Nile: Origin, environments, limnology and human use*. Dortrect: Springer Netherlands.

Chapter 3

The Water Resources of Ethiopia and Large-Scale Hydropower and Irrigation Development

Semu Moges, Helmut Kloos, Stuart McFeeters, and Worku Legesse

Introduction

Often referred to as the "Water Tower of Africa," Ethiopia is the country with the second largest hydropower potential in Africa. It would appear, therefore, that a resource base exists to help lift the country out of the poverty and low development level in which it currently finds itself. Further, while the country's irrigation potential does not match its hydropower potential, there exists substantial scope for augmenting the irrigated area through increased water storage. Despite Ethiopia's abundant water resources, it ranked 32nd among 48 African countries in regard to total renewable water resources per capita in 2005 (United Nations Educational, Scientific, and Cultural Organization, 2006). This discrepancy is mainly due to its large population, 73.9 million in 2007, which is twice the size of what it was in 1984 (Federal Democratic Republic of Ethiopia, 2008). Compounding matters is the fact that less than 5% of Ethiopia's abundant water resources are utilized in the country. The large population and low rates of utilization, therefore, explain many of the dire realities in Ethiopia rather than the country's natural water endowment. As explained in this book, however, availability and effective utilization of water resources is no simple endeavor, as interventions to achieve one objective can easily be undermined as a result of socioeconomic, environmental, and health externalities. This chapter examines the water resources conditions and potential in Ethiopia, with a focus on large-scale hydroelectric and irrigation development.

Topography and Climate

Prominent features of the Ethiopian topography are its extensive highlands with their undulating plateaus and deep river gorges, and the Great Rift Valley which divides the tectonically uplifted basaltic dome into the central/western and southern/eastern highlands. Altitudes range from 110 m below sea level in the northern part of the Ethiopian Rift Valley to more than 4,600 m in the Semien Mountains northeast of Lake Tana (figure 3.1). Seasonally high rainfall in the highlands above 1,500 m, where nearly 90% of the population lives on less than half the country's area, renders the mountains particularly vulnerable to erosion and degradation.

Despite Ethiopia's proximity to the equator, the highlands enjoy a temperate climate due to the moderating influence of high altitude, with a mean annual temperature rarely exceeding 20°C. The more sparsely populated lowlands have subtropical and tropical climates. Rainfall is most abundant in the western highlands (1,000–2,400 mm), with amounts increasing with altitude. The southern, eastern, and northern

FIGURE 3.1. The topography of Ethiopia.

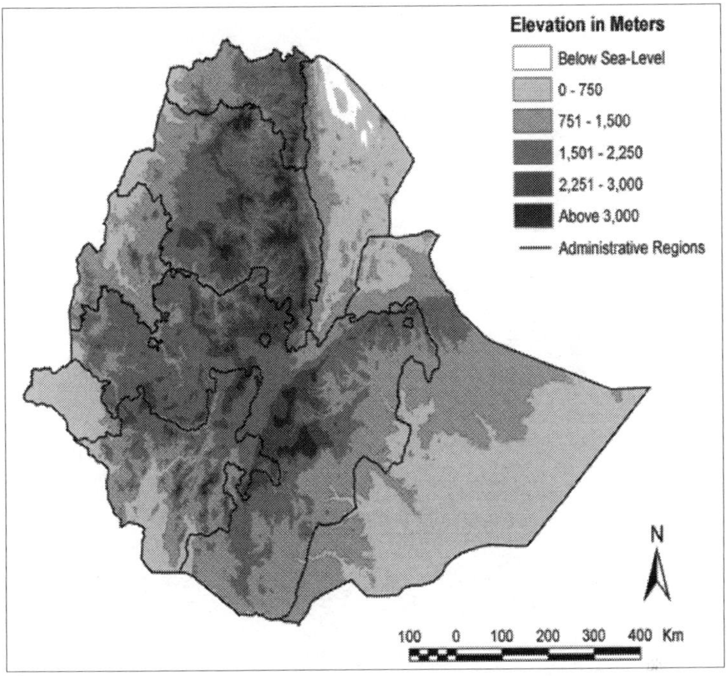

Source. Food and Agriculture Organization (n.d.).

highlands receive less precipitation (600–1,800 mm). The southern and eastern lowlands below about 1,000 m and those in the north below about 1,500 m receive between 200 and 600 mm (figures 3.1 & 3.2). Seasonal rainfall extends over much of the year in the western highlands, with a peak from around May to October, and is bimodal in most of the southern, central, and eastern parts of the country, with the small rains (*belg*) falling from February–March to May and the big rains (*keremt)* from June to September. These rainfall patterns are punctuated by periodic droughts, which have been traced back to the 9th century on the basis of written records (Pankhurst, 1984; Wood, 1977).

Droughts in Ethiopia, occurring about every 3 to 8 years in the northern part of the country and every 8 to 10 years in the more humid regions, are often associated with the El Niño-Southern Oscillation (ENSO), a coupled air and ocean phenomenon with global weather implications (Wolde-Giorgis, 2008). The relationship between climate change and water stress is poorly understood due to the lack of adequate data. There is considerable evidence, however, that rainfall has declined and become more erratic during the last four decades, characterized by longer dry spells and more flooding in both the

FIGURE 3.2. Mean annual rainfall in Ethiopia.

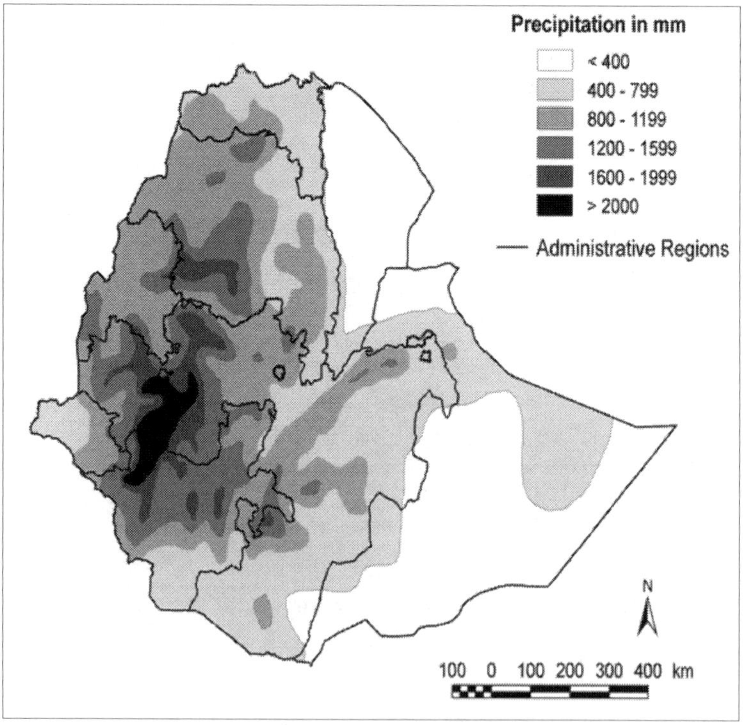

Source. Based on National Meteorological Service Authority (1996).

drought-prone southern, eastern, and northern parts of Ethiopia and the more humid southwestern parts (IGAD & ICPAC, 2007; Sileshi & Camberlin, 2006; Federal Democratic Republic of Ethiopia, 2001; Osman & Sauerborn, 2002). Similarly, both minimum and maximum temperatures increased at more than 40 stations throughout the country between 1952 and 1997 (IGAD & ICPAC, 2007). While the relationship between climate change and water stress is inadequately understood in Ethiopia due to lack of adequate data, one hydrological model predicted that a 10% reduction in rainfall would produce a 30% reduction in catchment runoff and a 1.5° C increase in air temperature a 15% reduction in simulated discharge (Legesse, Vallet-Coulomb, & Gasse, 2003).

Government, Socioeconomic, and Demographic Characteristics

Ethiopia is a federal republic under the 1994 constitution. The Ethiopian People's Revolutionary Democratic Front (EPDRF), the largest party, headed by Prime Minister Meles Zenawi, has promoted a policy of ethnic federalism, devolving significant powers to regional, ethnically based authorities. Ethiopia has a five-tiered administrative system, consisting of the federal administration, nine semiautonomous administrative regions (*kililoch* in Amharic) and two special city administrations (Addis Ababa and Dire Dawa, figure 3.3), 71 zones, more than 500 *woredas* (districts), and over 20,000 *kebeles* (peasant associations in rural areas and neighborhood associations in towns). Although the government has pursued a decentralization drive since 1991, with stepped up efforts under the present administration since 2001, the goals of regional governance and local self-rule have been elusive. A combination of lack of experience in decentralized governance, weak revenue bases, and administrative capacity, as well as an entrenched political culture of top-down decision making and governance structure has been associated with continuing federal interventions. Top-down decision making is reflected by the dominance of the ruling party at all levels of governance (Gebre-Egziabher & Berhanu, 2007; World Bank, 2001).

FIGURE 3.3. The administrative regions of Ethiopia.

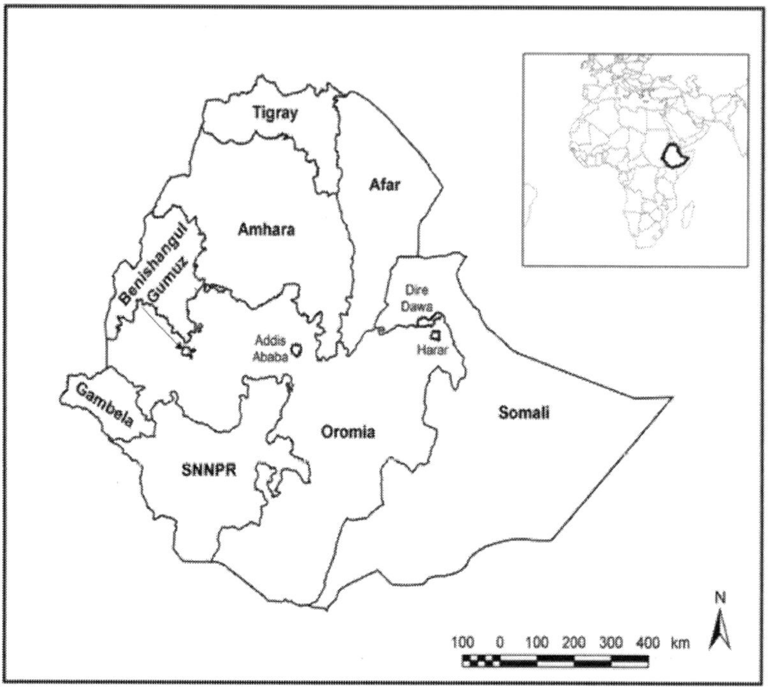

Source. Food and Agriculture Organization (n.d.).

Eighty-five percent of Ethiopia's population depends on subsistence agriculture and nearly exclusively on rain fed crop and livestock production. Low and highly variable rainfall, severe land degradation, and lack of irrigation are often identified as the major causes of food insecurity, although land tenure issues and lack of access to markets and technology are similarly important.

Even under normal rainfall conditions, more than 5 million Ethiopians are food insecure every year. According to the Food and Agriculture Organization and the World Food Program (2006), some 9.8 million people needed assistance in 2006, either through the government's

Productive Safety Net Program (PSNP) for the chronically food insecure or through humanitarian response mechanisms. Widespread poverty, unsustainable land use practices in the fragile Ethiopian highland ecosystem (Tefera, 2007), rapid population growth, low agricultural productivity, and lack of an industrial base contribute in multiple ways to the discrepancy between the large water resources and water scarcity.

The current government issued a number of policies and developed poverty reduction strategies to boost water and food security and meet the Millennium Development Goals (MDGs) (Federal Democratic Republic of Ethiopia, 1999, 2001, 2002; Ministry of Finance and Economic Development [MFED], 2006). In 2002, the government embarked on an ambitious 15-year (2002–2016) water sector development program, with the objective to achieve food security through irrigated agriculture, reduce flooding, and increase electric power production (Federal Democratic Republic of Ethiopia, 2002). All 11 administrative regions are developing water development master plans as part of the recently implemented decentralization drive. The eight rural regions were delineated by the current government on the basis of ethnic affiliation, introducing a cultural-political dimension to water resources development and management that is beyond the scope of this book.

Water Resources

Preliminary estimates reveal that Ethiopia's annual renewable water resources are 122 billion m^3, excluding groundwater resources (Halcrow, 1989). Ninety-seven percent of the surface water flows out of Ethiopia, greatly reducing the estimated 54.4 billion m^3 of surface runoff. An estimated 2.6 billion m^3 of groundwater may potentially be developed annually for various uses (United Nations Educational, Scientific, and Cultural Organization, 2006). About 94% of all utilized water in Ethiopia is used by agriculture, 6% for domestic and municipal needs, and 0.4% by industries (Food and Agriculture Organization, 2005).

Rivers

Ethiopia's resource potential has been identified and described in different master plans for integrated development of major river basins (BCEOM, 1999; Ministry of Water Resources, 1996, 1997, 1998). The country is subdivided into 12 river basins, of which 3 (Afar-Danakil, Ogaden, and Aysha) are dry basins with little flow of the drainage system (figure 3.4). Except for two internal basins (Rift Valley and Afar-Danakil), all basins are drained by rivers that flow across international boundaries or into lakes shared by Ethiopia with Kenya (Omo River into Lake Turkana) and Djibouti (Awash River into Lake Abe; figure 3.4). The four major rivers originating in the western highlands (Blue Nile, Baro-Akobo, Tekeze, and Omo-Gibe) account for about 83% of the total runoff from all catchments but their basins cover only 38% of the total area of the country. The Blue Nile River Basin generates nearly half of all surface water (52.6 billion m^3), followed by the Baro-Akobo and Omo-Gibe basins, largely reflecting the spatial distribution of rainfall. The main river basins in the central and eastern parts of the country (Awash, Wabe Shebele, and Genale-Dawa) occupy about 43% of the country's area, but their combined runoff constitutes only about 11% of the total runoff generated within the country (table 3.1; figure 3.4).

River flow throughout Ethiopia is highly seasonal, with far-reaching implications for water accessibility, the economy, dams, ecosystems, and flooding. The dry/wet season ratios of flow volume of the Blue Nile, Tekeze, Awash, Wabe Shebele, and Genale rivers vary from about 1:20 to 1:50, with peaks extending from June to October or November, after the end of the main rains. Even the Omo and other rivers originating in the most humid part of the southwestern highlands, although having relatively larger dry season flows and longer flood periods, are characterized by significant seasonal fluctuations. Interannual flows also fluctuate significantly with rainfall. During the 1972–1973 and 1984 droughts, for example, Blue Nile river flow was between half and two-thirds the volume during most other years in the 1970s and 1980s (Bekele, 2002; Conway, 2000).

TABLE 3.1. River basins, annual runoff, and specific discharge.

River basin	Catchment area (km²)	Annual runoff (BCM)*(%)	Specific discharge (L/second/km²)
Blue Nile	199,812	52.60 (43.1)	7.8
Baro-Akobo	74,100	23.60 (19.3)	9.7
Omo-Gibe	78,200	17.90 (14.7)	6.7
Tekeze	89,000	7.63 (6.0)	3.2
Genale-Dawa	171,050	5.80 (4.8)	1.2
Rift Valley	52,740	5.60 (4.6)	3.4
Awash	112,700	4.60 (3.8)	1.4
Wabe Shebele	200,214	3.15 (2.6)	0.5
Afar-Danakil	74,000	0.86 (0.7)	0.0
Mereb	5,700	0.26 (0.2)	3.2
Ogaden	77,100	0.00	0.0
Aysha	2,200	0.00	0.0
Total	1,136,816	122.00 (100)	

*Billion cubic meters.
Sources. Compiled from various river basin master plan studies and river basin surveys by the federal Ethiopian government.

Despite the continuous food and water insecurity, less than 5% of the potential irrigable land and 1% of the hydropower potential have been developed. Nevertheless, increasing utilization of streams, springs, and lakes in drought-prone areas for irrigation and domestic purposes and climatic change are already contributing to local water insecurity. Many springs dried up within the last few decades and in Hararge Highlands, lakes Haramaya and Adele disappeared in 2006 due to a combination of environmental degradation, siltation, and overpumping (Alemayehu, Furi, & Legesse, 2007).

Lakes

Nearly all of Ethiopia's lakes are located in internal basins, with important implications for their hydrology and survivability as pressure on water resources increases. The great majority of the more than 100 Ethiopian lakes are small (most of them crater lakes or shallow saline lakes in the lowlands) and relatively inaccessible or unsuitable for irrigation and domestic uses. With the exception of Lake Tana, the largest Ethiopian lakes are located in the Rift Valley (figure 3.4). Lake Tana has the lowest saline, conductivity, and fluoride concentrations of any of the large lakes in Ethiopia (Tekle-Haimanot et al., 2006).

FIGURE 3.4. River basins of Ethiopia and major rivers, lakes, and wetlands.

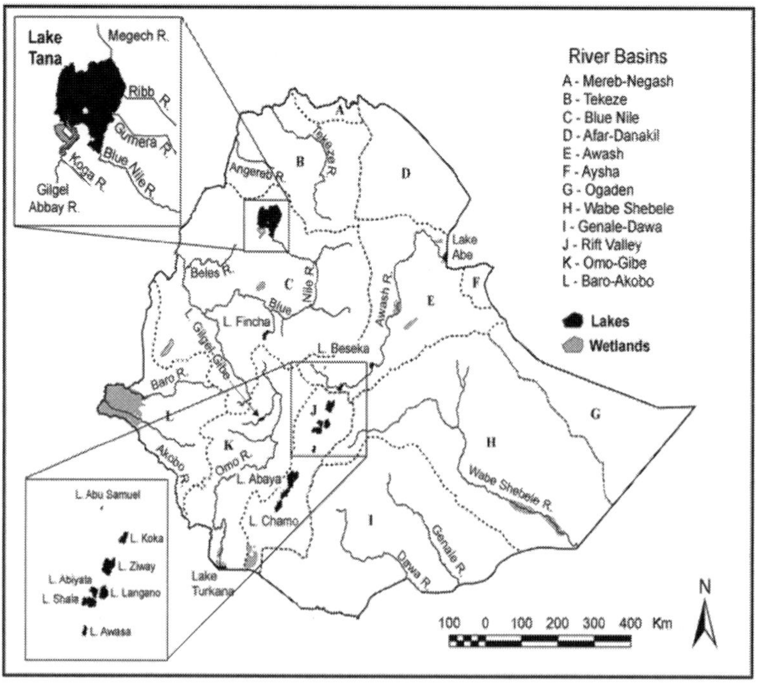

Source. The information on wetlands is based on Abebe (2003) and Hailu (2003).

Lake Tana

Lake Tana in the northwestern highlands, the largest lake in Ethiopia, is about 3,500 km^2, with an estimated storage volume of 28 km^3. The major tributaries, including the Gilgel-Abbay, Ribb, Gumera, and the Megech (figure 3.4), contribute more than 93% of the inflow into the lake (Kebede, Travi, Alemayehu, & Marca, 2006). The weir constructed from 1995 to 1997 regulates the water level and outflow of Lake Tana to increase generation of electricity through the Tis Abbay II hydropower plant on the Blue Nile. Although the use of Lake Tana was not, unlike other lakes, regulated by environmental laws until recently, the level of exploitation of its water resources, particularly for irrigation, was insignificant in the past. But Lake Tana and its tributaries have been selected for irrigation development and hydropower production. A notable development is the Tana-Beles growth pole, which is expected to stimulate integrated water resources development programs as part of the Nile Basin Initiative. The plan includes a basin water transfer scheme from Lake Tana to the Beles River for hydropower production and development of storage dams on the tributaries of the lake (Ministry of Water Resources, 2007). Although unofficial discussion with planners suggested that the Lake Tana ecosystem is not being undermined by the proposed water development projects, overbank flooding from the lake's tributaries draining unprotected watersheds is already threatening biodiversity through increased lake and wetland siltation and the socioeconomic wellbeing of the population (Tegegne, 2007).

Rift Valley Lakes

Of the 10 largest Rift Valley lakes (Turkana, Abaya, Chamo, Ziway, Shala, Abiyata, Abe, Langano, Awasa, and Beseka), four lakes (Shala, Abiyata, Abe, and Beseka) are salt lakes, four (Turkana, Langano, Awasa, and Chamo) have brackish water (between 1 and 2 g/L salt), and only lakes Ziway and Abaya have freshwater (Ayenew, 2007; International Lake Environmental Committee & United Nations Environmental Program, 2002). Fluoride levels show a similar pattern (Tekle-Haimanot et al., 2006).

Rising water levels in lakes Abaya, Chamo, and Awassa have been associated with increasing silt load deposited by their tributaries in the lake beds, a result of watershed degradation (Bekele & Horlacher, 2004; Gebreegzhiaber, 2005). Lake Beseka expanded more than 10-fold between the 1950s and 2005, largely due to a combination of increased groundwater inflow related to geotectonic activity and drainage water from nearby Abadir irrigation farm (Alemayehu, Ayenew, & Kebede, 2006; Ayenew, 2007; Belay, 2009). Increasing soda ash extraction from Lake Abiyata and ongoing large-scale irrigation development around Lake Ziway catchment are significantly reducing their volumes (Alemayehu et al., 2006; Ayenew & Becht, 2007). Lake Abe's water level also declined, by two thirds since the 1930s, due to large-scale irrigation development in the Awash Valley (Taddese, Sonder, & Peden, 2005). Its increasingly saline waters are not being used for irrigation and apparently also not being used for domestic purposes, but the lake provides habitat for local and migratory birds (see chapter 9).

Largely as a result of hydrological changes and abstractions, salinity levels have decreased in lakes Langano, Awasa, and Beseka, increased in lakes Abiyata, Chamo, and Ziway and remained largely unchanged in lakes Shala and Abaya between 1960 and 1990. Eutrophication and algal blooms caused by the heavy inflow of nutrient-rich silt originating in the adjacent degraded agricultural highlands are a growing threat to most Rift Valley lakes (Alemayehu et al., 2006; Awulachew, 2007; Institute of Biodiversity Conservation, n.d.; Zinabu, Kebede-Westhead, & Desta, 2004) and chemical and organic pollution from rapid urbanization threatens lakes Ziway, Awasa, and Abaya (Ayenew, 2007).

Reservoirs

Four large reservoirs have been created by Koka, Fincha, Melka Wakena, and Gilgel-Gibe I dams and at least five additional reservoirs (Tekeze in Tekeze Basin, Tendaho and Kessem in the Awash Basin, Koga near Lake Tana, and Gibe III in the Omo Basin) (figure 3.5) are expected to be filled by 2012, although delays are making this questionable. Their predicted longevity and socioeconomic and ecological impacts are discussed in chapters 8 and 9.

Wetlands

According to the wetland classification by Cowardin, Carter, Golet, and LaRoe (1979), which was adopted by the South African Wetlands Conservation Programme (South African Wetlands Conservation Programme, 1997), wetlands are “lands transitional between terrestrial and aquatic systems where the water table is usually at or near the surface or the land is covered by shallow water.” Various researchers and investigators have attempted to document the wetlands of Ethiopia but no comprehensive inventory has been made. Hillman and Abebe (cited in Abebe, 2003) identified 77 wetlands (including lakes) covering an area of about 13,700 km^2, or 1.14% of the country’s area. Wetlands are found in all agro-climatological and hydro-climatological regimes of the country. The 14 largest wetlands are shown in figure 3.4.

Wetlands provide many ecological and economic services, ranging from ‘water banks’ from where water may be drawn or groundwater replenished, flood control, shoreline stabilization, nutrient retention, water filtration, provision of a wide range of wild foods, fuel wood, grass and reeds for thatching houses, drinking water, and agricultural and forage resources. Moreover, wetlands are the most productive ecosystem in the world, producing greater quantities of herbaceous plants than sugar cane and maize grown in irrigated agriculture and supporting a great variety of wildlife and fish. Although little is known about the impact of agricultural and urban development on Ethiopia’s wetlands (Abebe, 2003), an estimated 35% of the largest wetland area of Ethiopia, in Ilubabor Region (figure 3.4), had been developed to agriculture by 1999 (Hailu, 2003), and pent-up demand for wetland cultivation in that region is threatening their sustainability (Mulugeta, 2004). Only one wetland area, in the Omo River Delta in Lake Turkana, has increased in recent years (by more than 500 km^2), due to silt inflow from the Omo River and decreasing lake levels, but its future is threatened by dam construction (Haak, 2005; see also chapter 10). Although some policies specifically address wetlands, no national wetlands policy exists and none of the country's wetlands are covered by a conservation and management plan (United Nations Educational, Scientific, and Cultural Organization, 2006), a situation Abebe (2003) attributes to a general lack of awareness of their beneficial services.

Groundwater

Little reliable information exists on the volume, location, and quality of groundwater resources, and estimation of the groundwater potential of Ethiopia is complicated by the complexity of its geology (Calow, McDonald, Nicol, Robins, & Kebede, 2002). According to the Ministry of Water Resources, Ethiopia's groundwater potential is estimated to be 2.6–13.5 billion m^3 per year (Dingamo, 2008). Groundwater is of paramount importance in areas affected by recurrent droughts or with limited or polluted surface water, in urban areas, and as a supplementary source of irrigation water. However, yield levels of many shallow water wells are low (less than 5 L per second), groundwater tables have been declining in recent years as a result of drought, and aquifers in the lowlands are generally too deep for economic exploitation for irrigation purposes. More than two thirds of the estimated utilizable groundwater potential is in the Blue Nile Basin and around 90% is in the western half of the country (Dingamo, 2008; WAPCOS, 1990). Groundwater has played a major role in reducing the social and economic consequences of drought in rural communities in Ethiopia and Southern Africa (Calow et al., 2002), and the Ethiopian government launched a national groundwater assessment program involving coordination of planning and management between the federal and regional levels and linkage with university research programs (Dingamo, 2008).

DEVELOPMENT POTENTIAL FOR HYDROPOWER PRODUCTION AND IRRIGATION

As pointed out in chapter 1, dam and large-scale irrigation development in African countries is an issue of increasing debate because of its apparent relationship with economic development and poverty alleviation on the one hand and its limited success in overcoming food insecurity in some drought-prone areas on the other. Its poor record in Ethiopia has often been attributed to sustainability issues and conflicts of interest with local populations, including failure to adapt imported technologies to local conditions, and ecological, health, and social impacts (Teshome, 2003). Nonetheless, the abundance of Ethiopia's untapped water resources and the burgeoning food and energy needs call for greater development of the water and land resources.

Hydropower Potential and Development

Ethiopia is endowed with vast, unexploited hydropower resources. The gross hydropower potential of the country has been estimated at 650,000 GWh (gigawatt hours) per year (CESEN & ENEC, 1986). Various sources rank Ethiopia second in hydropower potential in Africa. However, figures by different investigators for economically feasible energy development vary from 145,000 to 160,000 GWh, less than one quarter of the gross potential (CESEN & ENEC, 1986; Kyaruzi et al., 2005). Only about 1% of the feasible potential had been tapped by 2002. The economic hydropower potential energy was established through consideration of storage/diversion weirs of dam lengths below 750 m and dam heights below 120 m (Federal Democratic Republic of Ethiopia, 1999). Installed capacity corresponding to 145,000 to 160,000 GWh output per year, at an average plant utilization factor of 0.6, would generate about 27,000 to 30,000 MW (megawatts). Forty-nine percent of the country's hydropower potential and more than half of the 75% dependable surface water availability are in the Blue Nile Basin (table 3.2). Ethiopia's high rainfall in the highlands combined with high elevation drops (or "head"), provide for suitable conditions for hydropower development. Driven by local demand and encouraged in part by international financiers to develop electricity for export, Ethiopia has begun to put in motion a strategy by which more of its potential is to be utilized.

In recent years, Ethiopia's total energy production from electricity (including diesel) was only about 5%. The remaining 95% came from

TABLE 3.2. Hydropower (technical) potential of Ethiopia, 75% dependable surface water availability, and number of potential hydropower sites, by river basin.

River basin	Technical potential (GWh/yr)*	75% dependable surface water (BCM)**	Potential hydropower sites
Blue Nile	78,820	51.48	132
Wabe Shebele	5,440	2.34	18
Genale Dawa	9,270	4.58	23
Awash	4,470	4.10	43
Tekeze	5,930	5.73	15
Omo-Gibe	36,560	14.46	23
Ogaden	***	***	***
Baro-Akobo	18,880	8.51	39
Afar-Danakil	***	***	***
Rift Valley lakes	800	4.36	6
Aysha	***	0.57	***
Total	160,170	96.13	299

*Gigawatt hours
**Billion cubic meters
*** No data available
Source. NBCBN (2005).

traditional resources such as fuel wood, dung, crop residues, and human and animal power. Ethiopia's annual per capita production of electricity in 1996 (26 KWh) was the lowest even by regional standards, with Uganda using 40 KWh, Sudan 50 KWh, and Kenya 150 KWh. In comparison to Africa's average, Ethiopia's power production was also small. These low figures were reflected in the low per capita storage of water and the relatively low development of Ethiopia's hydropower potential. Ethiopia stores only an estimated 43 m^3 per capita and has developed only 1% of its hydropower potential, much less than in Sub-Saharan Africa overall, North Africa, the Middle East and the industrialized countries (Campagna per la riforma della Banco Mondiale [CRBM], 2008; Ethiopian Electric Power Corporation, 2006; see also chapter 1).

Eight large hydroelectric projects have been completed in Ethiopia since 1960 (figure 3.5). They generated a total of 663.6 MW in late 2006 (table 3.3). Tekeze Dam, designed to generate 300 MW, was put in operation in mid-2009, when Gibe III Dam, designed to create the second largest reservoir in Africa and generate 1,870 MW, was under construction. Koga and 13 additional dam projects were at the feasibility or construction stages. The planned Karadobi, Mabil, Mendaia, and Border dams on the Blue Nile (figure 3.5), with a combined storage capacity of 73.1 km^3 and installed capacity of 5,570 MW, may triple Ethiopia's power production and generate electricity for export to neighboring countries (Block, Strzepek, & Rajagopalan, 2007). About 300 additional potential sites have been identified for possible future hydropower generation, many of them for multipurpose projects (World Bank, 2006).

Table 3.3. Existing hydropower plants and installed capacity (in megawatts) in Ethiopia in late 2006.

Plant	Installed capacity in megawatts (MW)	Year of commission
Koka*	43.2	1960
Tis Abbay I	11.4	1964
Awash II*	32	1966
Fincha*	134	1973
Awash III*	32	1974
Melka Wakena*	153	1988
Tis Abbay II	74	2001
Gilgel Gibe I*	184	2004
Total	663.6	

*Projects with manmade lakes
Source. Ethiopian Electric Power Corporation (2006).

FIGURE 3.5. Hydroelectric, irrigation, and multipurpose dams.

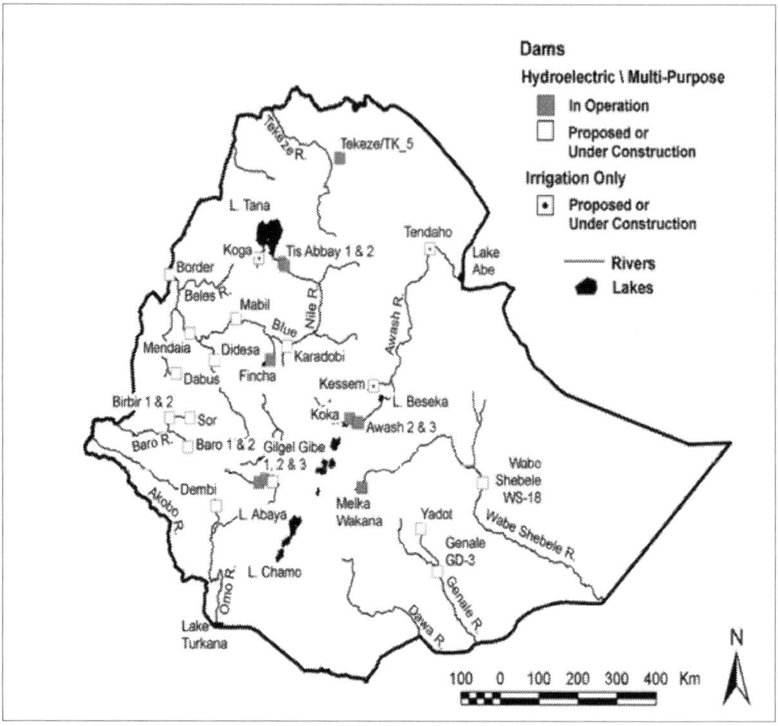

Source. Based on Abate (2009), CRBM (2008), and World Bank (2006).

After the completion of the Tana-Beles, Tekeze, and Gibe III projects, national energy production is expected to sharply increase and will constitute 96% of the country's electricity supply (Ethiopian Electric Power Corporation, 2006). Although the large size of these dams may provide sufficient water during periodic droughts, the effect of climate change are not known due to the uncertainties surrounding the few efforts in hydrological modeling of drought-risk. The ubiquitous siltation problem adds to the need to diversify energy sources toward achieving energy security.

There is also a need for a comprehensive cost-benefit analysis of the export of hydroelectricity versus using this energy toward energy security and poverty reduction in Ethiopia (CRBM, 2008). National power export is estimated to increase from 22.8% of total production in 2008 to 40.1% in 2012 (Ethiopian Electric Power Corporation, 2006).

In its push for rapid development of hydropower, the national utility corporation Ethiopian Electric Power Corporation and the Environmental Protection Agency occasionally side-stepped lengthy budget procurement and environmental impact assessment procedures. Examples of hastily planned and implemented projects date back to the construction of Koka Dam in the late 1950s in the Awash Valley and include the more recent Alwero, Tana Beles, Fincha, Koga, and the Omo/Gilgel Gibe projects (Eguavoen, 2009; Yarns & Fibers, 2008; see also chapters 8 and 9).

Irrigation Potential and Development

The availability of large areas of irrigable land and large water resources, together with recurring drought and chronic food insecurity prompted the government of Ethiopia and its neighbors to give high priority to water storage and irrigated agriculture. The estimates of the irrigation potential of Ethiopia range from 1.0 million ha to 4.3 million ha. These differences are largely due to the use of different criteria. The Ministry of Water Resources uses a 2.7 million ha estimate for planning purposes (Awulachew, Merry, Kamara, Van Koppen, de Vries, & Boelee, 2005). Twenty-eight percent of the irrigable land is located in the Blue Nile Basin, 67% in the five major basins in western Ethiopia, only 23% in the southern and eastern basins, and 10% in the northern basins (table 3.4; Halcrow, 1989). By 2002, Ethiopia was irrigating an estimated 197,000 ha, about 65,000 ha (33%) of which were under large-scale irrigation and the rest under medium- and small-scale irrigation. More than a third of all irrigated land was in the Awash River Basin (table 3.4), where Haile Selassie's regime had begun to establish large, export-oriented irrigation schemes in the 1960s. That initial commercial irrigation program paid

TABLE 3.4. Gross irrigation potential of the major river basins of Ethiopia and area irrigated in 2002.

River basin	Potentially irrigable area in ha	Irrigated area in 2002 in ha
Blue Nile	760,000 (28.0)	30,000
Baro-Akobo	600,000 (22.1)	*
Wabe Shebelle	355,000 (13.1)	*
Genale-Dawa	300,000 (11.0)	8,850
Omo-Gibe	248,000 (9.1)	*
Northern rivers	200,000 (7.4)	*
Awash	206,000 (7.6)	69,900
Rift Valley lakes	47,600 (1.8)	*
Total	2,716,600 (100)	197,000

*No data available.
Source. Based on Federal Democratic Republic of Ethiopia (2002).

little attention to peasant production and, in the mid-1970s, contributed to drought and famine. The total area under irrigation increased from 176,105 ha in 1991 to 197,250 ha in 1998. According to the Ministry of Water Resources, all of the increase (21,145 ha) was accounted for by modern small-scale irrigation schemes in various regions (Federal Democratic Republic of Ethiopia, 2002).

Another survey estimated that 75% of the total irrigated area comprised small-scale schemes (Awulachew et al., 2005). Little or no development occurred in medium- and large-scale irrigation in the 1990s (Federal Democratic Republic of Ethiopia, 2002), partly due to their generally poor performance, the current emphasis on small-scale, beneficiary operated irrigation schemes as ecologically and socially more sustainable, and high cost. Irrigation capital investment cost per hectare of US$18,000–25,000 in the 1990s was between two and four times higher than in Kenya and Tanzania (Appelgren, Klohn, & Alam, 2000). Other reasons include recurring drought and famine, poverty, social upheavals, and war, all of which impeded the development of large-scale irrigation infrastructure. Only 0.8% of all crop land nationwide was irrigated and 0.6% of the cereal staples were grown on irrigated fields in 2001–2002, mostly on small farms (Federal Democratic Republic of Ethiopia, 2003). On a per capita basis, Ethiopia had developed only 0.3 ha of irrigated land per 100 people by 2000, far below the 4 ha per 100 people of potentially irrigable land (Federal Democratic Republic of Ethiopia, 2002).

The underdevelopment of capacity to grow food and industrial crops prompted the government to allocate an additional 273,829 ha for irrigation under its 2002–2016 Water Sector Development Program, bringing the total acreage to 471,079 ha by 2016. Fifty-four percent of the additional acreage is to be developed by the federal government as large- and medium-sized schemes in all rural regions and 46% by the regional governments as small-scale schemes (Federal Democratic Republic of Ethiopia, 2002), indicating a renewed interest by the former in larger irrigation projects. The government is currently expanding the irrigated area of sugar cane production and increasing the milling capacities of existing sugar factories after reaching a favorable trade agreement with the European Union, under which Ethiopia can export sugar at higher prices than importing this commodity, and with the objective of increasing ethanol production from bagasse (Wudineh, 2009).

Interim monitoring and evaluation of irrigation works in progress revealed low levels of performance and implementation capacity, the result of shortage of heavy machinery and equipment, lack of experienced contractors, shortage of skilled manpower, shortages in building materials, and vagaries of the weather. Considering the dismal record of government rural development projects, including poor water management and salinization in at least 11 irrigation schemes, socioeconomic conflict with indigenous populations (Rahmato, 1999), negative ecological impacts on downstream communities (Ayenew, 2007) and the failure of several multipurpose dams and resettlement and conservation projects (see chapters 3 and 8), the new rush into large-scale irrigation is inviting a number of problems. Challenges include design issues, lack of implementation capacity, particularly lack of engineering and management experience, environmental problems (high silt load of rivers and salinization due to lack of watershed protection), health risks from vector-borne diseases, resettlement and compensation issues, and salinization. Salinization due to overirrigation and rising water tables affected more than 2,000 ha in 10 irrigation schemes in the Awash Valley in recent years, the result of lack of knowledge of crop water requirements and efficient water management (Ayenew, 2007). In medium and small schemes, many of them inadequately leveled, irrigation efficiency was estimated to be 20–30%. The practice of the government to levy low, subsidized water charges on the basis of hectares cropped rather than water volume further reduced irrigation efficiency (Abate, 1994).

Additional problems may arise out of over-optimism about benefits of irrigation. Although efforts were being made to mitigate social and environmental problems, these activities were carried out in an ad hoc fashion at various stages of the implementation process rather than from the start of projects. Many of the requirements and lessons to avert project failure that have been reported throughout Africa were being applied, albeit in haste and inadequately, during different stages of the implementation process (Aberra, 2004; Food and Agriculture Organization, 2004). In the Koga scheme near Lake Tana, for example, the first attempt by the Ethiopian government to develop a large-scale irrigation farm for peasants, less than half of the resettled farmers had been compensated for the loss of their homes and land to the reservoir. In addition, recommended watershed management to prevent siltation of the reservoir and public participation in the project were weak, environmental impact assessments were poorly implemented by the executing institution, and there was no follow-up by the regulatory agency (Abebe, 2007; African Development Fund, 2001).

Dam construction: Progress and challenges

Planning the damming of Ethiopia's rivers for irrigation, hydropower production, and flood control will have to consider not only downstream effects, periodic drought, and climate change, but also the rivers' high silt loads, landslides, and earthquakes, in addition to the resettlement of populations displaced by the reservoirs and possible negative hydrological, ecological, health, and socioeconomic impacts within the vicinity of reservoirs and downstream. Many dams and artificial lakes worldwide became inoperable or lost much of their capacities because of these and related problems (Evans, Mackey, Gottgens, & Gill, 2000). Hydrological models developed for the Blue Nile in Ethiopia have failed to comprehensively address these critical aspects in the absence of adequate river flow data, climate uncertainties, and the difficulty of assessing downstream impacts. The argument that Ethiopia provides ideal conditions for large dam development because of the relatively low temperatures and evaporation in the highlands (Block et al., 2007), must also consider the other abovementioned risks, particularly siltation of reservoirs and socioeconomic impacts, discussed in detail in chapters 8, 9, and 13. Silt reduced the capacity of Koka Reservoir from 1.8 billion m^3 to 1.0 billion m^3, causing frequent power shortages in Addis Ababa (Achamyeleh, 2003) and filled the Borkena Reservoir within a few years. Other dams expected to be seriously affected by heavy siltation attributed to soil erosion in the degraded highlands include the old Tekeze Dam in Tigray Region, the Angereb Dam on the Angereb River, and the Melka Wakena Dam on the upper Wabe Shebele River (Ayalew, 2007; Achamyeleh, 2003). A study of the recently filled Gilgel Gibe I Reservoir estimated that high silt loads and nutrient levels will cause the reservoir to be completely filled in and atrophied within 24 years (Devi, Tefahune, Legesse, Deboch, & Beyene, 2008), less than half the life period of 50 years predicted by the government's environmental impact assessment (Ethiopian Electric Light and Power Authority, 1997). The turbines of Aba Samuel Reservoir had to be shut down years ago due to eutrophication of the lake and the invasion of water hyacinth related to heavy river pollution and siltation originating in and around nearby Addis Ababa.

Additional potential ecological upsets, as well as malaria, schistosomiasis, and other health threats, resettlement and support of displaced communities, and hydrological changes affecting communities downstream of this dam will also have to be addressed if the intended economic objectives of this project are to be achieved (CRBM, 2008; Legesse, 2005).

Landslides and accelerated erosion are hazards in much of western Ethiopia, including in the Fincha Reservoir basin, because of the region's high rainfall, steep topography, and ongoing deforestation (Ayalew, 2007). Much of the increase in soil erosion in that basin was caused by the extension of crop land to steep parts of the water shed after the farms were inundated by the reservoir (Tefera & Sterk, 2006). These and other resettlement problems, as well as negative health impacts of large-scale dam development in Ethiopia and elsewhere in Africa are largely preventable through adequate planning and proper implementation of projects, further discussed in chapter 13.

Although ecological, health, and social problems related to dam construction and operation have been considered by the Water Sector Development Program 2002–2016 (Federal Democratic Republic of Ethiopia, 2002), the two regulatory agencies, the Ministry of Water Resources and the Environmental Protection Agency, have limited capacities and powers in monitoring projects and taking necessary actions (CRBM, 2008). In spite of anticipated high environmental risks and socioeconomic costs of the construction of multipurpose large dams, properly designed and operated hydropower plants can mitigate the effects of deforestation, floods, drought, and sediment transport. In addition, dams and sustainable reservoirs may contribute to national socioeconomic goals. Extension of the life of reservoirs and reduction of damage to the turbines of hydroelectric power plants through proper land use management can make dam projects more cost effective and economically viable. Integrated river basin development at the catchment level that promotes ecologically sustainable agriculture and urban and industrial use of water resources, vegetation, and water resources may be instrumental in achieving these objectives (Heathcote, 1998; Woube, 2005). However, this approach requires a longer time horizon and a better understanding of and massive inputs in agriculture and infrastructure development, which remain to be generated and coordinated (World Bank, 2006), a formidable task considering the state of the Ethiopian economy and lack of implementing capacity.

The prioritization of large hydropower has been questioned as a detriment to developing alternative energy sources, which have been neglected and underfunded. Wolde-Ghiorgis (2001) called for new energy policies promoting alternative energy technologies toward stimulating rural development. In addition to mini and microhydropower, coal, and gas (Asfaw, 2007; Mining exploration, 2008; Mulugetta, 2007), Ethiopia has considerable geothermal, wind, and solar potential. The renewable energy sources, in particular, are effective technical alternatives that may be deployed in widely dispersed rural communities (Mulugetta, 2007).

In 1969 Ethiopia began to develop its geothermal energy resources and is currently expanding its drilling program. The more than 1,000 MW potential, representing more than a quarter of all of East Africa's total, is adequate to meet all of Ethiopia's electric energy needs (Gawell & Greenberg, 2007; Teklemariam, 2005; United Nations Environmental Programme, 2008). Ethiopia is one of 39 countries, 12 of them in the East African Rift region, which could meet all their energy needs through geothermal sources. However, by 2005 Ethiopia obtained only 1.6% of its electric energy from this source, much less than the 15% in Kenya. High cost and risk of exploratory drilling and institutional, regulatory, and capacity barriers have impeded the development of this energy source (Earth Policy Institute, 2008). Recent developments of more efficient drilling technology in Kenya and the formation in 2003 of the African Rift Geothermal Development Facility Program by Ethiopia and five collaborating East African countries may facilitate the development of this dependable, nonpolluting energy source (United Nations Environmental Programme, 2008).

The total wind resource of Ethiopia is estimated at 20.1 million TJ (terra joules) per year (CESEN & ENEC, 1986) and wind speeds sufficiently high for wind energy production have been reported from various parts of the country (People's Democratic Republic of Ethiopia, 1988; Walelu, 2006). By 2008, 16 prospect areas had been identified for electricity production (Teklemariam, 2008). Due to continuing drought and the inability of hydropower plants to meet domestic demand, EEPCO signed a contract with a French firm in 2008 to construct a 150 MW wind farm, the largest in Africa (Ethiopia signs deal, 2008). Wind pumps were successfully demonstrated through donor funding and microhydropower-driven flour mills exist in several farming communities.

There is increasing evidence that solar cooking is sustainable and can result in a sharp decline in the use of fuel wood in Ethiopia and other developing countries (Drake & Mulugeta, 1996; Szulczewski, 2006). Solar photovoltaic systems were recently introduced to Ethiopia, and in 2008 the health and education ministries reached an agreement for their large-scale installation in schools. Training has been carried out for the installation, operation, and maintenance of solar systems, energy efficient wood stoves, and microhydropower plants for off-grid, community use, with the primary objective of reducing indoor air pollution and chronic respiratory diseases (World Bank, 2008). Another promising project is the establishment of a solar energy school by a private firm in a rural village in Amhara Region, where local people learn to install solar systems for power lighting, water pumps, radios, and the provision of warm water (Ehrlich, 2008).

In contrast to geothermal, wind, and solar energy, the development of biofuels is destined to contribute to water scarcity and environmental degradation. In 2006–2007, Ethiopia produced 8 million L of ethanol from sugar cane molasses and plans call for 129 million L by 2010–2011 (Fessehaie, 2009). In 2008, an estimated 5% of all irrigation water used in Ethiopia was for biofuel production, with significant increases expected in the future (Hoogeveen, Faures, & van de Giessen, 2009). Castor bean and Jatropha are increasingly planted by investors on arable land, forests, and woodlands, in violation of the new energy policy, which stipulates that these bioenergy crops be grown only on degraded or agriculturally marginal land (Mulugeta, 2008). There are concerns that in the absence of regulatory oversight, these companies, mostly foreign firms, are not acting in the best interest of local farmers and the environment (Lakew & Shiferaw, 2008; Patel, 2007).

CONCLUSION

Ethiopia possesses considerable water resources potential for hydropower and large-scale irrigation development. Based on accessibility, water resources development for hydroelectric power production and large-scale irrigated agriculture will probably have to focus more on surface water than groundwater, although the latter may assume a greater role in increasing water security during droughts. The Water Sector Development Program for 2002–2016, a useful road map to harnessing water resources for irrigation and hydropower using a comprehensive and integrated approach and the Plan for Accelerated and Sustained Development and Eradication of Poverty (Federal Democratic Republic of Ethiopia, 2006), can contribute to accelerated socioeconomic growth. Irrigation and hydropower development in Ethiopia can, if appropriately and efficiently implemented and managed, make major inroads into Ethiopia's food and water insecurity. Hydropower can be instrumental in reducing deforestation and land degradation and stimulate national economic activities if it can be generated reliably and affordably and pooled and distributed over the national grid to rural areas. However, alternative energy sources need to be given greater emphasis, as they may provide more reliable, accessible, environmentally friendly, and, in the long term, less expensive technology. Similarly, successful large-scale irrigation development is increasingly difficult for economic and environmental reasons and is likely to play a decreasing role in meeting the country's food needs compared to small-scale irrigation, discussed in chapter 4. The recent vigorous push by the Ethiopian government toward large-scale water resources development will also need to consider the national capacity, particularly integration of indigenous knowledge and institutions in the planning and implementation of programs, as demonstrated in the following chapters.

REFERENCES

Abate, G. (2009, July 10). Ethiopia's Tekezze dam starts generating power. *Ethiopolitics.com*. Retrieved from http://ethiopolitics.com/news_1/20090824998.html

Abate, Z. (1994). *Water resources development in Ethiopia: An evaluation of present experience and future planning concepts*. Reading, U.K.: Ithaca Press.

Abebe, W. B. (2007). EIA implementation and follow-up: A case study on Koga Irrigation Project, Ethiopia. Unpublished Master's thesis. UNESCO-IHE (Institute for Water Education), Delft, The Netherlands.

Abebe, Y. D. (2003). Wetlands of Ethiopia: An introduction. In Y. D. Abebe & K. Geheb (Eds.), *Wetlands in Ethiopia. Proceedings of a seminar on the resources and status of Ethiopia's wetlands* (pp. 1–11). Nairobi: International Union for Conservation of Nature and Natural Resources.

Aberra, Y. (2004). Problems of the solution: Intervention into small-scale irrigation for drought proofing in the Mekele Plateau of northern Ethiopia. *The Geographical Journal*, *170*, 226–237.

Achamyeleh, K. (2003). *Integrated flood management*. Report prepared for the World Meteorological Organization, Associated Programme on Flood Management, Geneva.

African Development Fund. (2001). *Ethiopia: Koga irrigation and watershed management project: Appraisal report*. Addis Ababa: Author.

Alemayehu, T., Ayenew, T., & Kebede, S. (2006). Hydro-geochemical and lake level changes in the Ethiopian Rift. *Journal of Hydrology*, *316*, 290–300.

Alemayehu, T., Furi, W., & Legesse, D. (2007). Impact of water overexploitation on highland lakes in eastern Ethiopia. *Environmental Geology*, *5*, 147–154.

Appelgren, B., Klohn, W., & Alam, U. (2000). *Water and agriculture in the Nile Basin* (Food and Agriculture Organization of the United Nations Rep. No. AGL/MISC/29/2000). Rome.

Asfaw, M. T. (2007, June). *Biofuels in Ethiopia*. Paper presented at the Eastern and Southern Africa Regional Workshop on Bio-fuels, Nairobi.

Awulachew, S. B. (2007, May). *Abaya-Chamo lakes physical and water resources characteristics, including scenarios and impacts*. Paper presented at the Lake Abaya Research Symposium 2007, Arba Minch University, Arba Minch, Ethiopia.

Awulachew, S. B., Merry, D., Kamara, B., Van Koppen, B., de Vries, F. P., & Boelee, E. (2005). *Experiences and opportunities for promoting small-scale/micro irrigation and water harvesting for food security in Ethiopia*. Colombo, Sri Lanka: International Water Management Institute.

Ayalew, L. (2007). Something that we need to know about our rivers' hydropower potential. Retrieved from http://www.mediaethiopia.com/Engineering/Lulseged_on_hydroelectric.htm.

Ayenew, T. (2007). Water management problems in the Ethiopia rift: Challenges for development. *Journal for African Earth Sciences, 48*, 222–236.

Ayenew, T. & Becht, R. (2007, May). *Comparative study of the hydrology and hydrogeology of selected Ethio-Kenyan rift lakes*. Paper presented at the Lake Abaya Research Symposium 2007, Arba Minch University, Arba Minch, Ethiopia.

BCEOM (French Engineering Consultants), in association with ISL and BRGM. (1999). *Blue Nile River Basin Integrated Development Master Plan Project* (Vol. I, *main report*). Montpellier, France: Author.

Bekele, S. (2002). Challenges and opportunities in capacity building for water resources development and research in Ethiopia: The AWTI experience. In P. G. McCormick, A. B. Kamara, & G. Tadesse (Eds.),

Integrated water and land management research and capacity building in Ethiopia. Proceedings of a MoWR/EARO/IWMI/ILRI International workshop (pp. 141–158). Addis Ababa: International Water Management Institute and Nairobi: International Livestock Research Center.

Bekele, S. & Horlacher, H.-B. (2004, May). *Water balance of the lakes Abaya and Chamo under limited data situation and impact scenario modeling*. Paper presented at the Lake Abaya Research Symposium 2004. Arba Minch University, Arba Minch, Ethiopia.

Belay, E. A. (2009). *Growing lake with growing problems: Hydrological investigation on Lake Beseka* (Unpublished doctoral dissertation). University of Bonn, Bonn, Germany. Retrieved from http://hss.ulb.uni-bonn.de/diss_online/math_nat_fak/2009/belay.eleni

Block, P. J., Strzepek, K., & Rajagopalan, B. (2007). *Integrated management of the Blue Nile Basin in Ethiopia: Hydropower and irrigation modeling*. Washington, DC: International Food Research Institute.

Calow, R., McDonald, A., Nicol, A., Robins, N., & Kebede, S. (2002). *The struggle for water: Drought, water security and rural livelihoods* (British Geological Survey Commissioned Report, CR/02/226N). Nottingham, U.K.: Keyworth.

Campagna per la riforma della Banco Mondiale. (2008). *The Gilgel Gibe affair: An analysis of the Gilgel Gibe hydroelectric projects in Ethiopia*. Retrieved from http://www.bankwatch.org/documents/gibe-study.pdf

CESEN & ENEC. (1986). *Cooperation agreement in energy sector between the Ministry of Mines and Energy of the provisional Military Government of Socialist Ethiopia, Ethiopian National Energy Committee (ENEC) and CESEN-Ansaldo/Finmecanica Group*. Addis Ababa: Author.

Conway, D. (2000). The climate and hydrology of the upper Blue Nile River. *Geographical Journal 2000, 166*, 46–62.

Cowardin, L. M., Carter, V., Golet, F. C., & LaRoe, E. T. (1979). *Classification of wetlands and deepwater habitats of the United States*. U.S. Department of the Interior, Fish and Wildlife Service, Washington, DC, Jamestown, ND: Northern Prairie Wildlife Research Center Home. Retrieved from http://www.npwrc.usgs.gov/resource/1998/classwet/classwet.htm

Devi, R., Tefahune, E., Legesse, W., Deboch, B., & Beyene, A. (2008). Assessment of siltation and nutrient enrichment of Gilgel Gibe dam, southwest Ethiopia. *Bioresource Technology, 9*, 975–979.

Dingamo, A. (2008, August). *Ethiopian groundwater resources management*. Paper presented at the 33rd International Geological Congress, Oslo. Retrieved from http://www.cprm.gov.br/331GC/1456705.html

Drake, F., & Mulugeta, Y. (1996). Assessment of solar energy and wind energy resources in Ethiopia. *Solar Energy, 57*, 205–217.

Earth Policy Institute. (2008). *Data for world geothermal power generation nearing completion*. Retrieved from http://www.earthpolicy.org/Updates/2008/Update74_data.htm

Eguavoen, I. (2009). *The acquisition of water storage facilities in the Abay River Basin, Ethiopia* (ZEF Working Paper No. 38). Bonn: Bonn University, Department of Political and Cultural Change.

Ehrlich, D. (2008, August 8). Ethiopia powers up for solar energy. *Cleantech Group News*. Retrieved from http://cleantech.com/news/3213/ethiopia-powers-solar-energy

Ethiopia signs deal for largest wind farm in Africa. (2008). *AFP*. Retrieved from http//:www.afp.google.com/articl/ALeqM5hnhy_9ZX7TD3X8iNqhONLasX5YSw

Ethiopian Electric Light and Power Authority. (1997). *Gilgel Gibe hydroelectric project: Environmental assessment* (Main report, Vol. 2). Addis Ababa: Author.

Ethiopian Electric Power Corporation. (2006). *Power system expansion update*. Addis Ababa: Author.

Evans, J. E., Mackey, S. D., Gottgens, J. F. & Gill, W. M. (2000). Lessons from a dam failure. *Ohio Journal of Science, 100*, 121–131.

Federal Democratic Republic of Ethiopia, Central Statistical Authority. (2003). *Statistical Abstract 2003*. Addis Ababa: Author.

Federal Democratic Republic of Ethiopia, Ministry of Finance and Economic Development. (2006). *Ethiopia: Building on progress, a plan for accelerated and sustained development to end poverty (PASDEP) (2005/06–2009/10)*. Addis Ababa: Author.

Federal Democratic Republic of Ethiopia, Ministry of Water Resources. (1999). *Ethiopian water resources management policy*. Addis Ababa: Author.

Federal Democratic Republic of Ethiopia, Ministry of Water Resources. (2002). *Water sector development programme.* Addis Ababa: Author.

Federal Democratic Republic of Ethiopia, Ministry of Water Resources. (2007). *Achieving water security: Lessons of Ethiopia country experience*. Addis Ababa: Author.

Federal Democratic Republic of Ethiopia, National Meteorological Services Agency. (2001). *Initial national communication of Ethiopia to the United Nations Framework Convention on Climate Change*. Addis Ababa: Author.

Federal Democratic Republic of Ethiopia, Population Census Commission. (2008). *Summary and statistical report of the 2007 population and housing census*. Addis Ababa: Author.

Fessehaie, R. (2009, June). *The status of bio-fuels in Ethiopia: Opportunities and challenges*. Paper prepared for the IUCN Regional Workshop on Bio-Fuel Production and Invasive Species. Retrieved from cmsdata.iucn.org/downloads/04_status_of_biofuels_in_Ethiopia_rezene_fessehaie.pdf

Food and Agriculture Organization of the United Nations. (n.d.). *Geo-Network.* Retrieved from http://www.fao.org/geonetwork/srv/en/main.home

Food and Agriculture Organization of the United Nations. (2004). *Experiences using water of various qualities.* Retrieved from http://www.fao.org/DOCREP/003/T0234E/T0234E09.htm

Food and Agriculture Organization of the United Nations. (2005). *Ethiopia: Aquastat-FAO's information system on water and agriculture, 2005 report.* Retrieved from http://www.fao.org/nr/water/aquastat/countries/ethiopia/index.stm

Food and Agriculture Organization of the United Nations and World Food Program. (2006, February). *Crop and food supply assessment mission to Ethiopia*, Rome: Author.

Gawell, K., & Greenberg, G. (2007). *Interim report: Update on world geothermal development* (Unpublished manuscript). Retrieved from www.geo-energy.org/publications/reports/GEA%20World%Update%202007.pdf

Gebreegziabher, Y. (2004). *Assessment of the water balance of Lake Awassa catchment* (Unpublished master's thesis). International Institute for Geo-Information Science and Earth Observation, Enschede, The Netherlands.

Gebre-Egziabher, T., & Berhanu, K. (2007). A literature review of decentralization in Ethiopia. In T. Assefa & T. Gebre-Egziabher (Eds.), *Decentralization in Ethiopia.* Addis Ababa: Forum for Social Studies.

Haak, B. (2005). Monitoring wetland changes with remote sensing: An East African example. *Environmental Management, 20*, 411–419.

Hailu, A. (2003). Wetlands research in south-western Ethiopia: The experience of the Ethiopian wetlands research programme. In Y. D. Abebe & K. Geheb (Eds.), *Wetlands of Ethiopia* (pp. 37–47). Nairobi: International Union for Conservation of Nature and Natural Resources.

Halcrow. (1989). *Master plan for the development of surface water resources in the Awash basin.* Report prepared for the Ministry of Water Resources, Addis Ababa.

Heathcote, I. W. (1998). *Integrated watershed management: Principles and practice*. New York: John Wiley and Sons.

Hoogeveen, J., Faures, J.-M., & van de Giessen, N. (2009). Increased biofuel production in the coming decade: To what extent will it affect global freshwater resources? *Irrigation and Drainage, 58*, S148–S160.

Institute of Biodiversity Conservation. (n.d.). *Site action plan for the conservation and sustainable use of the Lake Ziway biodiversity (Rift Valley Lakes Project)* (Unpublished manuscript). Addis Ababa.

Intergovernmental Authority on Development and International Climate Prediction and Application Centre. (2007). *Climate change and human development in Africa: Assessing the risks and vulnerability of climate change in Kenya, Malawi and Ethiopia* (Draft report). Retrieved from hdr.undp.org/en/reports/global/hdr2007-2008/papers/igad.pdf

International Lake Environmental Committee and United Nations Environmental Program. (2002). *World lake database. Survey of the state of world lakes*. Oroshimo-cho, Japan: Author.

Kebede, S., Travi, Y., Alemayehu, T., & Marca, V. (2006). Water balance of Lake Tana and its sensitivity to fluctuations in rainfall, Blue Nile Basin, Ethiopia. *Journal of Hydrology*, *316*, 233.

Kyaruzi, A., Hailu, Z., Mngodo, R. J., Mkandi, S. H., Ntungumburanye, G., Kizza, M., & Malisa, J. (2005). *Regional power integration in hydropower*. Nile Basin Capacity Building Network, Hydropower Development Research Cluster. Retrieved from http://www.nbcbn.com/Project_Documents/Progress_Reports/HP-G2.pdf

Lakew, H., & Shiferaw, Y. (2008). Rapid assessment of biofuels development in Ethiopia. In T. Anderson & M. Belay (Eds.), *Proceedings of the National Workshop on Environmental Impact Assessment and Biofuels* (pp. 1–3). Addis Ababa: MELCA Mahiber.

Legesse, D., Vallet-Coulomb, C. V., & Gasse, F. (2003). Hydrological response of a catchment to climate and land use changes in tropical Africa: Case study of South Central Ethiopia. *Journal of Hydrology, 275*, 67–85.

Legesse, W. (2005). *Physicochemical and biological characterization of the Gilgel Gibe River Basin: A tool for impact assessment study for the Gilgel Gibe Hydropower Dam construction.* Report prepared for the Ethiopian Science and Technology Agency, Addis Ababa.

Mining exploration company joint venture: Ethio-Pak Cola plc. begins coal mine production. (2008). Retrieved from http://paguntaka.org/2008/09/26/mining-exploration-company-joint-venture-ethio-pak-coal-plc-begins-coal-mine-production/

Ministry of Finance and Economic Development. (2006). *Ethiopia: Building on progress, a plan for accelerated and sustained development to end poverty (PASDEP) 2005/06–2009/10* (Vol. 1). Addis Ababa: Author.

Ministry of Water Resources. (1996). *Omo-Gibe River Basin integrated development master plan study* (Vol. 2). Prepared by Richard Woodroofe & Associates, Mascott Ltd. for MWR, Addis Ababa: Author.

Ministry of Water Resources. (1997). *Baro–Akobo River Basin integrated development master plan study* (Final report). Prepared by ULG Consultants. New York: Tams.

Ministry of Water Resources. (1998). *Tekeze River Basin integrated development master plan project.* Prepared by Netherlands Engineering Consultants for MWR, Addis Ababa: Author.

Ministry of Water Resources. (2007). *Ethiopia irrigation and drainage project: Environmental and social management.* Report prepared by Environmental Resources Management Ltd., Addis Ababa.

Mulugeta, A. (2008). *Companies derail biofuel policy, strategy*. Retrieved from http://www.capitalethiopia.com/archieve/2008/november/week1/local_news.htm

Mulugeta, S. (2004). Socio-economic determinants of wetland cultivation in Kemise, Illubabor Zone, southwestern Ethiopia. *Eastern Africa Social Science Research Review, 20*, 84–114.

Mulgetta, Y. (2007). Renewable energy technology and implementation mechanism for Ethiopia. *Energy Sources* (Part B), *2*, 3–17.

National Meteorological Service Authority. (1996). *Climate and agro-climatic resources of Ethiopia.* Addis Ababa: Author.

Osman, M., & Sauerborn. P. (2002, October). *Preliminary assessment of characteristics and long-term variability of rainfall in Ethiopia—Basis for sustainable land use and resource management.* Paper presented at the Conference on International Agricultural Research for Development, Witzenhausen, Germany.

Pankhurst, R. (1984). *The history of famines in Ethiopia prior to the twentieth century.* Addis Ababa: Relief and Rehabilitation Commission.

Patel, R. (2007). *Stuffed and starved: Markets, power and the hidden battle for the world's food system.* London: Black.

People's Democratic Republic of Ethiopia, Ethiopian Mapping Authority. (1988). *National atlas of Ethiopia.* Addis Ababa: Author.

Rahmato, D. (1999). *Water resources development in Ethiopia: Sustainability and participation.* Addis Ababa: Forum for Social Studies.

Sileshi, Y., & Camberlin, P. (2006). Recent changes in dry spell and extreme rainfall events in Ethiopia. *Theoretical and Applied Climatology, 83*, 181–191.

South African Wetlands Conservation Programme. (1997). *South African national wetland inventory.* Proceedings of a workshop held at the Department of Environmental Affairs and Tourism. Pretoria, South Africa: Author.

Szulczewski, M. (2006). Lasting impacts of solar cooker projects (Unpublished manuscript). Retrieved from http://solarcooking.org/lasting-impacts.htm

Taddese, G., Sonder, K., & Peden, D. (2005). *The water of the Awash River Basin: A future challenge to Ethiopia.* Retrieved from http://www.iwmi.cgiar.org/assessment/Research_Projects/irrigation_management_Ethiopia.htm

Tefera, B., & Sterk, G. (2006). Environmental impact of a hydropower dam in Fincha'a watershed, Ethiopia: Land use changes, erosion problems, and soil and water conservation adoption. In *Proceedings of the Second International Conference on Sustainable Sloping Lands and Watershed Management* (pp. 89–101). Luang Prabang, Lao PDR: National Agriculture and Forestry Research Institute.

Tefera, Z. (2007). The Ethiopian highland ecosystem. In S. Mengistu & S. De Stoop (Eds.), *Environment of survival: Taking stock of Ethiopia's environment* (pp. 75–96). Addis Ababa: Green Forum.

Tegegne, A. H. (2007). *Integrated assessment of the ecological and socio-economic importance of Lake Tana, Ethiopia* (Unpublished master's thesis). Environmental Systems Analysis Group, Wageningen University, The Netherlands.

Tekle-Haimanot, R., Melaku, Z., Kloos, H., Reimann, C., Fantaye, W., Zerihun, L., & Bjorvatn, K. (2006). The geographic distribution of fluoride in surface and groundwater in Ethiopia with an emphasis on the Rift Valley. *Science of the Total Environment, 367*, 182–190.

Teklemariam, M. (2005). *Geothermal exploration and development in Ethiopia*. Addis Ababa: Ethiopian Geological Survey.

Teshome, W. (2003). *Irrigation practice, state intervention and farmers' life-worlds in drought-prone Tigray, Ethiopia*, (Unpublished doctoral dissertation). Wageningen University, Wageningen, The Netherlands.

United Nations Educational, Scientific, and Cultural Organization. (2006). *Second UN World Water Development Report: Water, a shared responsibility*. Paris: Author.

United Nations Environmental Programme. (2008). *Hot prospect-geothermal electricity set for Rift Valley lift-off in 2009*. Retrieved from http://www.unep.org/Documents.Multilingual/Default.asp?DocumentID=553&ArticleID=6017&l=en

Walelu, K. (2006). *The wind energy projects in Ethiopia*. Paper prepared for the Ethiopian Electric Power Corporation, Addis Ababa.

WAPCOS. (1990). *Preliminary Water Resources Development Master Plan for Ethio*pia. Final report prepared for EVDSA, PDRE, Addis Ababa.

Wolde-Ghiorgis, W. (2001). *Renewable energy for rural development in Ethiopia: The case for new energy policies and institutional reform* (Unpublished manuscript). Retrieved from www.geocities.com/akababi/renewable.pdf

Wolde-Giorgis, T. (2008). *El Niño and drought early warning in Ethiopia*. Retrieved from http://ccb.colorado.edu/ijas/ijasno2/georgis.html

Wood, C. A. A. (1977). A preliminary chronology of Ethiopian droughts. In R. J. Dalby, H. Church, & F. Bezzaz (Eds.), *Drought in Africa*. London: International African Institute.

World Bank. (2001). Ethiopia: Woreda studies. Vol. 1. The main phase. Draft report. Addis Ababa: World Bank Country Office Ethiopia). Retrieved from http://info.worldbank.org/etools/docs/library/5687/Ethiopia%20Woreda%20Studies,20%WB.htm

World Bank. (2006). *Managing water resources to maximize sustainable growth: A World Bank water resources assistance strategy for Ethiopia*. Washington, DC: Author.

World Bank.(2008). *Status of projects in execution-FY08*. Washington, DC. Retrieved from http:www1.worldbank.org/operations/disclosure/sope.html

Woube, M. (2005). *Effects of resettlement schemes on the biophysical and human environments: The case of the Gambela Region, Ethiopia*. Boca Raton, FL: Universal Publishers.

Wudineh, Z. (2009, January 5). Ethiopian sugar exporters secure hold on EU market. *Addis Fortune*. Retrieved from http://tralac.org/cgi-bin/giga.cgi?cmd=cause_dir_news_item&news_id=1694news_id=58215&cat_id=0

Yarns and Fibers. (2008, March). Ethiopia-cotton projected to spin $500m from 2011. *News*. Retrieved from http://www.yarnsandfibers.com/news/print_article.php3?id=16209

Zinabu, G. M., Kebede-Westhead, E., & Desta, Z. (2004). Long-term changes in chemical features of waters of seven Ethiopian rift valley lakes. *Hydrobiologia, 477*, 81–91.

Chapter 4

Small-Scale Irrigation and Microdams

Helmut Kloos and Worku Legesse

Although small-scale irrigation and microdams have been used in mitigating droughts and boosting food security in Ethiopia since ancient times, they received little attention by most farmers and the Ethiopian government until recently (World Bank, 2005). Due to continuing food insecurity and poor performance of most large-scale irrigation schemes in Ethiopia, small-scale irrigation farming is now being promoted as part of a cautious move in the development paradigm to 'development from below.' The economic policy of Ethiopia calls for increased agricultural productivity in drought-prone areas through small-scale irrigation (Ministry of Water Resources, 2001). Microdams have similar advantages over large dams as small-scale irrigation systems vis-à-vis large schemes because of their relatively low unit cost and amenability to community participation (Aberra, 2004; Rahmato, 1999; Waterbury & Whittington, 1998). However, reports of poor performance of some irrigation systems (see, for example, Catterson et al., 1999; Awulachew et al., 2005) points out deficiencies that need to be addressed if small-scale irrigation is to contribute significantly to achieving food security in Ethiopia. This chapter examines recent developments, efficiency, institutional settings, and environmental impacts of small-scale irrigation schemes and storage facilities. A review of the country's experience of small-holder irrigated agriculture may contribute to evaluating the prospects of this intensive form of agriculture to achieve food security without major environmental and health risks. Small-scale irrigation, defined here as "irrigation, usually on small plots, in which small farmers have the major controlling influence, and using a level of technology which the farmers effectively operate and maintain" (Carter, 2004, p. 4), shares several characteristics of rainwater harvesting. In Africa, a focus on community initiatives and ownership are particularly important in successful capacity building and food security efforts of these two types of water use (see chapter 5). Microirrigation, used primarily in commercial flower and high-value backyard vegetable farming in Ethiopia and considered to have a relatively small impact on poverty and food security, is discussed by Awulachew et al. (2005) and the New Partnership for Africa's Development (2005).

The regional bureaus of agriculture and the federal Ministry of Water Resources classified small-scale irrigation as areas below 200–300 ha (Awulachew et al., 2005). This and other classifications based on size have been criticized for neglecting ownership, organization, and management of schemes, and physical and technical factors, which are considered to be more important in Sub-Saharan Africa (Carter, 2004; Rosegrant, 1997). The great majority of irrigated farm schemes in Ethiopia are very small, serving between 200 and 300 households (Tahal Consulting Engineers, 1988). This chapter examines the history, characteristics, and performance of both traditional irrigation systems (developed by farming communities without external assistance from the government, donors, or NGOs) and formal systems developed by the last two government administrations and other external support groups.

Traditional Institutions in Small-Scale Irrigation

Small-scale irrigation has a long history in Ethiopia that started before the Axum empire some 2,000 years ago (Fattovich, 1990). The Konso and several other groups in southwestern Ethiopia have traditionally used highly effective and sustainable irrigation systems, water conservation, and crop intensification systems (Beshah, 2003). Similar soil conservation and water harvesting techniques and irrigation management have been used for many years in central and northern Ethiopia (Teshome, 2003; see also chapter 8 on degradation). Irrigation was mostly seasonal and practiced by relatively few, favorably located communities in the mountain and valley landscape, usually with the objective of either growing one crop or supplementing the seasonal, typically unpredictable rainfall (Kloos, 1991; Teshome, 2003).

Access to land, crop, and livestock sharing, and indigenous institutions and mutual assistance formed an integral part of the cultural economy of Ethiopian farmers, and social networks, or social capital, supported the way farmers managed the land (Beshah, 2003; Conrad, 2007). Traditional irrigation systems in Ethiopia were developed and supported by indigenous institutions as part of local governance, including resource management, conflict resolution, and day-to-day mutual assistance. The highly evolved land use and irrigation management skills of the Konso, Wolaita, and other societies of western and southern Ethiopia owe much of their ecologically sustainable production systems supporting the highest rural population densities in the country to their reliance on the *ensete* staple or "false banana" *(Ensete ventricosum)*, their drought-tolerant staple, which conserves and restores soil in the erosion-prone highland areas. *Ensete* is also the most productive food plant in Ethiopia, producing over 5.6 tons per ha per year, ensuring a high degree of food security in most years. The agro-forestry system of the Gedeo and other ethnic groups within the *Ensete* Culture Complex Area (Shack, 1966) is part of this highly integrated agriculture/forestry system, which sustains about 600 persons per km^2. However, the call by the EPA for wider cultivation of *ensete* as "the cheapest remedy to the recurrent drought and food problems in Ethiopia" (Federal Democratic Republic of Ethiopia, 2004, p. 27) is unlikely to be heeded by the majority of the Ethiopian population, which considers *ensete* a food inferior to grain products. The future of the *ensete* culture is not certain due to overpopulation, labor shortage for maintaining terraces due to out-migration, and increasing competition with highland grain farmers for land use rights (Beshaw, 2003).

In many other highland communities currently pursuing irrigated agriculture, irrigation was introduced after the mid-20th century. A study in an area in eastern Amhara Region (South Wello Zone), which appears to be representative of small-scale development in the central and northern Ethiopian highlands during the last century, indicates that irrigation during the reign of Haile Selassie was usually introduced by external agents, particularly members of the nobility, landlords, and entrepreneurs from other countries. The small irrigated plots were typically confined to the river bottoms and landlords tended to have first choice. Only after their fields were irrigated could peasants compete over the surplus, or "hyena water." Landlords normally controlled the irrigation and in their absence a "father of water" (*Wuha Abbat*) ensured its fair distribution. Most peasants continued to grow their staple rain fed cereal crops under irrigation, barley and wheat in the highlands, and sorghum and maize in the lowlands. Only at the end of the 20th century were fruits and vegetables gradually introduced, mainly in response to increasing urban demands (Admassie, Pankhurst, & Yibabie, n.d.).

IRRIGATION UNDER SOCIALISM

During the socialist Derg period, small-scale irrigation declined together with agriculture overall until the 1984 famine, when the Derg government began to recognize surface water as a valuable, underutilized resource. Its hastily implemented program was hampered by the socialist land reform of 1975, disenchantment of peasants with cooperative farming, civil war, lack of market incentives, and shortages of technical, material, and skilled manpower farm inputs (Kloos, 1991). The functions of many indigenous water institutions, including the *Xela* in Konso, the *Idiriya* in Wolaita District, and the *Kire* in Wello, were sharply diminished with the establishment of peasant associations after the 1975 land reform. In line with socialist tenets, these organizations and local cultural dignitaries, which had been responsible for the distribution of access to and use of resources were now decried as exploitive landlords by the state, even though they usually managed resources more effectively than the new socialist associations (Beshah, 2003). According to Watson (2004, 2008), Christian missionaries contributed to the peripheralization of the traditional institutions on religious grounds, and Watson rejects the notion that population increases in Konso precipitated the recent famine conditions and food shortages. Watson (2008) noted that many of these institutions have started to function again and argued that government and development organizations interested in sustainable agricultural practices should engage more constructively with local populations in dealing with institutional and land management issues.

Traditional water users' associations or cooperatives and elected irrigation leaders (*Abo Mai*) and water committees play an important role in the administration and management of small-scale irrigation schemes in Tigray Region. A strong tradition of community mobilization and participation in resource management and irrigation development assured adequate manpower inputs in the construction and maintenance of microdams and irrigation systems. Nevertheless, one study in Tigray found that both management problems and sociocultural constraints impinged on irrigation activities, the former in the areas of operation, maintenance, water allocation, water distribution, decision making, and conflict resolution, and the latter in regard to land rights, labor shortage, shortage and inappropriate use of support services, market problems, and local institutions (Gum'a, 2004). The top-down approach used by the government, which did not permit for decision making by individual farmers, has exacerbated the management problems (Ayalneh, 2004). Water users' associations usually manage small-scale schemes, but even though they are legal entities the agricultural policy reportedly favors rain fed farmers in terms of providing credit and market information (Dinku, 2004).

GOVERNMENT PROGRAMS IN THE POSTREVOLUTION ERA

Small-scale irrigation had been largely neglected by the Haile Selassie regime, even though the well-documented study of famine occurrence by Wolde (1984) covering the period 1958–1977 demonstrated that famines had increased in frequency. The Derg government and NGOs began to promote small-scale irrigation after the 1975 land reform and actively encouraged it after the 1984 drought and famine as a strategy to eliminate food insecurity. Progress in small-scale irrigation development was slow, however, even after the establishment of the Irrigation Development Department within the Ministry of Agriculture in 1984 (Rahmato, 1999).

The pattern of small-scale irrigation management in Ethiopia has not changed substantially since Haile Selassie's regime. During his reign, it was the landlords and local governors, during the Derg period the Ministry of Agriculture and producer cooperatives, and under the current administration agricultural extension agents and "experts" who continue coercive strategies under the name of "participation." These various government agents and bureaucracies were involved in decision making of water allocation and conflict resolution, leaving farmers with little negotiating power and politicizing irrigation practices (Teshome, 2003).

The Socialist Period

The government's approach to small-scale irrigation development reflected the Derg's policies and practices. In spite of the Derg's plan to drought-proof the rural economy, the socialist government failed to establish viable irrigation institutions and constructed and rehabilitated few irrigation systems. Some traditional irrigation systems were incorporated into producer cooperatives, but relatively few farmers acquired irrigable land or water rights from the socialist government because of the requirement that they form producer cooperatives or, as in Tigray, water user associations. Local administrators and party officials controlled access to irrigation water, contributing to the decay of many hitherto sustainable systems (Rahmato, 1999; Teshome, 2003). The political objectives of irrigation were part and parcel of the state's drive to collectivize agriculture. However, the goal of collectivizing the majority of the farming population was not achieved due to the strong resistance by farmers to join state-sponsored institutions, as they saw this as a precursor to the demise of their family farms. Moreover, as the water rights of many traditional irrigators were eroded and access to water became problematic, many farmers leased out their land. This problem, together with the continued priority given by the Derg to large-scale commercial irrigation and technical inputs rather than social and management dimensions of irrigation development contributed to the failure of the government's goal to achieve food security (Rahmato, 1999; Teshome, 2003).

The Current Government's Small-Scale Irrigation Program

After 1993, the incoming EPDRF government gave high priority to small-scale irrigation as part of its Agricultural Development Through Industrialization (ADLI) policy and its efforts to meet Millennium Development Goals. The National Water Resources Policy of 2002 envisages the development of 127,000 ha to small-scale irrigation during the 15-year plan period (2002–2016) for an estimated 185,800 beneficiaries (Federal Democratic Republic of Ethiopia, 2002). The program includes water harvesting, crop production, and marketing components and is to be implemented in 315 drought-prone *weredas* and 3,500 *kebeles* in all nine rural administrative regions. The ambitious plan calls for the construction of bunds and ridges on 875,000 ha, 1.05 million water harvesting tanks and ponds for backyard watering and livestock watering, and 700,000 hand-dug wells for drinking water and gardening. The estimated cost of US$1.49 billion is expected to be financed by international donors (59%) and government contributions (23%), with beneficiaries (17%) contributing labor and local materials and paying water charges (Federal Democratic Republic of Ethiopia, 2005).

Whereas large- and medium-scale irrigation schemes are currently being managed mostly by public enterprises, the management of small-scale schemes is the responsibility of the farmers themselves, mainly through informal, traditional community groups and formal water users' associations the government promotes as the focal point for the development and management of irrigation schemes. The only government assistance farmers receive after the construction of schemes, mostly by external government agencies and NGOs, is in the form of extension and training services provided to water users' associations by the federal Ministry of Agriculture and Rural Development. The absence of appropriate local government organizations has left many farmers without guidance in irrigation operation and maintenance at the community level, further discussed below. In addition, institutional set-up and accountability issues vary among the different regions and are not clearly defined, resulting in confusion, inefficiencies, and occasional scheme failures. These deficiencies have become more problematic with the increasing complexity of and demands placed on small-scale irrigation in recent years. Planning, design, and construction of schemes are gradually being transferred to regional organizations as part of the decentralization drive, resulting in variable levels of sustainability due the variable capacities of the different regions (Awulachew et al., 2005; Food and Agriculture Organization, 2005; Federal Democratic Republic of Ethiopia, 2002; Rahmato, 1999).

Area Irrigated and Characteristics of Schemes

Of the 4.3 million ha identified as irrigable in Ethiopia, an estimated 247,470 ha (5.8%) were reportedly under irrigation in 2002, 55.9% of them on small-scale traditional schemes, 19.4% on small-scale modern schemes, and 24.6% on medium- and large-scale schemes (Awulachew et al., 2006). The existing information on the number and acreage of small-scale traditional irrigation schemes and the number of beneficiaries in different regions is unreliable and underreported due to regional differences in the definitions of small-scale irrigation and water harvesting and under registration of schemes at the regional Bureaus of Agriculture. According to the most comprehensive survey to date, 1,095 small schemes comprising 80,844 ha that benefited 323,376 households existed in Amhara Region in 2004, or about 15 times the acreage and number of beneficiaries of modern small-scale schemes. In Oromia Region, only 34,000 households benefited from traditional small irrigation systems, but 728,000 households benefitted from ponds and 370,000 households benefitted from rain and flood water harvesting. In Tigray Region, more than 15,000 ha were estimated to be under traditional irrigation and 4,989 ha under modern small systems, with no information on the number of users (Awulachew et al., 2005). No representative data are available for the other regions, two of which (Afar and Somali) are the most vulnerable to the effects of drought and climate change.

In 2006, 833,300 farmers irrigated a total area of 134,500 ha (Central Statistical Authority, 2007), twice the area under large-scale irrigation between 1986 and 2002 (McCornick, Kamara, & Tadesse, 2003; Kloos, 1991). By the end of 2004, 1,383 small-scale irrigation schemes had been developed in Tigray, Amhara, Oromia, and SNNPR regions. Their sizes ranged from less than 10 ha to 619 ha, with most of them around 20–150 ha (Awulachew et al., 2005).

Nearly four fifths (79.2%) of the 5,942 small-scale irrigation systems supported by six major Ethiopian and international NGOs in the late 1990s were spate systems spreading flood waters, 10.1% were diversion systems depending on gravity flow, 9.3% were storage systems using dams, and 0.1% were systems fed by springs, with no schemes involving mechanical or motorized lifting devices. Only diversion systems on perennial rivers and spring systems provided a dependable water supply throughout the year, permitting irrigation of dry-season crops in addition to providing supplemental irrigation during the rainy season. Spate and storage systems, on the other hand, typically operate intermittently during the year, the specific timing and amounts of water abstracted depending on the amount of rain received locally or in nearby highlands (Catterson et al., 1999). Microdams, introduced during the last few decades to Ethiopia, may also be affected by water shortages during the dry season and during droughts and are beset with siltation problems (Haregeweyn et al., 2006). Recently introduced technologies such as pumped systems have faced maintenance and fuel problems in the absence of long-term government, private sector, and NGO support (Carter, 2004).

Achievements and Challenges

Various studies of small-scale irrigation schemes have reported significant increases in crop production and food security (Ayalneh, 2004; Behailu, Abdulkadir, Mezgebu, & Yasin, 2004; Tesfaye, Bogale, Namara, & Bacha, 2008) and increased household income (Ayalneh, 2004; Dinku, 2004), including women renting irrigated plots (Behailu et al., 2004), contributing to reducing dependence on food aid (Mekonnen, 2003). New economic linkages were associated with irrigation development in two irrigation systems in eastern Oromia, where the production of cash crops created jobs for landless young people, including women, increased income, the demand for nonagricultural products, and investment in the cottage industry and small stores (Dinku, 2004). Even though these various studies did not assess the role of social capital in the development process, the above economic achievements were made possible to a great extent through social connections and relationships, including cooperation, reciprocity, neighborliness, trust, and civil involvement (Conrad, 2007).

Significant crop production increases were reported by the beneficiaries of irrigated plots fed by 15 earth dams in Tigray, but the livelihoods of only the more affluent farmers improved, with the poor not benefiting (Mekonnen, 2003). Several other farms in Tigray with adequate water supplies reported loss of soil fertility from double cropping in the absence of adequate soil fertilization. Another impediment to intensifying production was lack of market access and information on market economics, limiting the production of perishable and high-value crops (Behailu et al., 2004). A study of 1,785 irrigated plots in 100 villages in highland Tigray found that although mean yields of plots were significantly higher on the typically double-cropped, intensively cultivated irrigated plots than on nonirrigated plots, the value of crop production or household income was not higher. Increased farming intensity in more densely populated areas was not associated with significantly higher crop yields, also reported from other parts of the Ethiopian highlands (Pender & Gebremedhin, 2006). These findings from the overpopulated highlands contrast with Boserup's (1965) theory that density of population creates an environment conducive to innovation and progress in agricultural production.

Most deficiencies of small irrigation schemes were related to seasonal drought, the design and management of schemes, and the weakness of water management at the farm and catchment levels. Shortage of irrigation water during the dry season, when irrigation water is needed most, was experienced by some systems in Tigray Region, and wastage of excess water loss through seepage and overwatering was experienced in other regions (Ayalneh, 2004; Behailu, Abdulkadir, Mezgebu, & Yasin, 2005; Hassen, 2004). Salinity build-up, drainage problems, algal blooms, water logging, soil fertility declines, receding of river banks into irrigated land, and the deepening of river beds due to river erosion also limit the potential of small-scale irrigation in Tigray (Tesfaye, Haile, & Yazew, 2001). The use of pipes was suggested to reduce conveyance losses, and the imposition of water fees was suggested to reduce over-irrigation (Kahsay, 2003), neither of which has been implemented for small systems in Ethiopia. Prevention of overirrigation on two farms because of the cost of fuel further argues for the imposition of water fees. The same farms were able to improve household food security but were beset with weak management and lack of transparency (Assefa, 2008), two perennial problems which cannot be solved in the short term. Another culture-related problem—the low participation of females in decision making roles—was considered to impede efficient management of schemes in Tigray and Oromia regions, where existing female participation was associated with fewer conflicts and violation of rules (Deribe, 2008).

With increasing farmer demand for irrigation in recent years, existing irrigation areas were expanded and new ones developed, but there were also increases in conflict and competition over water. For instance, increased abstraction of water for irrigation in upstream locations is causing upstream/downstream tension in water-scarce areas in Tigray, Amhara, and other regions (Global Water Partnership, 2006). Competition for irrigation water with large commercial schemes was reported for two small farms in the upper Awash Valley (Hassen, 2004), and conflict between upstream and downstream water users has increased in many parts of Ethiopia (Admassie et al., n.d.; Food and Agriculture Organization, 2005). This problem arises from the tendency of individual farmers, communities, and NGOs to develop irrigation schemes oblivious to the needs of other users in the same catchments in the absence of water regulation (Carter, 2004).

Irrigation and Microdam Development in Tigray Region

Lowland farmers in drought-prone Tigray Region traditionally diverted river water with temporary stone and earthen weirs to their farm land, and highland farmers captured spring water (Teshome, 2003). Use of spring water harvested in ponds is generally preferred as demanding less maintenance and less likely to fail than diversion systems, but springs are scarce and many have been yielding less water in recent years. Manually dug wells used for irrigation on several valley plains with suitable soils and high water are of less importance in crop production (Kahsay, 2003).

The Tigray regional government introduced microdams in 1975 in favorable areas, but construction could not proceed on a large scale until after the end of the war in 1991. In 1994, the government implemented a 10-year plan to construct 500 microdams for irrigation as the main strategy of mitigating the problems of recurrent drought and reducing the reliance on rain-fed agriculture with the objective of achieving food security. However, due to land use, technical, and managerial problems, only 54 dams were constructed. A hydrological assessment of the poor performance of these dams identified insufficient inflow, excessive seepage, and rapid sediment deposition resulting from soil erosion in catchment areas that had not been rehabilitated prior to dam construction (Haregeweyn et al., 2006; Behailu et al., 2004). These deficiencies contributed to reducing the storage capacity and expected economic life of these reservoirs and the amount of water available for irrigation, a problem compounded during droughts when microdams filled only partly. Siltation was estimated to reduce the economic life of many reservoirs by about half. Only 30% of the reservoirs were expected to remain usable until the end of the entire design period. The reservoir sedimentation survey, the first one in Ethiopia, associated variations in reservoir sediment load with differences in land use, vegetation cover, the physical characteristics of rocks, and severity of gulley erosion in the respective catchments. The productivity of some of the irrigation schemes was unsatisfactory due to these and other dam- and catchment-related problems, causing beneficiaries to withdraw their support for the construction of additional dams (Haregeweyn et al., 2006), even though crop production could be increased over predam levels and well water production increased as a result of groundwater recharge from dam water (Behailu, Kifle, & Leulseged, n.d.). Additional on-site and off-site effects of poorly designed and managed microdams and irrigation schemes include overwatering of crops, salinization, water logging exacerbated by inadequate drainage works, and increased malaria risk (Behailu & Haile, 2002; Behailu et al., 2004). Similar water and production problems have been reported from Eritrea, where traditional water laws and technical features of spate irrigation systems similar to those in Tigray proved to be generally more sustainable than new approaches (Haile, 2007).

Conclusion

There is some evidence that the expansion of small-scale irrigation in Ethiopia has contributed to creating new opportunities for production and better income, reducing risk, and improving livelihoods and food security. But it is also evident that production and productivity levels are far from optimum, that equity and health issues need attention, that water management is deficient and environmental sustainability is not assured, and that more community participation is required to achieve targets. Four salient facts have significant policy and management implications. First, it is not the lack of water per se that constrains irrigation development in most areas, but lack of management strategies that can enable farmers to mitigate dry spells and droughts. Second, time-tested and locally adapted indigenous knowledge of water harvesting, conservation farming, and water conservation is not adequately integrated into the design and management of new schemes. Third, related to the second finding, simple irrigation systems that can be easily adopted and maintained by small farmers with limited outside support are preferable. Fourth, the multifaceted nature of small-scale irrigation, with its linked water, soil, fertility, and crop management, tillage, labor use, land ownership, watershed management, pest management, livestock interactions, and health components, requires that water system innovations be addressed in an integrated fashion. Clearly, overcoming the problem of physical and economic scarcity of irrigation water in Ethiopia cannot be a single-factor solution but must be consistent with the recently formulated environmental and economic policies and requires strengthening of infrastructure (roads, health, and communications), agricultural extension services, and greater access to land as part of the rural development program. The success of the program will ultimately hinge on the ability to build on the experiences of government organizations, community-based organizations, farmer organizations, and NGOs. Project planning, implementation, and evaluation from the perspective of the participant farmers will be essential for the optimization and sustainability of projects.

References

Aberra, Y. (2004). Problems of the solution: Intervention into small-scale irrigation for drought proofing in the Mekele Plateau of northern Ethiopia. *The Geographical Journal, 170*, 226–237.

Admassie, Y., Pankhurst, A, & Yibabie, T. (n.d.). Briefing, trends in irrigation development: Institutionalization of irrigation in South Wello. Brighton, U.K.: University of Sussex.

Assefa, M. (2008). Socio-economic assessment of two small-scale irrigation schemes in Adami Tullu Kombolcha Woreda, Central Rift Valley of Ethiopia (Unpublished master's thesis). Addis Ababa University, Addis Ababa.

Awulachew, S. B., Merry, D. J., Kamara, A. B., Van Koppen, B., de Vris, F. P., & Boelee, E. (2005). *Experiences and opportunities for promoting small-scale/micro irrigation and rainwater harvesting for food security in Ethiopia* (Working Paper 98). Colombo, Sri Lanka: International Water Management Institute.

Awulachew, S. B., Merry, D., van Koppen, B., Kamara, A., de Vries, F. P., & Boelee, E. (2006). *Roles, constraints and opportunities of small scale irrigation and water harvesting in Ethiopian agricultural development: Assessment of existing situation*. Colombo, Sri Lanka: International Water Management Institute.

Ayalneh, W. (2004). *Socio-economic and environmental impact of assessment of community based small-scale irrigation in the Upper Awash Basin: A case study of four community based irrigation schemes* (Unpublished master's thesis). Addis Ababa University, Addis Ababa.

Behailu, M., Abdulkadir, M., Mezgebu, A., & Yasin, M. (2004). *Preliminary report on community based irrigation management in the Tekeze Basin: Impact assessment. A case study of three small-scale schemes (micro dams)*. Report prepared for Mekele University and the International Livestock Research Institute, Addis Ababa.

Behailu, M., Abdulkadir, M., Mezgebu, A., & Yasin, M. (2005, March). *Report on community based irrigation management in the Tekeze Basin: Performance evaluation.* Report prepared for Mekele University, Mekele, Ethiopia.

Behailu, M., & Haile, M. (2002). *Water harvesting in northern Ethiopia: Environmental, health and socioeconomic impacts.* Paper presented at the Workshop on Integrated Water and Land Management Research and Capacity Building Priorities in Ethiopia. International Livestock Research Institute, Addis Ababa.

Behailu, M., Kifle, W., & Leulseged, T. (n.d.). *Fighting, famine and poverty through water harvesting in northern Ethiopia* (Unpublished manuscript). Mekele University, Mekele, Ethiopia.

Beshah, T. (2003). *Understanding farmers: explaining soil and water conservation in Konso, Wolaita and Wello, Ethiopia* (Unpublished doctoral dissertation). Wageningen University and Research Center, Wageningen, The Netherlands.

Boserup, E. (1965). *The conditions of agricultural growth.* New York: Aldine Publishing.

Carter, R. C. (2004, January). *Small-scale irrigation in Ethiopia. Part 1—Desk study* (Unpublished manuscript). Cranfield University, Silsoe, U.K.

Catterson, T., Worku, M., Endalew, M., Abate, C. G., Brockman, F., Amaneul, A. A., & Mamusha, A. K. (1999). *Programmatic assessment of small-scale irrigation in Ethiopia.* Report prepared for Catholic Relief Services, U.S. Catholic Conference, Baltimore.

Central Statistical Authority. (2007). *Agricultural Sample Survey 2006/2007 (1999 E.C.). Vol. 2. Report on farm management practices (private peasant holdings, meher season).* Addis Ababa.

Conrad, D. (2007). Defining social capital. *Electronic Journal of Sociology.* Retrieved from http://www.sociology.org

Deribe, R. (2008). *Institutional analysis of water management on communal irrigation systems in Ethiopia: The case of Atsbi Wemberta, Tigray Region and Ada'a Woreda, Oromiya Region* (Unpublished master's thesis). Addis Ababa University, Addis Ababa.

Dinku, L. (2004). *Smallholders' irrigation practices and issues of community management: The case of two irrigation systems in Eastern Oromia, Ethiopia* (Unpublished master's thesis). Addis Ababa University, Addis Ababa.

Fattovich, R. (1990). Remarks on the pre-Aksumite period in northern Ethiopia. *Journal of Ethiopian Studies, 23*, 1–33.

Federal Democratic Republic of Ethiopia. (2002). *Water Development Programme 2002–2016.* Addis Ababa: Ministry of Water Resources.

Federal Democratic Republic of Ethiopia. (2004). *The third annual report on the implementation of the UNCCD/NAP in Ethiopia.* Addis Ababa: Author.

Federal Democratic Republic of Ethiopia. (2005). *Support to NAPAD-CAADP implementation. Report TCP/ETH/2908 (I). Vol. 2. Water harvesting and small-scale irrigation*. Addis Ababa: Author.

Food and Agriculture Organization. (2005). *Ethiopia.* Rome: FAO, Aquastat Information System on Water and Agriculture.

Global Water Partnership. (2006). *Ethiopia's poverty reduction.* Entebbe, Uganda. Retrieved from htpp://csdwand.net/data/sheet.asp?cn=Uganda&fn=LA1762

Gum'a, D. (2004). *The socio-cultural aspect of irrigation management of two community-based small-scale irrigation schemes in the upper Tekeze Basin, Tigray Region* (Unpublished master's thesis). Addis Ababa University, Addis Ababa.

Haile, A. M. (2007). A tradition in transition, water management reforms and indigenous spate irrigation systems in Eritrea. Delft, The Netherlands: UNESCO/IHE Institute for Water Education.

Haregeweyn, N., Poesen, J., Nyssen, J., DeWit, J., Haile, M., Govers, G., & Deckers, S. (2006). Reservoirs in Tigray: Characteristics and sediment deposition problems. *Land Degradation and Development, 17*, 211–230.

Hassen, Y. K. (2004). *Assessment of small scale irrigation using comparative performance on two selected schemes in upper Awash River Valley* (Unpublished master's thesis). Alemaya University, Alemaya, Ethiopia.

Kahsay, L. (2003). *Design and performance of community based irrigation in Tigray*. Paper presented at the Workshop on Community Based Irrigation Management at ILRI, Addis Ababa.

Kloos, H. (1991). Peasant irrigation development and food production in Ethiopia. *The Geographical Journal, 157*, 295–306.

McCornick, P. G., Kamara, B., & Tadesse, G. (2003). Integrated water and land management research and capacity building priorities for Ethiopia. In P. G. McCormick, A. B. Kamara, & G. Tadesse (Eds.), *Integrated water and land management research and capacity building priorities for Ethiopia. Proceedings of a MWR/EARO/IWMI/ILRI international workshop* (pp. 25–47). Colombo, Sri Lanka: International Water Management Institute and Nairobi: International Livestock Research Institute.

Mekonnen, H. M. (2003). *Irrigation development in Tigray part of the Tekeze River: The experience of the Relief Society of Tigray (REST)*. Paper presented at the Workshop of Community-Based Irrigation Management in Ethiopia, Strategies to Enhance Human Health, Livestock and Crop Production, and Natural Resource Management organized by the International Livestock Research Institute, Mekele.

Ministry of Water Resources. (2001). *Irrigation development strategy (component of the water sector strategy)*. Draft Report. Addis Ababa: Author.

New Partnership for Africa's Development. (2005). *Support to NEPAD-CAADP implementation. Report TCP/ETH/2908(I). Vol. II. Water harvesting and small-scale irrigation*. Addis Ababa: Author.

Pender, J. & Gebremedhin, B. (2006). Land management, crop production, and household income in the highlands of Tigray, northern Ethiopia: An econometric analysis. In J. Pender, F. Place, & S. Ehui (Eds.), *Strategies for sustainable land management in the East African Highlands* (pp. 107–140). Washington, DC: International Food Policy Research Institute.

Rahmato, D. (1999). *Water resources development in Ethiopia: Issues of sustainability and participation.* Addis Ababa: Forum for Social Studies.

Rosegrant, M. W. (1997). *Water resources in the twenty-first century: Challenges and implications for action.* Washington, DC: International Food Policy Research Institute.

Shack, W. A. (1966). *The Gurage: A people of the ensete culture.* London: Oxford University Press.

Tahal Consulting Engineers. (1988). *Traditional organizational forms in small-scale irrigation schemes in Ethiopia*. Report prepared for the Ministry of Agriculture, Addis Ababa.

Tesfaye, A., Bogale, A., Namara, R. E., & Bacha, D. (2008). The impact of small-scale irrigation on household food security: The case of Filtino and Godino irrigation schemes in Ethiopia. *Irrigation and Drainage Systems*, *22*, 145–158.

Tesfaye, G., Haile, M., & Yazew, E. (2001). Land use and plot size determination issues in small-scale irrigation development in Tigray Region, Ethiopia. In W. Negatu, Y. Amare, & Y. Adal (Eds). *Current issues on land tenure in Ethiopia: Access, food production and natural resources management* (pp. 182–203). Addis Ababa: Institute of Development Research, Addis Ababa University.

Teshome, W. (2003). *Irrigation practices, state intervention and farmers' life-worlds in drought-prone Tigray, Ethiopia* (Unpublished doctoral dissertation). Wageningen University, Wageningen, The Netherlands.

Waterbury, J. & Whittington, D. (1998). Playing chicken on the Nile? The implications of microdam development in the Ethiopian highlands and Egypt's New Valley Project. *Natural Resources Forum, 22*, 155–163.

Watson, E. E. (2004). Agricultural intensification and social stratification: Konso contrasted with Marakwet. In M. Widgen & J. Sutton (Eds.), *Islands of intensive agriculture in Eastern Africa* (pp. 59–68). Oxford: James Currey.

Watson, E. E. (2008). *Landscape, culture and development: Konso, Ethiopia*. Retrieved from http://www.geog.cam.ac.uk/research/projects/konsoethnography/

Wolde, M. M. (1984). *Rural vulnerability to famine in Ethiopia: 1958–1977*. Addis Ababa: Addis Ababa University Press and Vicas Publishing Limited.

World Bank. (2005). *Water resources, growth and development.* A working paper for discussion prepared by for the Panel of Finance Ministers of the U.N. Commission on Sustainable Development. New York.

Chapter 5

Rainwater Harvesting

Sorssa Natea, Cherkos Tefera, and Helmut Kloos

Introduction

Rainwater harvesting (RWH) includes two broad application technologies: 1) the collection of ground surface runoff in ponds, reservoirs, and underground storage tanks for human and livestock consumption as well as for irrigation, the percolation of rainwater into the ground for recharge to be used for irrigation, and groundwater recharge, flood, and erosion control measures, and 2) roof water harvesting (Thomas & Martinson, 2007). Especially the former type of RWH has been practiced in the semiarid and arid parts of Africa, the Mediterranean, South Asia, and the Americas since ancient times (Gill, 2000; Pandey, Gupta, & Anderson, 2003; Wahlin, 1995). This increasingly utilized water technology may become a key climate adaptation opportunity and a means to boost food and water security efforts in Africa and other water-deficient areas. According to the United Nations Environmental Program (2006), Africa is not water scarce if the rainwater harvesting potential is considered. After being prolonged by NGOs, faith-based organizations, and water development networks, the use of RWH technologies in Africa has been increasingly recognized during the last two decades within the context of the Millennium Development Goals (MDGs) and other initiatives because of the need for managing rainwater resources for "drought proofing" communities subject to climatic variability and to facilitate social and economic wellbeing (United Nations Environmental Program, 2005).

RWH is potentially a pro-poor and gender-sensitive water development intervention and, as such, can significantly contribute to reaching the MDGs (Awulachew et al., 2005). This technology is deemed suitable in about one third of Africa, where rainfall is greater than 200 mm per year and where the great majority of the population lives. Worldwide, roof water harvesting accounts for 1–2% of domestic supply in some countries and up to 100% on several low tropical islands lacking suitable surface and ground water (Thomas & Martinson, 2007). In industrialized countries as well, rainwater harvesting is an appropriate method of storing water, as evidenced by the increasing popularity of roof water harvesting in recent years as a result of housing growth, changing lifestyles, climate change, and stretched water resources (Meera & Ahammed, 2006). With accelerated abstraction of surface and groundwater resources due to population pressure, urbanization, pollution, mismanagement of resources, poor long-term investment, lack of environmental research, and global warming, the need for alternative water resources is rapidly increasing worldwide. Thomas and Martinson (2007) even considered roof water harvesting a "postmodern" technology the whole world may be moving to in the quest for environmental sustainability. This chapter focuses on RWH in Ethiopia, particularly surface runoff-fed ponds, tanks, and roof water harvesting technologies that are currently prioritized by the Ethiopian government and NGOs. This information supplements the discussions on small-scale irrigation and microdams in chapter 4 and on runoff on-farm rainwater harvesting (contour catchment water harvesting, in-field rainwater harvesting, and water conservation tillage) in chapter 8.

A recent study by the United Nations Environmental Program and the World Agroforestry Center estimated that RWH could potentially meet the domestic and agricultural water needs of about 320 million people in Ethiopia, after allowing adequate river runoff for environmental needs (United Nations Environmental Program, 2006). Nearly all parts of the country may benefit from RWH (United Nations Environmental Program, 2005). Rainwater harvesting may, thus, contribute to ensuring national water and food security in a country where rainfall is increasingly unpredictable, potable water coverage by the formal sector is low in rural areas, and population growth is accelerating. By collecting surface runoff in rural areas, RWH can complement water and soil conservation efforts aiming to retain rainwater in situ (Falkenmark & Rockström, 2004).

History of Rainwater Harvesting in Ethiopia

Rainwater harvesting in Ethiopia dates back to the pre-Axumite period in the 6th century B.C. (Fattovich, 1990). There is much evidence (the large urban population, dams, ponds, and cisterns) that the Axumites had developed an elaborate water harvesting technology in their capital city (Butzer, 1981; McCann, 1999). This technology was later used in the medieval Ethiopian capital of Gonder for agricultural and domestic needs as well as religious rituals (Alem, 1999). Some of the indigenous water harvesting technologies and practices surrounding irrigation and domestic water use are described in chapters 4 and 8. They include stone and earth dams as well as ground and rock catchments, terraces, soil bunds, stone-lined canal systems, and storage tanks. Their survival has been put in jeopardy in recent years because of the prioritization of modern schemes by the Ministry of Agriculture and Rural Development (Alamerew, 2006) and the weakening of traditional institutions after the 1975 land reform (see chapter 8). Nevertheless, attempts have been made to promote some of these culturally entrenched practices, knowledge, and institutions in new small-scale irrigation schemes. The slow development of RWH in Ethiopia in recent years has been attributed primarily to lack of information on their characteristics and constraints

(Kamara & McCornick, 2002) and high cost of building materials (Alamerew, 2006; Alamerew & Natea, 2004), but inadequately planned and implemented government programs are also a major factor.

The Period Prior to 1990

Periodic famines during historic times and in recent decades, including those in the early 1960s and in 1974–1975, the latter of which contributed to the demise of the Haile Selassie regime, did not generate much interest in RWH in Ethiopia. After the 1984–1985 famine, the Derg regime started to promote RWH pond construction for irrigation and household needs through the then Ministry of Agriculture. All *woredas* were instructed to coordinate the planning and construction of at least two ponds per peasant association, although these directives were not closely adhered to, largely due to the top-down planning approach employed by the government and a general lack of experience with pond construction in Ethiopia. Improved rainfall after the 1984–1985 famine further contributed to obliterate the pond construction initiative.

Rainwater Harvesting Since 1990

The need for alternative water sources in Ethiopia received new impetus in the 1990s, when some NGOs began to introduce various RWH technologies to rural communities facing water shortages, mostly in the lowlands. Most of the technologies introduced proved to be viable alternatives for water production, particularly for domestic consumption. In 2002, the Ethiopian government started to promote RWH ponds in the four regions with the largest populations (Oromia, SNNPR, Amhara, and Tigray) on a large scale. The strategy is based on constructing water harvesting ponds near households for close management by family members. The stored water is used for supplementary irrigation during the growing season to reduce the risk of moisture stress. Many households use RWH ponds to meet some of their domestic water needs, thereby also reducing the time involved in fetching water at distant water sites (Oromia Regional Government Irrigation Development Authority, 2004a, 2004b). In 2005, the Ethiopian government launched the Comprehensive

Food Security Program (CFSP), which comprises the Productive Safety Net Program (PSNP) and related food security interventions. The major components of the PSNP are public works and the creation of communal assets, including RWH ponds. By 2007, about 65,000 communal ponds had been constructed through the PSNP program, but little information is available on their operation and maintenance. Unlike the previous program, the PSNP program constructed ponds primarily for livestock and occasionally for backyard supplementary irrigation when rainfall was inadequate (Food Security Coordination Bureau, 2007). A sand dam, an underground water storage technology first developed in Kenya and introduced to Ethiopia by a NGO in 2007 in the form of a pilot project, is a promising RWH technology in arid and semiarid regions (Rainwater Harvesting Implementation Network, 2008).

Absence of a legal framework for RWH has hampered its development in Ethiopia, a situation prevailing in all African countries except South Africa, where water legislation recognized RWH as a possible supply for private households (Hartung & Patschull, 2001). Although the Federal Water Resources Policy of Ethiopia provides for the need to promote and enhance rainwater harvesting, it does not provide for the independent planning and development of water works by municipalities and towns or the use of a second household source (Federal Democratic Republic of Ethiopia, 2000). The idea of one water source per household originated in water-rich countries in the North and was adopted by African countries but contrasts with the traditional practice of using multiple sources in water-scarce communities and also runs counter to the concept of water as a basic human right. Ethiopia and Uganda have been moving toward developing a comprehensive policy and legislative framework for implementing RWH (Government of the Republic of Rwanda, 2007; Hartung & Patschull, 2001).

High-Need Areas for Rainwater Harvesting Development

Needs for RWH are greatest in the drought-prone northeastern, eastern, and southern lowlands and in the Ethiopian Rift. Mean annual rainfall ranges from less than 400 mm (in the northeastern part of the Rift

Valley) to over 2,000 mm in the western mountains of Oromia and SNNPR, with most of the highlands receiving between 800 mm and 1,600 mm of precipitation. In the western half of the country, rain usually falls for 4–10 months during the summer (*keremt* in Amharic), with erratic rainfall or a distinct dry season (*bega*) between October and April, the length of which increases from south to north. In the eastern part, the main rainy season extends from June to September and is followed by a dry season *(bega)* from October to January and the small rains *(belg)* from February to May. In the southern lowlands, precipitation falls in relatively small amounts during spring (March to May) and fall (October to November). The length of the dry periods is generally inversely correlated with the annual rainfall totals in this and other parts of the country (figure 5.1). Annual and spatial variations in rainfall are considerable, often within short distances, resulting in many microclimates and highly unpredictable precipitation in most years. Rainfall unpredictability has increased since the mid-1980s in most parts of Ethiopia (Seleshi & Camberlin, 2005; United States Agency for International Development, 2003), further discussed in chapter 3. Assessments of the need for RWH in a given area will also have to consider population density, type of subsistence economy, particularly the mix of farming and livestock husbandry, conservation methods, and cultural preferences. According to the 1994 national census, nearly all *woredas* in the lowlands had densities below 44 persons/km^2, about half of the *woredas* in the central and northern highlands had densities of 95–231 persons/km^2, 18 *woredas* had densities of 232–412 persons per km^2, and 7 *woredas* had 413–725 persons/km^2, nearly all of them in the highlands in south-central Ethiopia (SNNPR and Oromia regions). Most districts in the remaining areas of the highlands above 1,500 m altitude had intermediate densities (95–231 persons/km^2), and those in nearly all the peripheral lowlands had only 2–43 persons/km^2 (figure 5.2; see also Kloos, Haile Mariam, & Adugna, 2006).

The Rift Valley is another area that may greatly benefit from RWH. The major water sources in the Rift Valley include groundwater, springs (including hot springs), lakes, and rivers, many of which are unusable

FIGURE 5.1. Seasonal distribution of rainfall in Ethiopia and Eritrea for 38 stations and by rainfall region.

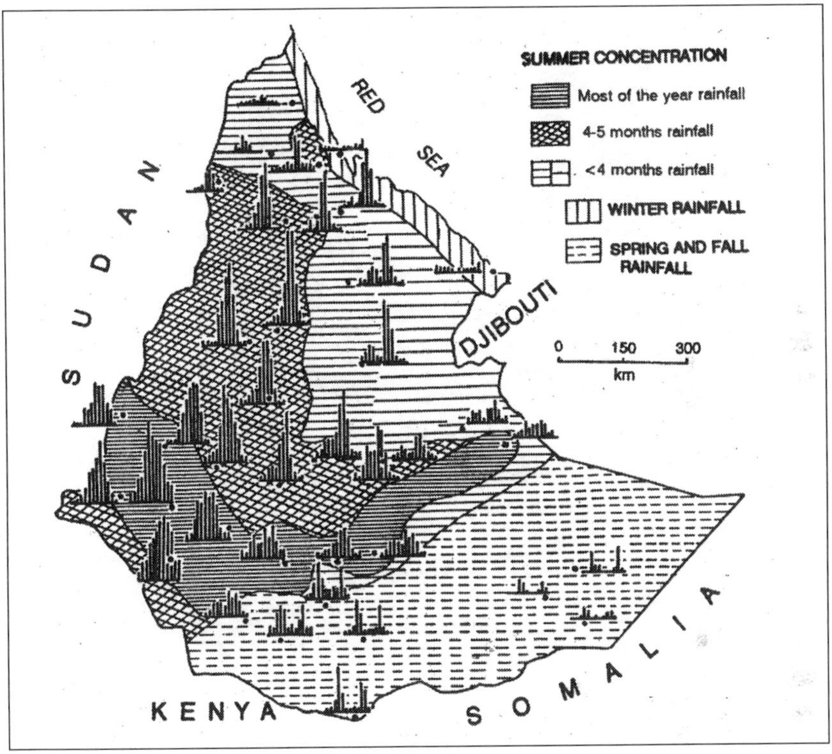

Source. Ethiopian Mapping Authority (1988), National Meteorological Services Agency (1996), and Dr. Bekure Wolde Semait, unpublished map.

for human and livestock consumption because of high levels of fluoride, salt, and other geochemicals. Utilization of surface water is limited around some towns due to the constraint of treatment costs. Most of the rural water supply is dependent on small springs and boreholes. Water quality problems associated with groundwater, dominantly high fluoride concentrations, have made difficult the effort toward utilization of this source to meet the demand for potable water supplies. Water supply and

FIGURE 5.2. Population density, by *woreda*, 1994.

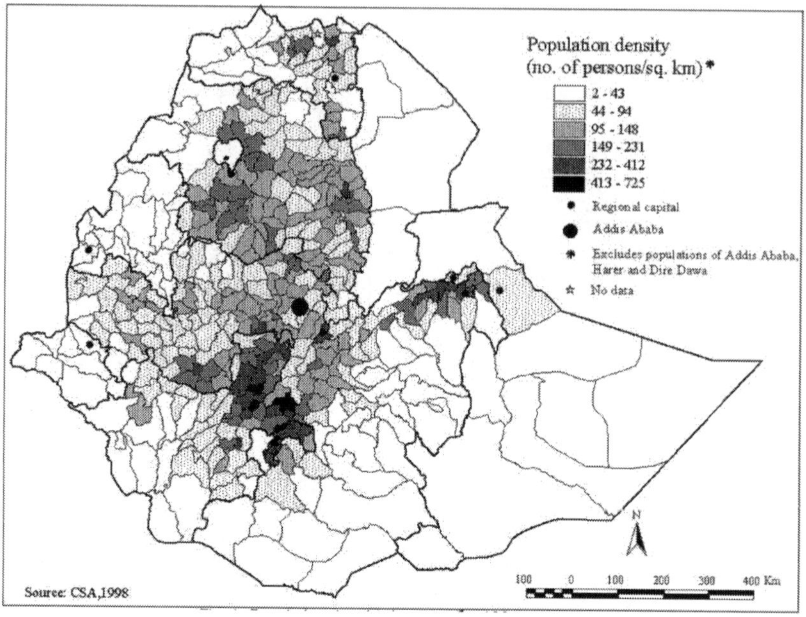

Source. Based on 1994 census data (Central Statistical Authority, 1998).

sanitation coverage in the Rift Valley are low. Only 60% and 88% of urban areas and 45% and 40% of rural areas have access to potable water supply in SNNPR and Oromia Region, respectively (Halcrow Group, 2007). Water, sanitation, and sewerage coverage of the Rift Valley communities is similarly low, especially among pastoralists whose seasonal mobility and fine-tuned water resources management in their once extensive areas traditionally minimized the need for these services. Population growth and increasing sedentariness, urbanization, and reduction of the size of their rangelands (see chapter 9) increase the need for stationary water supplies and sanitary facilities. RWH is an appropriate technology to relieve acute water shortages in the Rift Valley, where most surface waters are alkaline/saline, especially if soil conditions of the area can be dealt with. The dominant soil in the Rift Valley is silt loam with low water holding capacity, discouraging the construction of ponds because of seepage.

Pastoralists and agro-pastoralists inhabiting the Ethiopian lowlands outside of the Rift Valley have traditionally constructed different RWH systems, mostly communal ponds and cisterns, to store rainwater for use during the dry season. No statistics are available, but it is believed that several thousand traditional RWH structures exist in the semiarid lowlands (DCG, 2002). In the Ogaden area of Somali Region, ground runoff-fed *birkas* (underground tanks) and *haffirs* (low earth dams) are water harvesting structures used by Somali pastoralists (Development Technology Unit, 2002; Gebre Michael, 2006). The Borana Oromo pastoralists in the southern part of Oromo Region traditionally relied on deep wells (*ellas*) for all human and livestock water needs. These wells are no longer able to meet these needs in many communities due to population increase and permanent settlement, pointing out the need for water harvesting alternatives in areas lacking streams and rivers.

The Government's Water Harvesting Program

Droughts and crop failure, chronic food insecurity, and increasingly erratic rainfall in the highland area prompted the Ethiopian government to promote RWH and start an intensive pond construction program as a strategy to overcoming acute food insecurity within a short time. The large-scale campaign was aimed at popularizing RWH among rural populations using the mass media, public meetings, and training at various levels (Mills, 2004), with the objective of storing runoff during the rainy season for use during the dry season. The government launched the program in the four regions with the largest populations (Tigray, Amhara, Oromia, and SNNPR). Within each of these regions, particular emphasis was given to areas where rainfall distribution has become increasingly erratic in recent years, many of them highland areas where farmers generally have less experience with RWH than lowland populations (Ministry of Agriculture and Rural Development, unpublished data).

Various RWH technologies were introduced to the farmers, awareness was created, and skill training provided. Farmers were expected to adopt these technologies on their private holdings using their own resources. But because RWH activities were not an integral part of *woreda* annual plans, implementation of the program faced implementation difficulties, and *woreda* technical support to communal pond construction activities was limited (Rami, 2003; S. Natea, unpublished data). Other difficulties encountered during this hastily planned and implemented program are described below.

Under the Comprehensive Food Security Program, the government is providing both financial and technical support for communal pond construction. After a slow start in the 1990s that was hampered by top-down approaches, ponds are now being planned and implemented by communities based on the Community Based Participatory Watershed Development Approach using the existing administrative structure. At the federal level, the Ministry of Agriculture and Rural Development is responsible for overall support, including technical assistance and coordination. The implementation of RWH activities is undertaken by the regions, *woredas,* and *kebeles,* with the support of the Bureau of Water Resources and some NGOs (Oromia Regional Government, 2004b). The typical hierarchical implementation structure established in different parts of the country consists of technical support committees at the regional, zonal, and *woreda* levels and construction teams in the *kebeles*.

In addition to ponds, ferro/cement and stone tanks were demonstrated to farmers and constructed in selected areas. Both systems were built under the technical supervision of government technical staff for demonstration purposes. Four standard trapezoid-shaped pond designs were made available to farmers to select from. The storage capacity of the five types of ponds constructed ranged from 81 m^3 to 306 m^3 (Mills, 2004). While the standard design of ponds is simple, their management and maintenance by farmers and development agents has been difficult.

Between 2003 and 2006, 952,120 RWH ponds were reportedly built in the four regions of Amhara (327,670 ponds), Oromia (267,039),

Table 5.1. RWH storage structures constructed by the Ministry of Agriculture and Rural Development in 2003–2004.

Year	Types of water harvesting storage structures				Total
	Dome-shaped tanks	Hemispherical tanks	Household ponds	Community ponds	
2003	266	5,116	82,274	1,318	88,974
2004	779	9,858	216,101	27,134	253,872
Total	1,045	14,974	298,375	28,452	342,846

Source. Ministry of Agriculture and Rural Development, unpublished data. www.moard.gov.et

SNNPR (242,511), and Tigray (114,900; Ministry of Agriculture and Rural Development, unpublished data). According to online information and personal communication by one of the authors, Sorssa Natea, with officials of the Ministry of Agriculture and Rural Development involved in program implementation, the government constructed 342,846 water harvesting storage structures in 2003 and 2004 alone, of which 87.0% were household ponds, 8.3% community ponds, and 4.7% covered storage tanks (table 5.1). Water harvesting became the main activity of *woreda* agriculture offices between 2003 and 2006. Subsequently, government support for RWH shifted to the broader National Food Security Program. No information is available on the proportion of ponds that could be successfully maintained, but implementation difficulties, briefly described below, indicate that a considerable number were deficient or became unusable shortly after their completion.

Implementation Challenges and Evaluation of the Pond Construction Program

Of the various RWH schemes initiated by the government, only the pond construction program has been evaluated to date, although inadequately.

There is considerable evidence that while it contributed toward reducing water scarcity, this large-scale program was characterized by a number of shortcomings at the planning and implementation stages and the failure to achieve full and effective participation of stakeholders.

Program goals were too ambitious and construction targets set by the regions too high, beyond the capacity of the *woredas* and *kebeles*. In an attempt to meet government construction quota, implementing institutions focused on physical quantity, particularly at the *kebele* level, at the expense of quality of design and construction (Oromia Regional Government, 2004a). Particularly in areas where pond construction was new, site selection was often inappropriate. For example, in some cases, catchment sizes were too small for adequate runoff for ponds. In other cases, ponds were located on sandy soil, rock outcrops, and in marshy areas (Mills, 2004; Rami, 2003). Most of the staff involved in pond construction and supervision lacked adequate experience in RWH and pond design, and construction procedures were not closely followed. As a result, many ponds did not meet construction standards, reducing their storage capacity and discouraging many farmers to adopt RWH technology. Nevertheless, satisfactory outputs were achieved in communities where farmers had traditionally constructed ponds. There, farmers had sufficient RWH management skills in constructing and maintaining ponds and in efficiently using the stored water, permitting them to alleviate their water shortages (Mills, 2004; Rami, 2003).

Water loss through seepage was a major problem, particularly in the Rift Valley and other lowland parts of the country where the soil has low water holding capacity. To control seepage, ponds were lined with clay soil and plastic sheets, with variable results (Rami, 2003). In areas of vertisols (black cotton soils), many cement ponds cracked. Lack of irrigation knowledge further reduced the potential benefits of ponds. Although the storage capacity of ponds allowed for the irrigation of only small plots of land, some farmers cultivated crops with high water demands, such as rice, bananas, and sugar cane or irrigated larger areas than recommended, resulting in under-irrigation and in small crops or crop failure (Mills, 2004).

Level of farmers' experience in water harvesting was found to be associated with their willingness to accept RWH technology. However, quotas for the construction of RWH units at the *kebele* and *woreda* levels did not always take into account public awareness of the nature and potential of RWH and farmers' knowledge and experience in managing schemes. The top-down approach used did not foster a sense of local ownership and the establishment and strengthening of water committees and other community-based organizations, further reducing the participation of communities in the planning and implementation processes and, thus, the acceptance rate. Most of these mishaps might have been prevented if the program had been guided by consideration of indigenous knowledge, best practices in RWH, the improvement of traditional practices, and the adaptation of new technologies to local conditions and needs.

Rainwater Harvesting in Tigray and Oromia Regions

Evaluation of the effectiveness and impact of pond construction programs was carried out in Tigray and Oromia regions after the second year of implementation, briefly summarized below. In Oromia Region, evaluation was conducted by a team of government staff, while the Tigray Region evaluation was done by an international consulting firm. Both evaluation missions concluded that RWH ponds can effectively contribute to food security if well planned and implemented (S. Natea, unpublished data).

The RWH program in Tigray, initiated to improve agriculture productivity, was supported by the Ethiopian government, the European Commission (EC), and other donors. Farmers participating in RWH pond construction were paid relief grain and cash. Over 3 million Birr of the EC Fund were utilized for the construction of 30,450 RWH clay and plastic-lined ponds in 16 food insecure *woredas* (Mills, 2004). Half of the ponds visited by the evaluation mission were either empty or had little water. The stored water in the ponds was used by the farmers for different purposes, including household consumption, livestock watering, construction, and irrigation. Irrigation of vegetables proved to be more profitable than of cereal crops, making RWH ponds more economical for the cultivation of high-value crops (Mills, 2004). Siltation reduced the capacity and lifetime of ponds, also noted among microdams in the Tigray Region, discussed further in chapter 3.

Out of Oromia's 255 *woredas*, 88 were selected for water harvesting program implementation, based on their vulnerability to drought. To support the construction of ponds, training was provided on site selection and construction techniques of ponds at different levels in the *woredas*. During the first year, 27 *woreda* experts, 179 development agents, and 760 model farmers were selected and trained. Model farmers were selected from each *kebele*. After returning to their *kebeles,* the model farmers trained fellow farmers in pond construction. Except for the provision of hand tools, farmers received no compensation, with all construction work being carried out by volunteered community labor (Oromia Regional Government, 2004a). By the end of 2004, more than 31,000 domestic and 20,887 irrigation ponds had been constructed, as well as a small number of underground concrete tanks for demonstration purpose. Seventy-five percent of the irrigation ponds and 67.2% of the domestic storage ponds were in drought-prone highland zones.

Utilization patterns of stored rainwater for food production were strongly influenced by regional differences in agricultural systems, experience with RWH, and rainfall. In the highland zones in eastern Oromia Region, where rainfall is often inadequate for crop production and farmers have developed efficient water use and conservation methods, the stored water was used primarily for crop production. In pastoralist and agropastoralist areas in southern Oromia Region, harvested water was used mostly for livestock production. In the more humid Jimma and Illubabor zones of western Oromia Region (1,200–2,200 mm annual precipitation), most farmers have no experience in RWH and did not use the stored water for crop production (Oromia Regional Government, 2004b).

The few studies undertaken to date on health impacts of RWH do not provide conclusive information. A study in Amhara and Oromia regions reported increased malaria risk to pond owners placing their ponds within close proximity of houses due to tenure insecurity and convenience (Waktola, 2008). Information from Tigray Region, on the other hand, indicated that pond water warmed by black plastic in ponds prevents mosquito breeding (Rami, 2003).

NGO-Supported Projects

A number of NGOs (nongovernment organizations) have promoted and supported RWH projects in all administrative regions since the mid-1990s. No reliable statistics exist on RWH systems supported by NGOs, but field observations and discussions with officials by S. Natea indicate that three major NGOs had constructed by 2007 about 4,700 to 8,200 RWH units, mostly ponds, followed by roof catchments, ground catchments, *haffir* dams, and rock catchments (table 5.2). Community participation, training of local workers, and the formation of water committees minimized construction costs and increased the sustainability of systems. Nevertheless, dependence on outside funding has been an impediment to the construction of stone/cement and other relatively expensive RWH systems (Alamerew, 2006). New impetus to develop RWH systems has come from small sand dams, first introduced by a NGO in Ethiopia in 2007 and described in the section "Sand Dams."

Table 5.2. Estimated number of RWH harvesting systems constructed in Ethiopia by NGOs by 2007.

Organization	Roof catchments	Ground catchments	Rock catchments	*Haffir* dams	Ponds
CARE	200–300	500–600	20–30		1,000–1,500
World Vision	100–200				1,000–1,500
Water Action	40–50				50–60
Others	300–400		30–40	400–500	2,000–3,000
Total	640–950	500–600	50–70	400–500	3,050–6,060

Source. S. Natea, unpublished data.

Unlike the government, most NGOs have involved target communities in the planning, design, and implementation of RWH projects. Considerable efforts were also made to strengthen community organizations and social networks toward these objectives, mostly with encouraging results. Nevertheless, the technologies that were introduced were not widely replicated since some of these require costly material inputs and technical skills which are beyond the capacity of communities (Alameraw, 2006).

The establishment of the Ethiopian Rainwater Harvesting Association (ERHA) in 1999 as part of a regional movement in Eastern and Southern Africa to promote RWH has been crucial in increasing awareness, establishing networks with international NGOs and local civil society organizations and NGOs, and building capacity toward developing, adapting, and disseminating the concept and techniques of RWH. Ethiopian partnership initiatives involving sprouting local organizations, recipient communities, and international donors have successfully coordinated the implementation of appropriate and sustainable RWH projects in a number of Rift Valley and highland communities. Inclusion of women in elected water and sanitation committees in several projects has been associated with efficiency and water conservation in the management of several community-owned RWH systems using ferro/cement tanks (Alamerew, 2006, 2007; Alamerew & Natea, 2004) and of privately owned *birkas* (Guleid, 2002). Nevertheless, effective participation of Ethiopian women in decision-making processes, urgently needed in view of their experience in managing household water supplies and environmental sanitation, remains unsatisfactory because of cultural and social constraints (Natea, 2002). Our review of the NGO literature indicates that the crucial role of women in RWH development is rarely recognized by program managers preoccupied with technical details and funding issues. Cultural and social restrictions further limit the active participation of women in decision making regarding the management and sustainability of water supply systems through water and sanitation committees. Thus, while NGOs generally require that women be represented at these committees, which are responsible for distributing harvested water, maintaining structures, preventing water contamination, and educating others in the community about improved sanitation and hygiene, age-old customs prohibiting women from speaking out in public limit their participation (Natea, 2002).

Roof Water Harvesting

Roof water harvesting systems provide access to relatively clean water close to home, thus reducing the time spent by household members, mostly females, to carry water from distant sources. Collecting household water from distant stream and river sources often takes up to 5 hours a day in Ethiopia, especially during drought periods; reducing this time could contribute to overall socioeconomic development by reducing opportunity costs and freeing time for personal and household pursuits (Calow, MacDonald, Nicol, & Kebede, 2002). It is not known how soon after the commencement of modern urbanization in Ethiopia in the late 19th century and the construction of houses with corrugated iron roofs (Pankhurst, 1968) people began to collect water from roofs. Teka (1977) and Kloos et al. (1980/1981) first reported the collection of roof water in barrels and *enseras* (clay jars) in small Ethiopian towns four decades ago, noting a general preference of this water over river water.

Relatively easy management of RWH systems by home owners and their effectiveness in nearly all climates regardless of local geology or topography are additional advantages over competitor technologies. However, the high cost of roof water harvesting systems, discussed below, their relative ineffectiveness in meeting water needs during longer droughts, lack of experienced installers, and the unsuitability of thatched roofs for roof water harvesting limits their use in poorer rural regions with scarce and erratic rainfall (Thomas & Martinson, 2007). No negative attitudes toward roof water harvesting have been reported from Ethiopia, unlike in other sub-African countries, where this technology has been associated with traditional or pagan cultures and where some church groups consider it backward and pre-Christian (Development Technology Unit, 1999). The predominance (nearly 90%) of thatched roofs in rural areas limits the potential use of roof water in Ethiopia (Haile & Natea Merga, 2002). The requisite corrugated sheet metal or another type of clean, hard-surface roofs for the production of clean water, the greater availability of construction materials, and higher incomes in towns make the installation of RWH systems particularly suitable for urban areas, although the need for them is greater in rural areas. The harvesting potential in larger towns is considerable. According to GIS-based analyses, an estimated 11.44 million m^3 can be harvested annually from roofs in Addis Ababa (United Nations Environmental Program, 2005). The potential problems of chemical contamination of roof water from environmental sources, microbial contamination from animal and human sources, and mosquito breeding in tanks can be prevented by proper construction and maintenance of RWH systems (Lye, 2009; Thomas & Martinson, 2007).

Selected buildings in rural areas (schools, clinics, churches, mosques, office buildings, stores, and sufficiently large residential buildings) where there is acute shortage of domestic water supplies without alternatives or where a supplementary water source is needed by households during the dry season may also benefit from this technology. The government's plan to provide universal access to water supply and sanitation by 2012 under the Ethiopian Universal Access Plan for WASH Program, described in chapter 6, may have to consider the use of roof rainwater systems as an appropriate technology in areas lacking adequate water resources (Alamerew, 2007).

Rainfall distribution and roof size of individual buildings, two critical parameters in the selection and efficiency of roof rainwater systems, vary considerably in Ethiopia. Three seasonal rainfall regions have been identified: 1) a region with summer rainfall concentration (peaks from June–September) covering the Ethiopian highlands, the western lowlands, and most of the northeastern lowlands, 2) a smaller region of spring and fall rainfall (bimodal, peaking between March–May and October–November) in the southern and southeastern lowlands, and 3) a region receiving scarce winter rainfall mostly from December–January along the Red Sea coast (figures 5.1 and 5.2). Mean annual rainfall ranges from about 600–2,400 mm in the highlands, with the lowest values in the northern and eastern parts, and declines with altitude, reaching less than 200 mm in the northern part of the Rift Valley and in the southeastern lowlands (figure 5.1).

Roof catchment areas typically range from 30–60 m^2 in Ethiopia; they can provide 20,000–40,000 L of water in areas receiving 800 mm of precipitation. Households with five members, the average size of households in Ethiopia (Central Statistical Authority, 2001), would obtain enough water in areas receiving between 1,000 and 2,500 mm precipitation if they depended on rainwater harvesting as their only water source and if their roofs were 55 m^2 or larger. Households with that roof size in areas receiving at least 700 mm would obtain enough water only if they used it as their main or supplementary, not their sole source (Thomas & Martinson, 2003). Quantitative data on intensity, duration, and frequency of rainfall under different rainfall regimes, storage volume required per person in households with different water demands, and seasonal supply and demand analysis for households using roof water as a sole and supplementary source will also have to be considered for the installation of systems with appropriate tank sizes (Tefera, 2008). Preliminary studies in several Ethiopian communities indicate that per capita water use from roof water harvesting systems significantly decreased during the dry season and that households using these systems as their sole source of water reporting experienced more water shortages than those using them to supplement their main supply (Development Technology Unit, 2003).

Preliminary evaluation of the durability of roof water harvesting systems installed on Ethiopian buildings indicated that gutters and down pipes that are being installed on public buildings have a short operating life because most schools, mosques, churches, and other community buildings in rural areas are not planned and constructed for the installation of RWH systems. While some roof rainwater harvesting structures on public buildings have been functioning for more than two decades, operation and maintenance requirements are high. The smaller RWH systems installed on residential houses tend to be more durable, and RWH systems with underground storage tanks have been functioned for several decades without major problems (S. Natea, unpublished data).

Sand Dams

Seven sand dams were first constructed in Ethiopia as a pilot project in Borana District in 2007, and 10 NGOs were trained in dam construction (Rainwater Harvesting Implementation Network, 2009). Sand dams are a cost-effective, ecologically and environmentally friendly technology having several advantages over surface dams. They use simple, inexpensive technology amenable to community participation, have a long operating life, benefit from rather than being obliterated by siltation, do not lose water through evaporation, provide clean water by protecting against contamination by livestock and other sources through soil filtration, and prevent breeding habitats for mosquitoes and aquatic snails transmitting diseases. These characteristics make sand dams cost effective, environmentally appropriate, and safe water sources for domestic, livestock, and agricultural needs in semiarid and arid areas lacking permanent surface and groundwater sources (Rainwater Harvesting Implementation Network, 2008).

While the sand dams recently constructed in Ethiopia had not been evaluated by mid-2009, the social and economic impacts of 500 sand dams constructed in the Kitui region of Kenya are very encouraging. In two catchments in Kitui, one with and the other without sand dams, access to and use of water for all needs, as well as crop production, household income, local small business capacity, vegetation density, and biodiversity increased, vector-borne diseases decreased, and exposure to drought was controlled in the former during the first seven years of operation, with no significant changes observed in the latter. The reduction in the time women walked to the water sources was achieved because of year-round availability of water from the sand dams that eliminates trips to more distant river sites during the dry season and decreases in the queues and, thus, waiting time. Moreover, the recharge of aquifers, effective soil filtration, and protection and storage of river water makes sand dams an adaptive technology to climate extremes, particularly drought, and thus climate change (Lasage, Aerts, Mutiso, & Vries, 2008). As with all other water storage systems, hydrological, socioeconomic, and environmental impacts and limits of up-scaling sand dams need to be evaluated at the river basin level to equitably share water with downstream users (Pachpute, Tumbo, Sally & Mui, 2009).

Management Under Different Ownership Systems

Type of ownership of RWH systems, reflecting culturally based preferences and social factors, has greatly influenced the effectiveness and sustainability of RWH systems. In pastoral areas, all natural resources (land, pasture, and water) have traditionally been owned communally. Communities cooperate in digging wells and constructing runoff water harvesting structures, such as ponds and storage tanks. The relatively large structures are constructed by the communities, as the resources required for their construction are typically beyond the capacity of individual households. Although numerous communal water harvesting schemes have been in use for several decades, their management has proven to be challenging. Although some communal water harvesting schemes were abandoned within a few years of their construction, they have a better chance of functioning sustainably when they are initiated and owned by the beneficiaries. Most government and NGO-sponsored communal water harvesting schemes failing to fully consider the preferences and priorities of the users and local organizations failed due to lack of effective community participation.

Private ownership of water harvesting schemes is a recent phenomenon, introduced with the ownership of *birkas*. Somali pastoralist households privately constructed *birkas* for human and livestock uses after the 1950s, when they were introduced from Somalia. Fewer *birkas* have been built in recent years, and 30–40% of these systems have become nonfunctional, partly due to a combination of increasing insecurity in the area, high labor requirements to maintain older structures, increasing poverty, and the introduction of other technologies, including modified *haffir* dams (Gebre Michael, 2006). In some other Ethiopian communities, private ownership has been found to be most appropriate for domestic ponds and roof water harvesting systems and has been associated with a propensity of users, mostly homeowners, to conserve water and maintain the operational efficiency of structures (Alamerew, 2007).

COST OF RAINWATER HARVESTING SYSTEMS

Availability and cost of construction materials, construction skills and knowledge of local people, and environmental parameters are key factors in the cost and sustainability of RWH systems. The cost of constructing 1 m^3 storage capacity of nine different water abstraction and groundwater systems used in Ethiopia was estimated for comparative purposes (table 5.3). These estimates are based on an analysis of costs incurred by projects carried out by different institutions and involve free labor, sand, and wood provided by communities. Costs are considered to be fairly representative of costs nationwide in the 1990s, but have more than doubled since then. Springs, the least expensive system, is the first choice where they are available, reflecting a long Ethiopian tradition of using this highly valued water source. Hand-dug wells and earth dams are also widely used because of their low cost and ease of abstraction. Earth dams are feasible where there is sufficient rainfall (at least 500 mm) and where enough runoff can be harvested from the catchment area. High evaporation and siltation rates and the generally flat topography of the lowlands render surface dam construction less cost effective than in the highland environment but provide ideal conditions for moderately priced sand dams. Clay-lined ponds are also affordable but lose water through seepage. Plastic-lined ponds and especially concrete-lined tanks are expensive (Sileshi & Kedir, 2005). Overall construction and maintenance costs of each cement-lined tank has been even higher than the data indicate in table 5.3 because of their tendency to crack in the seasonally expanding and shrinking vertisol soils. For example, 14 out of 16 tanks built in the highlands of eastern Amhara Region were found to leak heavily within 10 months of construction. Under pressure to fill quotas, *woreda* officials had ignored manual guidelines, which recommended not building cement tanks in black clay (vertisol soil). Similarly, indirect costs of tank construction were incurred when construction

TABLE 5.3. Cost of constructing RWH and groundwater systems for 1 m³ of water storage capacity.

Methods of water abstraction	**Cost per m³***
Spring development	0.004
Hand dug well	0.01
Earth dam	0.02
Pond (clay-lined)	0.05
Borehole	0.39
Sand dam**	0.63
Rock catchments	3.41
Birka	4.15
Roof water catchments	6.20***

Sources. Habtamu & Natea (1993) and Lasage et al. (2008).
*in US$
**sand dams in Kitui District, Kenya
***This figure is for 1992. Costs have increased to over $30 per m³ since then. See also Development Technology Unit (2002).

materials were delayed for months and seasonal rains caused erosion and caving in of numerous excavated holes that resulted in considerable loss of topsoil and impacted crop production (Rami, 2003).

Roof catchments, *birkas*, and other underground tanks are the most expensive systems, and their costs have risen sharply as a result of increases in the cost of cement and steel (the most expensive items) since the early 1990s, making them unaffordable without government or NGO support (table 5.3; Martinson, 2007; Natea, 2002). The cost of roof harvesting systems in three Ethiopian communities in 2000 ranged from US$690 to $1,164 (CARE Ethiopia, unpublished data; WaterAction, unpublished data). Nevertheless, the construction of *birka* underground storage tanks by Somali households for decades without outside assistance (Food and Agriculture Organization, 2007; Gebre Michael,

2006) and the elaborate drainage systems and reservoirs of the Konso (see chapter 8) suggests that larger structures may be feasible using local materials and manpower. *Birka* development also indicates that individuals seek new ways of exploiting available resources as cost of their construction in other pastoral areas in East Africa is often recovered within the commons by the sale of water. These new initiatives contribute to the commercialization of the pastoral economy but tend to exclude the poor (Helland, 2006). Hemispherical underground tanks with a capacity of about 100 m^3 have been installed by some farmers obtaining government and NGO credit (ERHA, unpublished data).

Research on improving existing designs, the development of low-cost tanks constructed by communities using local materials, and refined calculations of determining appropriate size of tanks are expected to make RWH systems more affordable (Rainwater Harvesting Implementation Network & Ethiopian Rainwater Harvesting Association, 2006; Tefera, 2008; Thomas & Martinson, 2007). Costs may be further reduced through lower management cost, greater efficiency, and further innovation of this evolving technology, including its adaptation for multiple household water uses such as sanitation, hygiene, and horticulture (Alamerew, 2006). Studies of farmer-initiated activities outside the mainstream government-promoted water harvesting approaches, including gulley rehabilitation and check dams using local materials (Rami, 2003) may provide valuable information for the design, implementation, and operation of appropriate, low-cost systems in the future.

CONCLUSION

Rainwater harvesting provides an ecologically and technologically appropriate approach to improving water security and the livelihood of Ethiopian populations facing inadequate and diminishing surface water resources, moisture stress, frequent crop failures, and difficulties of accessing domestic water sources in rural areas. Thus, RWH can contribute to achieving the objectives of relieving shortage of water for domestic supply and food insecurity provided that it is well planned, coordinated, and implemented involving the beneficiaries and receives sufficient technical and management assistance, especially in communities lacking experience with RWH. Awareness of RWH is still low from the policy makers' level down to the farmers and urban residents. More coordinated efforts are required to educate communities and other stakeholders about RWH. A particularly serious impediment has been the lack of a long-term vision and user-oriented planning and implementation by key stakeholders, particularly the government. Follow-up to consolidate results achieved and experience gained is lacking, impeding the scaling up of RWH. Similarly, the hastily implemented government-sponsored RWH program in Ethiopia resulted in oversights and deficiencies that affected the acceptability, viability, and sustainability of systems from technical, social, and economic points of view, a problem which needs to be examined at all administrative levels. Assessment of program impact on household livelihood is particularly needed for the planning and implementation of affordable, culturally acceptable, technically and environmentally appropriate, and, thus, sustainable RWH systems.

Studies also need to be carried out on the feasibility, acceptability, and sustainability of various RWH systems in different climatic zones, terrains, and cultural areas in Ethiopia to gauge needs, preferences, intended uses, affordability, and maintenance of systems. Research may guide the selection of appropriate technologies in different environments, terrain, and on different buildings, inform on the availability and affordability of materials or their substitution with local materials, and on the viability of social structures and community participation needed to assure the sustainability of RWH systems.

REFERENCES

Alem, G. (1999). Rainwater harvesting in Ethiopia: An overview. In J. Pickford (Ed.), *Proceedings of the 25th WEDC Conference: Integrated development for water supply and sanitation* (pp. 387–390). Laughborough, U.K.: Water, Laughborough University, Engineering and Development Center.

Alamerew, E. (2006). *Towards widespread rainwater harvesting in Ethiopia: ERHA's experience as a successful Rainwater Harvesting Capacity Centre (RHCC)*. Paper presented at the 10th SEARNET (Southern and Eastern Africa Rainwater Network) Conference, Mombassa, Kenya.

Alamerew, E. (2007). *Rainwater harvesting for augmenting the rural wash and urban WDM programmes: Experiences of ERHA and its partners in Ethiopia*. Paper presented at the First International WASH Practitioners' Knowledge Fair and the 19th Regional Sanitation Seminar, Mombassa, Kenya.

Alamerew, E., & Natea, S. (2004). *Participatory assessment of rainwater harvesting systems providing domestic water supply: Case studies in the Rift Valley area, central Ethiopia* (Unpublished manuscript). Ethiopian Rainwater Association, Addis Ababa.

Awulachew, S. B., Merrey, D. J., Kamara, A. B., Van Koppen, B., de Vries, F. P., & Boelee, E. (2005). *Experiences and opportunities for promoting small-scale/micro irrigation and rainwater harvesting.* Colombo, Sri Lanka: Water Management Institute.

Butzer, K. (1981). Rise and fall of Axum, Ethiopia: A geoarchaeological interpretation. *American Antiquity, 3*, 471–495.

Calow, R., MacDonald, A., Nicol, A., & Kebede, S. (2002). *The struggle for water: Drought, water security and rural livelihoods*. Keyworth, Nottingham: British Geological Society.

Central Statistical Authority. (1998). *The 1994 population and housing census of Ethiopia: Results at country level*. (Vol. 1, Statistical report). Addis Ababa: Author.

Central Statistical Authority. (2001). *Report of the 2000 welfare monitoring survey*. Addis Ababa: Author.

Development Technology Unit. (1999). *Attitudes towards rainwater harvesting* (Domestic Roofwater Harvesting Program, Report ERB IC18-CT98-0276). Warwick, U.K.: University of Warwick, School of Engineering.

Development Technology Unit. (2002). *New technology for very-low-cost domestic roofwater harvesting* (Domestic Roofwater Harvesting Program, DFID KaR Contract R78833). Warwick, U.K.: University of Warwick, School of Engineering.

Development Technology Unit. (2003). *Very-low-cost domestic roofwater harvesting in the humid tropics: User trials* (Domestic Roofwater Harvesting Program, Report R3). Warwick, U.K.: University of Warwick, School of Engineering.

Ethiopian Mapping Authority. (1988). *National atlas of Ethiopia*. Addis Ababa: Author.

Falkenmark, M., & Rockström, J. (2004). *Balancing water for humans and nature: The new approach in ecohydrology*. London: Earthscan.

Fattovich, R. (1990). Remarks on the pre-Aksumite period in northern Ethiopia. *Journal of Ethiopian Studies*, *23*, 1–33.

Federal Democratic Republic of Ethiopia. (2000). *Proclamation No. 197/2000. Ethiopian Water Resources Management Proclamation. Federal Negarit Gazeta* (6th Year, No. 25, pp. 1250–1258). Addis Ababa.

Food and Agriculture Organization. (2007). *Potential of rainwater harvesting in Somalia: A planning, design, implementation and monitoring framework* (Somali Water and Land Information Management Project, Project Report No. W-09). Nairobi: Author.

Food Security Coordination Bureau. (2007). *2007 Progress Report.* Addis Ababa: Author.

Gebre Michael, Y. (2006). Water harvesting technologies in assuring food security: Lessons from the pastoral areas of Somali Region, Ethiopia. In G. Alemu, B. Simane, & M. Feseha (Eds.), *Proceedings of the Conference on Poverty and Development in Ethiopia: Challenges and options* (pp. 90–101). Addis Ababa: Addis Ababa University, Institute of Development Research.

Gill, R. D. (2000). *The great Maya droughts: Water, life and death.* Albuquerque: University of New Mexico Press.

Government of the Republic of Rwanda. (2007). *Legal aspects of rainwater harvesting in Sub-Saharan Africa.* Report prepared by the Ministry of Agriculture and Animal Resources in collaboration with the World Agroforestry Centre (ICRAF), Kigali.

Guleid, A. A. (2002). Water harvesting in the Somali National Regional State of Ethiopia. In M. Hailu & S. Natea Merga (Eds.), *Report of the Workshop on the Experience of Water Harvesting in the Drylands of Ethiopia: Principles and practices* (pp. 28–30). Ås, Norway: Drylands Coordination Group.

Habtamu, G., & Natea, S. (1993). *Area based integrated development: Water resource development component.* Report prepared for the Disaster Prevention and Preparedness Commission/UNDP, Addis Ababa.

Haile, M., & Natea Merga, S. (Eds.) (2002). *Workshop on the Experience of Water Harvesting in Dry Lands of Ethiopia: Principles and practices* (DCG Report No. 19). Ås, Norway: Dryland Coordination Group.

Halcrow Group. (2007). *Limited and generation integrated rural development: Rift Valley Lakes Basin integrated resources development master plan.* Study project inception report prepared for the FDRE Ministry of Water Resources, Addis Ababa.

Hartung, H., & Patschull, C. (2001). *Framework for the inclusion of rainwater harvesting in national water legislation in Sub-Saharan Africa*. Stuttgart: FAKT Consult for Management, Training and Technologies.

Helland, J. (2006). *Pastoral land tenure in Ethiopia*. Paper presented at the Colloque International, Les Frontiéres de la Question Fonciére-At the Frontier of Land Issues, Montpellier, France.

Kamara, A., & McCornick, P. (2002). Synthesis of research issues and capacity building in water and land resources management in Ethiopia. In P. G. McCornick, A. B. Kamara, & G. Tadesse (Eds.), *Integrated water and land management research and capacity building priorities for Ethiopia: Proceedings of the MoWR/EARO/IWMI/ILR International Workshop* (pp. 204–215). Addis Ababa: Ministry of Water Resources.

Kloos, H., Haile Mariam, A., & Adugna, A. (2006). Population. In Y. Berhane, D. Haile Mariam, & H. Kloos (Eds.), *The epidemiology and ecology of health and disease in Ethiopia*. (pp. 31–51). Addis Ababa: Shama Books.

Kloos, H., Higashi, G. I., Schinski, V. D., Mansour, N. S., Polderman, A. M., Lemma, A., & Murrell, K. D. (1980/81). Human behavior and schistosomiasis in an Egyptian village and an Ethiopian town: Tensae Berhan and El Ayaisha. *Rural Africana, 8/9*, 35–64.

Lasage, R., Aerts, J., Mutiso, G.-C. M., & Vries, A. (2008). Potential for community based adaptation to droughts: Sand dams in Kitui, Kenya. *Physics and Chemistry of the Earth, 33*, 67–73.

Lye, D. J. (2009). Rooftop runoff as a source of contamination: A review. *Science of the Total Environment, 407*, 5429–5434.

Martinson, D. B. (2007). *Improving the viability of roofwater harvesting in low-income countries* (Unpublished doctoral dissertation). University of Warwick, Warwick, U.K.

McCann, J. C. (1999). *Green land, brown land, black land: An environmental history of Africa, 1800–1990*. Portsmouth: Heinemann and Oxford: James Curry.

Meera, V., & Ahammed, M. M. (2006). Water quality of rooftop rainwater harvesting systems: A review. *Journal of Water Supply Research and Technology-Aqua, 55*, 257–268.

Mills, L. (2004). *Evaluation of the water harvesting schemes component of the EC funded programmes IFSP 1998 and IFSP 2000 in Tigray Regional State*. Mekele.

Natea, S. (2002). Rainwater harvesting systems for domestic consumption. In M. Hailu & S. Natea Merga (Eds.), *Report of the workshop on the experience of water harvesting in the drylands of Ethiopia: Principles and practices* (pp. 40–45). Ås, Norway: Drylands Coordination Group.

National Meteorological Services Agency. (1996). *Climatic and agroclimatic resources of Ethiopia*. Addis Ababa: Author.

Oromia Regional Government Irrigation Development Authority. (2004a). *Annual implementation report for 2003–2004*. Addis Ababa: Author.

Oromia Regional Government Irrigation Development Authority. (2004b). *Evaluation report on benefits of water harvesting schemes implemented in 2002–2003 in nine zones of the region*. Addis Ababa: Author.

Pachpute, J. S., Tumbo, S. D., Sally, J., & Mui, M. L. (2009). Sustainability of rainwater harvesting systems in rural catchment of Sub-Saharan Africa. *Water Resources Management*, doi 10.1007/s11269-009-9411-8

Pandey, D. N., Gupta, A. K., & Anderson, D. M. (2003). Rainwater harvesting as an adaptation to climate change. *Current Science, 85*, 46–59.

Pankhurst, R. (1968). *Economic history of Ethiopia: 1880–1935*. Addis Ababa: Haile Sellassie I University Press.

Rainwater Harvesting Implementation Network. (2008). *A practical guide to sand dam implementation*. Amsterdam: Author.

Rainwater Harvesting Implementation Network. (2009). *Second sand dam training in Ethiopia*. Retrieved from http://www.rainfoundation.org/index.php?id=184

Rainwater Harvesting Implementation Network & Ethiopian Rainwater Harvesting Association. (2006). *Towards widespread rainwater harvesting in Ethiopia*. Paper presented at the fourth World Water Forum, Mexico City.

Rami, H. (2003). *Ponds filled with challenges: Water harvesting experiences in Amhara and Tigray*. Addis Ababa: Office for the Coordination of Humanitarian Affairs, United Nations Ethiopia Office.

Sileshi, Y., & Camberlin, P. (2005). Recent changes in dry spell and extreme rainfall events in Ethiopia. *Theoretical and Applied Climatology, 83*, 181–191.

Sileshi, Y., & Kedir, Y. (2005). *Water harvesting technologies: A challenge to Ethiopia in environmental, ecological, health conditions and economic sustainability*. Report prepared for the International Livestock Research Institute, Addis Ababa.

Tefera, C. (2008, May). *Enhanced performance and design of roof rainwater harvesting based on rainfall regimes: The case of Ethiopia* (Unpublished manuscript). Addis Ababa.

Teka, G. A. (1977). *Water supply in Ethiopia: An introduction to environmental health practice*. Addis Ababa: Addis Ababa University Press.

Thomas, T. H., & Martinson, D. B. (2003). *Roofwater harvesting: A handbook for practitioners*. Coventry, U.K.: University of Warwick Press.

Thomas, T. H., & Martinson, D. B. (2007). *Roofwater harvesting: A handbook for practitioners*. Delft, Netherlands: IRC International Water and Sanitation Center.

United Nations Environmental Program. (2005). *Potential for rainwater harvesting in Africa: A GIS overview*. Retrieved from www.unep.org/pdf/RWH-AFRICAN-CITIES.pdf

United Nations Environmental Program. (2006, November 13). *Harvesting rainfall a key climatic adaptation opportunity for Africa*. UNEP News Center Nairobi. Retrieved from http://www.unep.org/Documents.Multilingual/Default.asp?DocumentID=485&ArticleID=5420&J=en

United States Agency for International Development. (2003). *Rainfall in Ethiopia is becoming increasingly erratic* (Unpublished manuscript). Retrieved from http://www.reliefweb.int.rw/rwb.nsf/AllDocsByUNID/66e4686182a84da49256d5100071

Wahlin, L. (1995). *3000 years of household water collection in Jordan*. Paper presented at the Third Nordic Conference on Middle Eastern Studies: Ethnic Encounter and Culture Change, Joensuu, Finland.

Waktola, D. K. (2008). Malaria and pond-based rainwater harvesting linkages in the fringes of central highland Ethiopia. *Rural and Remote Health, 8*, 956.

Chapter 6

Water Supply and Sanitation

Mark Harvey

Introduction

Worldwide, the percentage of people served with some form of improved water supply increased from 79% (4.1 billion) in 1990 to 82% (4.9 billion) in 2000 to 84% in 2004 and 87% in 2006. Over the same period, the proportion of the world's population with access to excreta disposal facilities increased from 54% to 62%. In 2006, 884 million people were without access to improved water supply, down from 2.4 billion in 2000. It is estimated that the Millennium Development Goal target for drinking water can be achieved at current trends (United Nations Development Program, 2006; WHO/UNICEF, 2008; World Bank, 2008). However, large regional disparities persist. Whereas many regions have water supply coverage rates of 80% or higher, Sub-Saharan Africa has a coverage rate below 60%. Great disparities within the Sub-Saharan region exist, and some rural and urban poor people have even worse access to water and sanitation facilities than average figures suggest (WHO/UNICEF,

2008; World Bank, 2008). The economic cost and health impact of unsafe water supply and poor sanitation are enormous, as reflected by the fact that water- and sanitation-related diseases account for half of the world's hospital bed occupancy (WHO/UNICEF, 2000). The majority of these people were reported as living in Asia and Africa, where 2.2 million people, mostly children, die every year from diseases associated with lack of safe drinking water, inadequate sanitation, and poor hygiene (Department for International Development, 2001).

Safe water supplies and adequate sanitation facilities are central to raising health levels and, hence, quality of life in the developing world. Although estimates are widely diverging, it is generally agreed that 70–80% of communicable disease episodes are due to contaminated water or lack of sanitation (Tebbut, 1992). Some authors have argued that it makes little economic sense to invest large amounts of money in building hospitals and training doctors if the people they intend to serve face continued risk of water-borne diseases (Nadakavukaren, 1992). A broad range of issues in poverty reduction, including economic well being, human development, and economic sustainability relate to the availability and adequacy of safe water supplies (Ministry of Finance and Economic Development & United Nations Country Team, 2004). It was for this reason that the United Nations underscored the importance of safe water and sanitation at the World Summit for Sustainable Development held in Johannesburg in 2002 by calling for the proportion of people without clean water and proper sanitation to be halved by the year 2015 as one of the Millennium Development Goals (MDG) (United Nations, 2002). However, in developing countries, populations and their demands are growing rapidly beyond the sustainable yield of local resources (Cunningham & Saigo, 1995) and at the current rate of progress, it will take Africa many years to achieve the MDG water target, and the sanitation target may take as long as 70 more years (House of Commons, International Development Committee, 2007). In Ethiopia, the proportion of households with access to hygienic sanitation facilities declined during the 1990s as new construction fell behind population growth and the maintenance of existing facilities was neglected

(Terefe & Edstrom, 1999). Ethiopia's sanitation levels are among the lowest worldwide.

In this chapter, I will review aspects of coverage, access, and use of water supply and sanitation services in Ethiopia as well as sector plans, performance, and priorities. I will also discuss associated health aspects and current efforts and programs to increase the coverage and sustainability of facilities. I will present a sector management and development perspective as distinct from a technical one because implementation and operational challenges constitute the overriding constraints in scaling up water and sanitation coverage. Major sources of information for this chapter include the WASH Sector Review Report prepared under the auspices of the European Union Water Initiative (EUWI) Country Dialogue for Ethiopia (MWR/MH/ME/CRDA/EUWI, 2006a) and available in the electronic library held originally at www.itacaddis.org/italy, and other recent government reports. Readers are referred to Legesse, Haile Mariam, & Kloos (2006) for water supply and sanitation developments around the turn of the millennium.

The Government's Response: Recent Legislative Actions and Policy Initiatives

The government of Ethiopia's response to this situation of poor service coverage is founded in its sector policy framework which comprises the Ethiopian Water Resources Management Policy which was issued in 1999, the Ethiopian Water Sector Strategy which followed in November 2001, the Water Sector Development Programme 2002–2016 (WSDP) in 2002, and the National Water Supply and Sanitation Master Plan Framework which was completed in January 2003 (Ministry of Water Resources, 1999, 2000, 2002, 2003a). The WSDP and the National Water Supply and Sanitation Master Plan Framework provided plans and a great deal of background data, but neither of them culminated in major sector investment programmes. A stronger response by the sector has been made in the government of Ethiopia's overall poverty reduction strategies of the past several years (Ministry of Finance and Economic

Development, 2001, 2005). This response arose in 2005 as national and international political commitments to address water supply and sanitation reached a peak around the United Nations MDG Needs Assessment Initiative. As a result, the government of Ethiopia committed to launch a major sector programme—the Universal Access Plan (UAP) (Ministry of Water Resources, 2006). Universal coverage targets were approved by Parliament in August 2005.

In addition to the sector policy framework mentioned above, which largely focused on water, the Ministry of Health issued in 2006 and 2007 key policy documents that set out how it aimed to achieve universal access to improved hygiene and sanitation (Ministry of Health, 2005, 2006, Ministry of Water Resources, Ministry of Health, & Ministry of Education, 2006b). Following the establishment by proclamation of the Ethiopian Environmental Protection Authority (EPA) in 1995 to coordinate and regulate activities in environmental management, various environmental health proclamations were issued, including one on environmental pollution control (Federal Democratic Republic of Ethiopia, 2002). These various programs and legislative actions are important initiatives and can draw attention to the role of water as an essential resource for both economic development and human welfare. Any progress in water resources and sanitation development also requires promotion of conservation practices and water pollution abatement. The development of alternative water sources, particularly rainwater harvesting, is encouraged by the Ministry of Water Resources in its Universal Access Plan (UAP), but the actual means to promote this approach are less evident. This is discussed further in chapter 5.

Low levels of sanitation coverage and high prevalence of communicable diseases account for three fourths of all health problems in Ethiopia (Ministry of Water Resources, 2003b) and are symptoms that promotion of public awareness on hygienic practices is essential. Sanitation can be defined as the quality of living that is expressed in clean homes, farms, businesses, industries, neighborhoods, and communities. As a way of life, sanitation must originate with the people, since it is promoted by knowledge and grows as an obligation and an ideal in human relations (WaterAid, 2003).

Past Performance and Baselines

Most studies of basic minimum water requirements recommend 20–50 L per person per day for different climates, cultures, traditions, diets, technological levels, wealth, and whether or not households use flush toilets or pit latrines (Gleick, 1996). The most extensive longitudinal water use survey in Africa, carried out in two phases in the late 1960s and the late 1990s in 34 rural and urban study sites in Kenya, Uganda, and Tanzania, revealed that while average daily per capita water use declined by 30% (from 61.4 L to 39.6 L) during that 30-year period, water use by households without piped supplies almost doubled (from 11.0 L to 19.7 L). Major factors in the decline in piped water use included the inability of municipal authorities to operate and maintain effective and efficient water services and population growth resulting in greater demand (Thompson et al., 2001). The UAP takes a pragmatic approach and, on the principle of "some for all rather than all for some," adopts standards of 15 L per person per day (lpd) in rural areas and 20 lpd in urban areas—these being typical current practice. The UAP expects that in future phases, these volumes will increase.

Baseline data against which to report progress in water supply and sanitation in Ethiopia vary. According to the MDG baseline data, drawn from the global UNICEF/WHO Joint Monitoring Programme (WHO/UNICEF, 2000), water supply coverage in 1990 was 15% in rural areas, 81% in urban areas, and 23% nationally, and in 2004 was 11% in rural areas, 81% in urban areas, and 22% nationally. The Water and Sanitation Program (WSP, 2006) reported water supply coverage in 1990 as 11% in rural areas, 70% in urban areas, and 19% nationally; in 2000 as 23%, 74%, and 31%; and in 2004 as 31.4%, 83.1%, and 39.4%, respectively. The same WSP report put sanitation coverage at 3.9% in rural areas, 49.7% in urban areas, and 11.5% nationally.

Baseline data on rural access to safe water shows the variation across the regions and puts into context the recent trends in service coverage discussed below (table 6.1). While these various baseline figures are useful to show longer term trends, the actual numbers are less relevant. The overall picture shows some progress in urban areas but with a recent

TABLE 6.1. Proportion of population with access to safe water in rural areas, in 1994, 1998, and 2000, by region.

Regions	1994	1998	2000
Amhara	15.2	11.2	11.3
Oromia	15.8	14.2	17.4
SNNPR	15.5	14.6	20.4
Tigray	10.6	17.5	25.6
Afar	5.2	8.7	7.7
Somali	9.0	14.3	22.5
Benishangul-Gumuz	14.8	13.3	22.1
Gambela	21.3	19.8	24.6
Harari	11.2	34.5	47.7
Dire Dawa	23.6	64.7	57.2
Addis Ababa			
National	14.7	13.7	17.1

Source. Ministry of Water Resources (2003a).

stagnation and very slow progress in rural areas; the sanitation sector is performing particularly poorly.

Recent Sector Performance

Official recent performance is reported in the Ethiopian government's two annual reports on its Poverty Reduction Startegy (PASDEP) for 2005/06 and 2006/07 (Ministry of Finance and Economic Development, 2007a, 2007b). Draft progress figures for 2007/08 are also available from the Ministry of Water Resources. According to these reports, coverage of clean water supply increased in 2005–2006, 2006–2007, and 2007–2008 to 41.2%, 46.4%, and 53.9% in rural areas, to 78.8%, 82.0%, and 86.2% in urban areas, and to 47.3%, 52.5%, and 59.5% nationally (data for 2006–2007 and 2007–2008 shown in table 6.2). The reports

TABLE 6.2. Coverage of water supply (percent of population), 2005/06 and 2007/08 by region.

Regions	Rural		Urban		Total	
	2005–2006 pleas	2007–2008	2005–2006	2007–2008	2005–2006	2007–2008
Amhara	36.6	49.0	80.0	87.8	41.5	**53.7**
Oromia	40.2	52.0	87.6	97.9	46.5	**58.3**
SNNPR	53.0	63.0	64.5	72.1	54.0	**63.6**
Tigray	42.8	56.0	50.9	72.0	44.3	**59.1**
Afar	41.1	53.1	73.0	77.4	44.0	**55.4**
Somali	21.5	32.9	60.0	61.6	28.0	**37.9**
Benishangul-Gumuz	46.0	44.3	66.2	93.1	48.0	**49.3**
Gambela	41.4	43.9	37.0	98.6	40.6	**54.7**
Harari	29.0	41.0	21.0	27.5	24.0	**32.5**
Dire Dawa	57.0	75.8	72.0	72.0	68.2	**73.0**
Addis Ababa	–	–	90.1	95.0	90.1	**95.0**
National	**41.2**	**53.0**	**78.8**	**822.0**	**47.3**	**59.5**

Sources. Ministry of Finance and Economic Development, 2007a, 2007b. Data for 2007–2008 are unpublished Ministry of Finance and Economic Development data made available to the author in the course of his work.

also highlight that maintenance and rehabilitation of nonfunctional or poorly functioning rural water supply schemes has resulted in their proportion falling from 30% in 2004–2005 to 25% in 2005–2006, to 20% in 2006–2007, and maintained at 20% in 2007–2008. These very recent trends are encouraging, though less has been revealed about the surveys undertaken to gather this information, as it is widely accepted that the water and sanitation sector lacks an adequate monitoring and evaluation system with supporting management information system. Also lacking is an analysis of how these increases in coverage have been achieved—some donor-supported projects have been slow to disburse, while other efforts, including under the government's Food Security Programme, are not routinely captured in sector reporting. Regional variations are presented in table 6.2, and the report noted that in some regions urban coverage has declined in 2006–2007 owing to lack of investment. The coverage data for the period 2005–2006 to 2007–2008 can be compared with those of the National Water Supply and Sanitation Master Plan, according to which all regions (except the predominantly urban regions of Harari and Dire Dawa) had rural water supply coverage of less than 30% even as recently as 1998. The low coverage in Harari in 2005–2006 may be explained by the shortage of piped water in Harar Town experienced as a result of the decline and drying up in 2006 of its main water source—Lake Haramaya. The sharp increase in Somali (53.0%), on the other hand, is difficult to explain in this area of pastoral nomadism. The notorious challenge of measuring sector performance is evident here, and this will be compounded in the coming years by the changes in Ethiopian definitions of service standards brought about under the recently approved UAP. Progress on sanitation is less well known, although the recent annual review meeting of the health sector reported that sanitation coverage in 2006–2007 was 44.0% nationally, 8.0% in Gambela, 28.1% in Oromia, and 64.6% in Harari regions (Ministry of Health, 2007).

Future Demand and Targets

Best estimate projections of future demand are presented in the government's Universal Access Program (Ministry of Water Resources, 2006) for water supply and sanitation. The figures presented in this ambitious plan are not particularly precise in terms of demand projections but do provide a comprehensive set of sector targets to which the government of Ethiopia has given the highest political commitment. The UAP aims nationally to reach 100% service coverage in urban areas and 98% service coverage in rural areas by 2012 (the rural target is weighted between 100% in most regions and 70% coverage in regions with large pastoralist populations). Considering the prevailing climatic vagaries, the large lowland areas with scarce potable water resources, the large number of people seasonally migrating with their livestock among pastures or living in widely scattered highland farm homesteads, and the difficulty of providing and maintaining potable water supplies in these isolated rural areas, these targets will be extremely difficult to achieve even under the most favorable scenario.

For urban coverage, Addis Ababa is key, and the city states of Harar and Dire Dawa are important. Harar's serious source problems mentioned above are currently being addressed through a major project supported by the African Development Bank. Addis Ababa is about to embark on a major project also to enhance source capacity. Demand management approaches are yet to be attempted, but behind the figures mentioned above it can be seen that in the coming years such approaches will have to be introduced.

Very recently in 2008 the government reviewed progress on the UAP, which was felt to be making progress too slowly. The Ministry of Water Resources used this opportunity to reaffirm its sector targets and also strengthen promotion of "self-supply" approaches where individual households will be encouraged to make their own investments in improving their water supplies.

Monitoring and Evaluating Water Supply and Hygiene Programs

There are few effective monitoring and evaluation (M&E) systems in the water supply and sanitation sector in developing countries. The best known example is Uganda. Ethiopia is currently in the very early stages of a process to establish a WASH (water, sanitation, and hygiene) M&E framework (Ministry of Water Resources, 2007a). Until such a framework is in place, there will continue to be doubts about the accuracy and reliability of data. To compound this problem, the figures reported above for progress, baselines, and targets lack consistency because of the recent introduction of the UAP strategy which has adopted relaxed service coverage standards compared to earlier years. The UAP now defines access to water supply in rural areas as households being within 1.5 km of an improved source using 15 L per person per day (lpd) and in urban areas being within 0.5 km of an improved source using 20 lpd. The adoption of self-supply approaches (as previously mentioned) will compound this problem further. The results of the national census, which was started in 2007, should in time provide a more reliable denominator population for the calculation of current water supply and sanitation coverage and, thus, a more robust comparison with data gathered around 1994—the time of the previous census.

Accessibility Problems

Accessibility problems can be categorized as physical, social, financial, and institutional problems—many of which are typical and common across countries at similar levels of economic development and sector performance as Ethiopia. Physical constraints include the basic availability of water—both in time and space. Much is written elsewhere (World Bank, 2006) and not repeated here about the spatial and seasonal availability of water across Ethiopia's varied landscape (see chapters 2 and 3). It is recognized also that there needs to be a deeper and more broadly shared knowledge and understanding of groundwater resources (Ministry of Water Resources, 2007b). Ethiopia's terrain and low level of travel and transport infrastructure also limit access—with people, women and children inevitably, having to walk many hours per day in some places to their source of water. Social constraints to access include fractures within communities between richer and poorer groups, the socially excluded, and the physically disabled. Financial constraints to access arise from weak ability to pay for an adequate service—in urban areas in terms of regular user charges and in rural areas in terms of being able to maintain a scheme. Institutional constraints compound all of the others when it comes to implementation capacity to scale up service coverage. But these generalities do not provide specific evidence of rationing during water scarcity or water stress episodes. Such episodes, of course, exist but are not necessarily identified and managed as such, and certainly not at the federal level. When such episodes can be explicitly identified is during periods of response to emergencies, but during these times, certainly in 2006 and 2007, the priority has been to tackle matters of water quality during outbreaks of acute watery diarrhea. Provisions to deal with such situations are handled by the Disaster Prevention and Preparedness Commission and its regional equivalents which have helped set up an Emergency Water and Environmental Sanitation Task Force in the Ministry of Water Resources, supported by UNICEF.

Water Sector Performance

The government of Ethiopia's approach to improve sector performance has a medium term history. Following the fall of the Derg in 1991, Ethiopia adopted a new federal constitution in 1994 which gives decentralization high priority. Policies, strategies, and plans for the water sector followed by the end of that decade—as described in the introduction to this chapter. As previously mentioned, neither the WSDP nor the National Water Supply and Sanitation Master Plan culminated in major sector investment programs. It was not until 2005, when national and international political commitment reached a peak around the United Nations MDG Needs Assessment Initiative and the U.K. government's Commission for Africa, on which the Prime Minister of Ethiopia, H.E. Meles Zenawi, sat, that a concerted effort was made by the government to launch a major sector program—the Universal Access Plan (UAP). At the same time as the government was preparing its plans, donors (who provide a significant proportion of development funds) and government were redefining how they were providing and managing aid—harmonization is the name of the game.

Starting in 2006, donors began to rally around the UAP as a framework for support to implementation but, as is mentioned above, the sector lacks any real performance measurement system. The sector is also at an early stage in many areas—in terms of shifting from a project approach to one of a sector-wide way of working, grappling with the challenges of decentralization and a reducing role in implementation for the federal Ministry of Water Resources, adopting more of a demand-driven approach, opening up space for private sector involvement, and establishing the environment for genuine community management of service delivery.

The early part of this chapter has described recent trends in service coverage which sets the scene for the sector. However, this only captures a few measures of sector performance. A deeper analysis of sector performance is provided in the WASH Sector Review Report (Ministry of Water Resources, Ministry of Health, & Ministry of Education, 2006a). The report was prepared under the auspices of the European Union Water Initiative (EUWI) Country Dialogue for Ethiopia through an iterative, participatory consultation through multistakeholder working groups comprising government at all levels, civil society, the private sector, and donors; and presented to the first annual WASH Multi-Stakeholder Forum in October 2006. The report proposed a number of scorecards for policy; strategy; regulations; institutions; environmental issues; hygiene practices; planning; financing, expenditure, and innovation; sector performance (according to PASDEP indicators as mentioned above); and coordination. While the scores in these scorecards are subjective and qualitative, they provide a useful baseline against which to review sector performance. At the moment the report stands as a snapshot of 2006.

By 2008, government and donors had made good progress in working in a sector-wide manner rather than through separate projects. Regions had also begun to work in this way, and there was strong momentum in translating the UAP into a National WASH Program for villages, towns, and cities. With support from the World Bank in rural areas, the program had been introduced in more than 200 *woredas*—a total of 801 water supply systems, 195 institutional latrines, and 12,674 improved household latrines had been constructed; planning, capacity building, and service improvements were underway in 87 small towns; and preparations were being made for planning and capacity building in 30 larger towns (World Bank, 2008).

These new ways of working demand that the sector develop, in addition to a sector monitoring and evaluation system, a more sophisticated approach to sector performance—progress against PASDEP and UAP targets needs to be measured and reported as a minimum, and a policy monitoring system needs to be introduced.

Water Use Efficiency and Wastage

The recently approved Universal Access Plan has introduced somewhat relaxed service standards as mentioned above. The rationale behind this approach is that government is prioritizing getting services to those currently not served before increasing service standards to higher levels (a "some for all," rather than "all for some" approach). Designs are based on rural households using as little as 15 L per capita per day and urban households using 20 L per capita per day. These are very small volumes, but are reported to be typically what people will manage with. It would be hard to argue that such low volumes were not an efficient use of water—many people will manage with less.

The problem of broken down or partially working schemes is a common one in rural water supply (the reasons do not need to be repeated here), but reversing this trend has been recognized by government. The PASDEP has a high level indicator that aims to measure functionality of rural water supplies. National targets for the proportion of malfunctioning water facilities have been set to improve from 30% in 2005–2006 to 10% in 2009–2010 (by 5% per year). Some recent data (Ministry of Water Resources, 2007c) show variation across regions from 17% to 30%. Efforts toward rehabilitation are being taken. An initiative in SNNP Region with support from Partners for Water and Sanitation (PAWS) (an industry network based in the U.K.) is to develop a region-wide rehabilitation strategy.

Leakage rates in urban areas are not well reported; it is revealing that in a word search of the UAP, nothing was found for leakage, losses, leak, or UFW (unaccounted for water). The Project Appraisal Document of the recently launched Urban Water Supply and Sanitation Project, with financing from the World Bank, reports the rate of nonrevenue water in Addis Ababa as more than 35%. No data is offered for the other four towns in this new project (Jimma in Oromia, Mekelle in Tigray, Awasa in SNNPR, and Gondar in Amhara), but a search of the National Water Supply and Sanitation Master Plan Framework (Ministry of Water Resources, 2003a) revealed that rates of UFW range from as low as 17.3% in some towns with populations of 20,000–30,000 up to 73.9% in very small towns below 5,000 population, with a weighted average of 30.1%.

SANITATION IMPROVEMENTS

Access to sanitation services is very low. The UAP estimates "coverage of toilet users" for the year 2004–2005, ranging from 1.6% to 65.0% in rural areas and from 31.7% to 63.0% in urban areas in all regions except Addis Ababa (shown in table 6.3). The surprisingly high user rate in rural SNNPR appears to be due to an intensive sanitation program, but there are doubts about the quality, actual use, and sustainability of the recently installed facilities.

In 2006 and 2007, there was a strong effort led by the Environmental Health Department of the Federal Ministry of Health to promote the importance of improved hygiene and sanitation. Evidence for this can be found in the sanitation strategy and protocol (Ministry of Health, 2005, 2006) and the accelerated Health Extension Program, which planned to train 30,000 female health extension workers by 2008, two for each of the 15,000 rural *kebeles* nationwide. About 2,700 of them had been trained and assigned to communities by 2004, and more than half by 2007. Health extension workers are being trained at 34 training centers to provide health education and promote family and environmental health, particularly enabling communities to safely dispose of solid waste using local resources (Kitaw, Ye-Ebiyo, Said, Desta, & Teklehaimanot, 2008; Interregional Information Network, 2007). The program was expanded in late 2007 to provide mobile health services in the areas of pastoral nomadism in Afar Region (Afar State, 2007).

TABLE 6.3. Estimated percent coverage of toilet users in urban and rural areas, by region.

	Rural	Urban
National	17.5	57.0
Afar	5.0	31.7
Amhara	2.6	47.5
Benishangul Gumuz	24.6	58.8
Dire Dawa	18.0	66.3
Gambela	3.8	54.5
Harari	1.6	63.0
Oromia	3.8	58.2
Somali	7.0	55.6
SNNPR	65.0	60.6
Tigray	11.9	48.0

Source. Ministry of Water Resources (2006, table 2.6).

There is also good practical experience from government, in SNNP Region, and from NGOs such as PLAN International, with an approach adapted from Community Led Total Sanitation. Also, success is reported by Catholic Relief Services in their promotion of the Arberloo—an approach to simple latrines whereby the waste pit, once filled, is planted with a sapling as the latrine superstructure is relocated over a newly dug pit. The expansion of the health extension program may contribute to the sanitation and hygiene efforts. An urban sanitation strategy is under development, and it is expected that a cadre of urban health extension workers will be established.

Water Sector Priorities and Challenges

The water sector is moving from a project-centric and, to a large extent, donor-project-centric approach to a sector-wide approach, it is grappling with decentralization and a changing role of the federal Ministry of Water Resources, it is adopting new approaches to increase private sector participation, community management, and capacity building, it is working with donors who are increasingly harmonizing their support, and it is adopting new aid-effectiveness goals. These drivers set the priorities for the sector.

Priorities for the sector have been, of course, set out in earlier government policies and plans, and the UAP is the latest incarnation. Unlike previous efforts, the UAP was conceived at a time when donors were also looking to scale up efforts in water supply and sanitation, with aid-effectiveness being high on their agenda. These combine well into what can best be described as a programmatic or sector-wide approach. At the same time, at the World Summit for Sustainable Development in Johannesburg in 2002, the European Union launched its Water Initiative and sought partner countries in which to pilot "Country Dialogues." Ethiopia embraced this opportunity and, since November 2005, has successfully led the Ethiopia WASH Country Dialogue and held two multistakeholder forums. These forums gathered over 200 people from federal and regional governments, sector ministries and bureaus, finance and economic development, parliament, NGOs, academia and researchers, and private sector and donors. They are an event and a process. They built on a number of other building blocks within the sector as well as other initiatives such as WSP-Africa's MDG Road Maps. What they delivered was a consensus among key sector stakeholders on the priority issues facing the sector and the policy means and strategy to tackle them.

In the past two years, the multistakeholder forum (MSF) has gained high level commitment to tackle important "undertakings" over the subsequent 12 months. These, therefore, are the priorities for the sector. The October 2006 MSFs listed the following 10 undertakings and, at the December 2007 MSF, the forum prioritized establishing a WASH sector monitoring and evaluation system and a capacity building pooled fund:

1. Popularize and Operationalize the Universal Access Plan

While the UAP had been officially approved in Parliament, at that time it was yet to be translated into English and had not been shared with civil society and, thus, was not a widely known plan. Since 2006, the English version has been issued, but it appears that many stakeholders may only be aware of the overall targets of universal coverage and know little of the strategies and approaches within the UAP. It remains important therefore to disseminate the UAP further. A recent Ministry of Water Resources review of progress found similar weaknesses and recommended an increased effort to address them.

2. Disseminate and Communicate Water, Sanitation, and Hygiene Sector Policies

The stakeholders generally felt that the government needed to do a better job of communicating its policies, strategies, and plans. Donors, too, need to improve in this area, especially regarding some of the decisions they make.

3. Implement Policy and Regulatory Measures to Increase Private Sector Participation

The Ethiopian government recognizes that the achievement of the stated water and sanitation targets cannot be achieved with public investment alone. Although the private sector in Ethiopia, unlike in other African countries, does not independently supply water to consumers, private firms (domestic and foreign) are increasingly involved in the supply of potable water, especially in urban areas. Specific areas include the construction of springs, hand-dug wells and dams, the supply of free labor and local materials, consulting in project design, minor maintenance and installation of private water lines, and retail and wholesale of water supply pipes, chemicals, and water meters (Legesse et al., 2006).

Furthermore, under the World Bank-supported project, which is the foundation for the national WASH program, decentralized service delivery is being facilitated by embryonic private sector consultants or *woreda* support groups and town support groups. While these are still at the stage of being project specific technical assistance, it is hoped that some may evolve into private sector service providers. Certainly, their existence in what was formerly a government-only arena has changed the picture.

We can also expect to see increasing competition for the state-owned enterprises, the Water Works Design and Supervision Enterprise, and the Water Works Construction Enterprise.

While the involvement of the private sector in service delivery is covered in existing policies and plans, translation into action was proving to be a challenge. A nascent private sector forum had been established in 2006, some initial work has been done to look into the constraints and opportunities for increasing private sector participation in service delivery, but government needs to take a stronger lead in establishing an enabling environment. In December 2007, this received an additional emphasis in the form of a push to document and scale up models of sustainable service delivery.

NGOs are heavily involved in water and sanitation development. A recent report (Christian Relief and Development Association [CRDA], 2007) presents an overview of NGO contributions to the sector. The study looked at the work of some 47 NGOs (20 international and 27 indigenous) across eight regions and those engaged at the federal level. It also collected some data from government organizations (GONGOs). The 47 NGOs had 262 projects under implementation between 2002 and 2006, building 4,531 water supply schemes and 45,581 latrines, maintaining and rehabilitating some 486 schemes, serving about 3.6 million people, around 5% of the population in 2006, at the cost of Birr 258 million.

The government acknowledges this contribution and indeed expects more (a doubling of effort). Roles can be as direct implementers or as *woreda* support groups (see above). Niche areas, such as working with remote, very low capacity *woredas*, with pastoralists, and with the physically disabled and excluded groups, are seen as important. A key challenge is coordination, however, both *within* the civil society sector and *by* government. The recent upgrading of status of the CRDA Water Working Group to Water and Sanitation Forum should help.

4. Establish and Improve Regulations for Community Management

A national task team has made good progress, in particular working with regional authorities and the legal department of the Ministry of Water

Resources, but it was felt that it needed more teeth. In 2007, further progress was made and in December this received a reinforced recommendation from the MSF to finalize legislation at the regional level.

5. Enhance and Harmonize Financing Mechanisms

In the lead up to the MSF in 2006, the EUWI task force (the temporary multistakeholder group of government, donors, civil society, and private sector) had commissioned the development of a financing strategy for the sector (Ministry of Water Resources, Ministry of Health, & Ministry of Education, 2007). One of its main recommendations was to move the sector from a project-centric, fragmented arrangement with high transaction costs to a programmatic sector-wide way of working, in particular to address the array of different project financing arrangements. Work on this has made good progress. Important progress is being made toward setting up a WASH capacity building pool—a pooled fund for sector capacity building and technical assistance (see below). In the medium term, there is an intention to work toward a pooled fund for investment. Until that stage is reached, the development partners will be at least harmonizing their sector financing through the Channel 1B (a sector earmarked grant) of the Ministry of Finance and Economic Development—the decision to do this was taken in mid-2007. Before a pooled investment fund can be considered, however, more effort must be made on improving financial data and its analysis in the sector and harmonizing procurement systems aligned with those of the government.

Government's policy on community and private efforts on investment in rural areas has been adapted and adopted from the World Bank-supported project, which expects up to 10% of scheme capital costs to be contributed by communities, with all operation and maintenance costs to be met. This policy is inconsistently applied. In urban areas, full cost recovery is expected. Further work on the financing strategy is needed and this may emerge in 2009.

6. Implement the Monitoring and Evaluation System

This was recognized as an important undertaking in 2006 and received the highest priority in 2007. It is essential that the sector gets better at reporting what it achieves with the funds it receives. If it does this, then it can demonstrate that it is a sector worth investing in. Improving sector M&E will not be an easy task and will not be quick, but it is fundamental to resource allocation and decision making. At the end of 2008, the Ministry of Water Resources was about to tender for consultants to establish a supporting management information system—this is a significant step, and although it will take up to 18 months to put it in place, it is probably the most important initiative in the sector at the current time.

7. Plan, Budget, and Scale Up Existing Human Resource Development Activities

This priority was set in October 2006. Little progress was made in any systematic manner, but in 2007 the sector was on the verge of signing a memorandum of understanding between government and development partners to launch and operate a pooled fund for sector technical assistance. Recent problems on the pooling arrangements mean that individual bilateral arrangements may be easier to launch in the short term, but irrespective of that, this is a significant achievement and should enable sector managers, in particular at the federal level but also in the regions, to develop the skills and capacity for sector management.

8. Implement and Monitor the WASH Memorandum of Understanding

Ethiopia took a unique step in early 2006 when the ministries of Water Resources, Health, and Education signed a memorandum of understanding for the implementation of an integrated approach to water supply, sanitation, and hygiene—known as the "WASH MOU" (Ministry of Water Resources, Ministry of Health, & Ministry of Education, 2006). Putting the intersecting committees and teams into effect has been a challenge, though some regions have made progress. It is essential that these core governance and management arrangements become established and effective.

9. Establish Supply Chain Outlets

Poor supply of equipment and spare parts is a brake on progress, thus, regional governments need to be more flexible in their ways to enable entrepreneurs and the private sector to become involved. Decentralized and localized procurement will be an important change and help accelerate progress and strengthen community ownership.

10. Other Priorities and Challenges

A key step in the near future will see the Project Implementation Manual, originally drafted for the World Bank-supported Water Supply and Sanitation Project, revised and relaunched as the National WASH Program Implementation Manual (PIM). This will be more than just a cosmetic change and will see the PIM as an important instrument for scaling up service delivery. Remote *woredas* with limited capacity and poor government reach may need to be addressed through different approaches, with NGOs acting as service providers under performance contracts with their local governments.

Decentralization and the federal/regional roles and relationships are perhaps the most difficult challenges facing the sector—the new approach expects local governments with limited capacity and financial resources to be responsible and accountable for service delivery, while regional bureaus and federal ministry provide capacity building support. Research needs to be better coordinated and linked to policy and implementation, and this is emerging. The Department for International Development (DFID) of the U.K. is financing Research Inspired Policy and Practice Learning in Ethiopia and the Nile Region (RiPPLE), which is a five-year research project with three main themes of finance, growth, and governance and planning, with crosscutting themes of mapping as well as sanitation. Other important research projects include PLAN International looking at scaling up. Research coordination is limited, however, and the capacity in the Ministry of Water Resources is low, but 2008 saw the establishment of the Forum for Learning on Water and Sanitation (FLoWS), led by the Research and Development Coordination Department of the Ministry of Water Resources. In the future, we may see an institutional strengthening process to establish an Ethiopia Water Research Institute. Lastly, models of sustainable service delivery including private sector and NGO participation, legalized community management, effective supply chains, cost comparisons, and technologies need to be better documented and shared.

CONCLUSION

Ethiopia's water supply and sanitation subsector is poised to scale up and achieve impressive gains in service coverage. Government, donor, and civil society commitment is strong. A clear framework is in place, even if it is not drawn together into a single strategy document, in line with the government's water resources policy and strategy, as follows:

1. The Universal Access Program sets ambitious targets for urban and rural water supply scheme construction and rehabilitation; human, financial, and material resource needs; water source choice and costs; and implementation approaches;
2. the WASH sector financing strategy stresses the need to improve data and information availability, analysis, and reporting and to work in a programmatic, sector-wide manner; it also describes short, medium, and long-term phases to sustainable sector financing;
3. the sanitation and hygiene needs assessment sets out the huge human resource and promotion needs;
4. the WASH capacity building study provides a strategic plan for improving sector human resources to complement investment-related capacity development in communities;
5. the conceptual framework for the development of WASH monitoring and evaluation is being implemented;
6. WASH subsector coordination arrangements are evolving for the integrated approach across the three ministries/sectors;
7. the World Bank project implementation manual is being revised to become a framework national WASH Program Implementation Manual;

8. research and lesson learning is being linked better to policy and practice; and
9. an annual multistakeholder sector forum for proposing, debating, and setting priorities is becoming established.

These building blocks are impressive and comprehensive. The main challenges continue to be those that bedevil many less developed countries—weak capacity at local government and implementation levels, typical public sector problems of low motivated staff and staff turnover, weak service culture, and poor flow and sharing of information and overall communication. Greater clarity on budget allocation and expenditure is needed, including improved reporting of commitments and disbursements by donors and the links between financial inputs and sector outputs and outcomes. But perhaps a key challenge overshadows these: sector leadership and management need to be strengthened to take forward the above agendas in a coherent and well-communicated manner. Moreover, the federal ministries need to find better ways of implementing national sector policies and plans through autonomous regional bodies which set their own priorities.

REFERENCES

Afar State expanding pilot project health extension package compatible with life style of pastoralists (2007, October 4). *Walta Information Center News*. Retrieved from http://www.waltainfo.com/EnNews/2007/oct04Oct07/36897.htm

Christian Relief and Development Association. (2007, February). *Assessment of civil society contributions to the water sector in Ethiopia, 2002–2006* (Draft report submitted to WaterAid Ethiopia).

Cunningham, W. P., & Saigo, B. W. (1995). *Environmental science: A global concern* (3rd ed.). Boston: Brown Publisher.

Department for International Development. (2001). *Addressing the water crisis: Healthier and more productive lives for poor people: Strategies for achieving the international development targets*. London: Author.

Federal Democratic Republic of Ethiopia. (2002, December 3). Proclamation No. 300/2002. Environmental pollution control. *Negarit Gazeta* (9th year, No. 12). Addis Ababa: Author. See also Environmental Protection Agency-Ethiopia. (1999). *National profile to assess the national infrastructure for the management of chemicals in Ethiopia*. Addis Ababa: Author.

Gleick, P. H. (1996). Basic water requirements for human activities: meeting basic needs. *Water International, 21*, 83–92.

House of Commons, International Development Committee. (2007). *Sanitation and water, sixth report of session 2006–07* (Vol. 1). London: The Stationery Office Limited.

Interregional Information Network. (2007, May 25). Ethiopia: New programme boosts village health service delivery. *IRIN Newsletter*, UN Office for the Coordination of Humanitarian Affairs.

Kitaw, Y., Ye-Ebiyo, Y., Said, A., Desta, H., & Teklehaimanot, A. (2008). Assessment of the training of the first intake of health extension workers. *Ethiopian Journal of Health Development, 21*, 232–239.

Legesse, W., Haile Mariam, D., & Kloos, H. (2006). Water supply and sanitation. In Y. Berhane, D. Haile Mariam, & H. Kloos (Eds), *The epidemiology and ecology of health and disease in Ethiopia* (pp. 127–151). Addis Ababa: Shama Books.

Ministry of Finance and Economic Development. (2001). *Sustainable development and poverty reduction program (SDPRP*). Addis Ababa: Author.

Ministry of Finance and Economic Development. (2005). *A plan for accelerated and sustained development to end poverty (PASDEP).* Addis Ababa: Author.

Ministry of Finance and Economic Development. (2007a). *Ethiopia: Building on progress: A plan for accelerated and sustained development to end poverty (PASDEP), Annual Progress Report 2005/06.* Addis Ababa: Author.

Ministry of Finance and Economic Development. (2007b). *Ethiopia: Building on progress: A Plan for Accelerated and Sustained Development to End Poverty (PASDEP), Annual Progress Report 2006/07.* Addis Ababa: Author.

Ministry of Finance and Economic Development & United Nations Country Team. (2004). *Millennium development goals report: Challenges and prospects for Ethiopia*. Addis Ababa: Author.

Ministry of Health. (2005). *National hygiene and sanitation strategy*. Addis Ababa: Author.

Ministry of Health. (2006). *National protocol for hygiene and "on-site" sanitation*. Addis Ababa: Author.

Ministry of Health. (2007). *Annual performance report of HSDP-III: EFY 1999 (2006/2007)*. Addis Ababa: Author.

Ministry of Water Resources. (1999). *Ethiopian water resources management policy*. Addis Ababa: Author.

Ministry of Water Resources. (2000). *Ethiopian water sector strategy*. Addis Ababa: Author.

Ministry of Water Resources. (2002). *Water sector development programme, 2002–2016*. Addis Ababa: Author.

Ministry of Water Resources. (2003a). *National water supply and sanitation master plan framework, environmental support project component 3* (DHV Consultants, 6 vols.) Addis Ababa: Author.

Ministry of Water Resources. (2003b). *Sanitation and hygiene issue paper* (Vol. II, Annexes). Addis Ababa: Author.

Ministry of Water Resources. (2006). *Universal access program for water supply and sanitation* (First English ed.). Addis Ababa: Author.

Ministry of Water Resources. (2007a). *A conceptual framework and work plan for the development of WASH monitoring and evaluation* (Unpublished draft). Addis Ababa: Author.

Ministry of Water Resources. (2007b). *Assessment, development and management of groundwater in Ethiopia (EGRAP).* Addis Ababa: Author.

Ministry of Water Resources. (2007c). *Water sector, 2006/07 PASDEP implementation report.* Addis Ababa: Author.

Ministry of Water Resources, Ministry of Health, Ministry of Education. (2006a). Memorandum of understanding between Ministry of Water Resources. Addis Ababa: Author.

Ministry of Water Resources, Ministry of Health, & Ministry of Education. (2006b). *CRDA Water Working Group, EUWI Ethiopia, CRDA Working Group, and EUWI Ethiopia, needs assessment to achieve universal access to improved hygiene and sanitation by 2012* (Final draft). Addis Ababa: Author.

Ministry of Water Resources, Ministry of Health, & Ministry of Education. (2007). *CRDA Water Working Group, EUWI Ethiopia, CRDA Working Group, and EUWI Ethiopia.* Final Draft, WASH sector financing strategy. Addis Ababa: Author.

Nadakavukaren, A. (1992). *Man and environment.* Prospect Heights, IL: Waveland Press.

Tebbut, T. H. Y. (1992). *Principles of water quality control.* Oxford: Pergamon Press.

Terefe, A., & Edstrom, G. (1999). ECOSAN-ecological sanitation. In J. Pickford (Ed.), *Proceedings of the 25th Water Engineering and Development Center Conference: Integrated development for water supply and sanitation* (pp. 16–17). Loughborough, U.K.: Loughborough University.

Thompson, J. K., Porras, I. T., Tumwine, J. K., Mujwahuzi, M. R., Katui-Katua, M., Johnstone, N., & Wood, L. (2001). *Drawers of water II. 30 years of change in domestic water use & environmental health in East Africa.* London: International Institute for Environment and Development.

United Nations. (2002). *With a sense of urgency. Johannesburg Summit sets an action agenda.* Retrieved from www.johannesburgsummit.org-new/feature-story38

United Nations Development Program. (2006). *Summary: Human development report 2006: Beyond scarcity: Power, poverty and the global water crisis.* New York: Author.

WaterAid. (2003). *WaterAid's strategy 2005–2010.* Retrieved from www.wateraid.org/strategy

Water and Sanitation Program. (2006). *Getting Africa on track to meet the MDGs on water and sanitation: A status overview of 16 countries.* New York: Author.

WHO/UNICEF. (2000). *Joint monitoring program for water supply and sanitation*. Geneva: Author.

WHO/UNICEF. (in press). *Joint monitoring programme for water supply and sanitation.* Geneva and New York: Author.

World Bank. (2006). *Ethiopia: Managing water resources to maximize sustainable growth*. New York: Author.

World Bank. (2008). *FY 2008 Report on the status of projects in execution-FY08*. New York: Author.

Chapter 7

Hydrological and Related Aspects of Deforestation and Degradation of Woody Vegetation

Badege Bishaw and Zebene Asfaw

The diverse climate, topography, and soils of Ethiopia support a variety of forests and woodlands, ranging from dense tropical forests in the humid and mountainous areas in the western part of the country, to the dry savannas in lowland Ethiopia's forests. Trees have played an important role in providing many economic goods and services and regulation of hydrological and climatic patterns. However, most of the forested area has been cleared over the centuries due to many factors related to agricultural land use practices, fuel needs, and inadequate forest policies (Tadesse, 2007). This chapter examines the history, causes, impacts, and possible remediation of deforestation and woodland degradation on water resources in Ethiopia, with an emphasis on the highlands.

A SHORT HISTORY OF DEFORESTATION IN ETHIOPIA: MISCONCEPTIONS AND RECENT EVIDENCE

Although forest decline is identified as an important problem in Ethiopia (Ethiopian Forestry Action Program, 1994), relevant data are scarce and their interpretation is influenced by different perceptions of the prevailing environmental crisis in the country (Hoben, 1995; Rahmato, 2001). The deforestation debate, particularly over historical changes in vegetation cover, has been fueled in the absence of sufficient empirical evidence of rapid deforestation. A number of academics and development organizations subscribed to the view that 40% of Ethiopia had been forested at the beginning of the 19th century. A paper presented at the First Green Forum Conference in Addis Ababa in 2006 alleged that "the country's forests were estimated to cover 40% of the country's surface in the 1940s, but they now cover less than 3%" (Tefera, 2007, p. 88). Even semiarid Eritrea was reported to have been covered on 30–40% of its area by forests in 1900 (Robinson, MacKay, Afwerki, & Yirgalem, 1995).

These unfounded assertions run counter to photographic records, the descriptions of European travelers in the 19th century, recent pollen analyses, and soil stratigraphy, which reveal a much longer history of deforestation. Considerable evidence points to the predominance of open grasslands and wood savannah in the central, northern, and southern highlands and dense hardwood forests in the west during historic times, the result of extensive land use practices and widespread water logging of highland soils (McCann, 1999; Pankhurst, 1995; Ritler, 1997). There is increasing evidence that forest vegetation cover changes due to human impacts predate the Axumite kingdom in northern Ethiopia (Butzer, 1981; McCann, 1999), and that deforestation has been a gradual, cyclical process rather than a linear process, accompanying and possibly even predating the southward expansion of the Ethiopian empire and plough/grain agriculture (Hagos, Pender, & Nega, 2002; McCann, 1999). Pollen analysis of the substrate of Lake Turkana, for example, indicates that vegetation changes from primary *Podocarpus* forests to secondary forest and shrub, a strong indication of extensive forest destruction, began as early as 150 A.D. in the catchment of the Omo River in western Ethiopia (Bonnefille & Hamilton, 1986). Soil accumulations overlying burnt strata and pollen analysis in various parts of the highlands indicate that extensive forest fires, either by climatic or human agency, occurred even earlier, more than two millennia ago (Nyssen et al., 2004).

Recent studies of forest change cast further doubt on the alleged confinement of massive deforestation to the last century. The Ethiopian Forestry Action Program (1994) estimated forest loss per year at 150,000–200,000 ha, the Food and Agriculture Organization (2007) estimated it at 141,000 ha, and Reusing (1998) put the annual reduction at 163,000 ha of high forests alone and concluded that between 1973 and 1990 the area of closed forest stands was reduced from 2.6% to 2.0%. Studies in three natural forests in the western mountains and in the Rift Valley revealed mean annual deforestation rates of 2.5% to 3.5% between 1957 and 2005 (Dessie & Kleman, 2007; Sherefa, 2007; Zeleke & Hurni, 2001), and a study in the semiarid mountains of eastern Amhara Region reported 1.8% mean annual loss of shrub land (Tekle & Hedlund, 2000). Considering that the first three rates were reported for closed forests subject to rapid incursion and deforestation by settlers during the political turmoil of the revolution of the 1970s, they are unlikely to be representative of long-term deforestation patterns, considering that there are areas of minimum deforestation and even with expanding forests in the form of wood lots (McCann, 1999). Although deforestation has undoubtedly accelerated during the 20th century and resulted in increased soil erosion, habitat destruction, and land degradation, the magnitude of these impacts have been overestimated for that short period. Most areas experienced a gradual change in vegetation cover due to land-use change rather than destruction of large forest tracts in recent years, including the removal of trees and shrubs between fields and on steep slopes (Nyssen et al., 2004). The unsupported deforestation claims, together with the widely accepted conventional wisdom that forests invariably act as giant "sponges" that release water during the critical dry season (Food and Agriculture Organization, 2005) provided the impetus for a massive tree planting and soil conservation program which sought to reverse land degradation processes and contribute to famine prevention. This coercive program failed to achieve its goal, caused production shortfalls, and even exacerbated land degradation through water erosion in some areas (see chapter 8).

The direct contribution of forests to the national economy of Ethiopia has not been examined in depth but appears to be substantial. The value of the annual coffee harvest alone, most of it destined for foreign markets, has been estimated as high as US$1.5 billion (Hein & Gatzweiler, 2006). Few data exist on the value of forest products traded in the informal market and the contribution of forests in nonwood products and its environmental and ecosystem services (Ministry of Economic Development and Cooperation [MEDC], 1999). One of the few economic studies of high forests in Ethiopia estimated that the combined value from economic use and environmental services of a high forest in western Ethiopia was approximately US$1,200 per ha per year (Seyoum, 2007).

A review of postreform development by the Ministry of Economic Development and Cooperation (MEDC, 1999) for the period 1990–1991 to 1996–1997 shows that, on average, the agriculture and forestry sectors accounted for 53.3% and 6.7% of the total national GDP, respectively. The contribution of forestry to the agriculture GDP over the same six-year period ranged from 11.4% to 13.2%. However, since Ethiopian national accounting largely considers recorded transactions in the formal market, these figures underestimate the total contribution of forests to the country's economy in the form of fuel wood and nontimber forest products.

FOREST AND WOODLAND TYPES

Natural Forest

Ethiopia's remaining natural forest cover includes various montane forest types concentrated in the less populated southern and southwestern mountain regions. Until the mid-20th century, surveys and descriptions of the forest vegetation were based on the observation of a few foreign travelers. Von Breitenbach (1963), the founder of Ethiopian forestry studies, classified the forest vegetation of Ethiopia and studied the distribution of indigenous tree species in relation to climatic and edaphic conditions. Unfortunately, he contributed to the deforestation controversy by stating that 37% of Ethiopia's land surface was covered by forests at the beginning of the 19th century, a hypothetical estimate solely based on rainfall, soil, and altitude data (Von Breitenbach, 1962). Friis (1992) identified seven forest vegetation types: (1) Dry Peripheral Semi-Deciduous Guineo-Congolian Forest, which includes dominant species such as *Diospyros abyssinica* and *Celtis toca*, (2) Transitional Rain Forest, with *Aningeria altissima* and *Morus mesozygia*, (3) Afromontane Rain Forest, including *Croton macrostachys, Eckebergia capensis* and *Olea welwitchii*, (4) Undifferentiated Afromontane Rain Forest of the Ethiopian Highlands, including *Podocarpus facatus*, *Juniperus excelsa*, and *Olea europaea*, (5) Dry Single-Dominant Afromontane Forest, with *Cordia africana* and *Acacia abyssinica*, (6) Riverine Forest, including *Tamarix nilotica*, *Tamarindus indica*, and *Acacia albida*, and (7) Mangroves.

Woodlands and Shrub Land

Woodlands refer to vegetation types characterized by an upper story of 5–12 m high trees forming a more or less closed canopy which admits sufficient sunlight to promote dense 1–3 m high shrub vegetation. Of Ethiopia's land area, approximately 20 million ha are covered by woodlands and shrub lands (Ethiopian Forestry Action Program, 1994), mostly in the agro-pastoral and pastoral areas below about 1,500 m altitude. Dominant lowland woodlands include *Combretum-Terminalia* woodlands, *Acacia* woodlands, *Boswellia-Comiphora-Acacia* woodlands, and mountain woodlands (*Hagenia* woodland).

Planted Forests and Remnant Trees

Plantation forestry in Ethiopia dates back to the turn of the 19th century, when Emperor Menelik requested that fast-growing tree species be planted to overcome the fuel wood shortage he faced at that time.

Subsequently, *Eucalyptus globulus* has been planted as the main tree species in the central highlands of Ethiopia since 1895 (Bishaw, 2001; Pohjonen & Pukkala, 1990). In the mid-1960s, plantations were initiated on a large scale in a few national forest priority areas to rehabilitate formerly forested areas and to produce construction and energy wood supply. The emphasis was on fast-growing exotic species, especially of eucalyptus, cypresses, and pines, and short rotation plantation development programs (Tadesse, 2007). Today, most Ethiopian rural and urban communities depend on eucalyptus plantations for fuels, construction purposes, and extra household income and little is done to grow indigenous trees because of their slow growth rate, limited silvicultural knowledge, and low economic return (Jagger & Pender, 2003). Plantations include industrial and peri-urban manmade forests established and operated by the government, as well as communal woodlots and plantations for watershed protection and fuel. The total area of planted forests in Ethiopia is estimated to be about 255,000 ha (Bekele, 2000). Agro-forestry and maintenance of indigenous trees on farmland are two other forms of forestry in Ethiopia. Although there is no information on the area covered, these plantations form a prominent part of the Ethiopian landscape.

CAUSES OF DEFORESTATION

Ecological and historical studies have demonstrated the dramatic human impacts on the forests of Ethiopia, where expansion of agricultural land, unsustainable use of forest resources, overgrazing, and establishment of new settlements on forest land in response to population growth are major driving forces (Bishaw, 2001; Senbeta, & Denich, 2006). Thus, deforestation rates are influenced by various environmental, socioeconomic, cultural, and political forces, which tend to interact synergistically. The mix of these forces varies among communities and over time, as indicated by case studies, rendering generalizations difficult. In most cases in Ethiopia, deforestation is a process involving competition amongst different land users for scarce resources, a process exacerbated by counterproductive policies and weak management and regulatory institutions. Most deforestation is due to the growing number of poor people, open access, and unclear property rights, also posited by two of the three schools of thought regarding deforestation described by Potter (2006). There is some evidence, however, that the relationship between population growth and deforestation is not linear. For example, scarcity of fuel wood in several villages in northwestern Ethiopia prompted households to plant trees, resulting in a slightly larger forest cover during the last four decades (Bewket, 2003).

According to a broad definition of deforestation, forest is gradually converted to nonforest, resulting in degradation that reduces forest quality, including changes in density and structure of trees, ecological services supplied, biomass of plants, species diversity, and genetic diversity (Food and Agriculture Organization, 2000). This definition describes the gradual degradation of forests observed in Ethiopia and most other developing countries. Definitions referring to changes in land cover or land use, based on empirical on-the-ground measurements and remote sensing, can promote more quantitative studies, but scarcity of detailed surveys and technical difficulties with land use definitions limit data bases in Ethiopia.

Direct Causes

Population pressure and poverty

While the Ethiopian population doubled between 1984 and 2007, the proportion of people living in rural areas decreased only slightly (from 86% to 84%) (Population Census Commission, 2008), exerting increasing pressure on agricultural and forest resources. The effect of population pressure in deforestation is influenced by the carrying capacity of the land, prevailing land use practices, the importance of forest-derived products and services to the local people, and the relative strength of the institutional framework in place. Increasing population pressure and rural poverty were associated with deforestation in several areas with high forests in southern Ethiopia (Sherefa, 2007). Poverty contributes to deforestation by limiting peoples' economic options, the formation of rural capital, and institutional and infrastructure development. With few economic alternatives available to them, poor rural people look to short-term gains, often exploiting water and soil resources in an unplanned, unsustainable manner, as described further in chapter 8.

Subsistence farming and animal husbandry

By far the most important agents of deforestation throughout the country are the farmers who live inside or at the margins of natural forests. Deforestation may take various forms, ranging from short- and long-fallow shifting cultivation to forest pioneer farming (Sherefa, 2007) and complete conversion to crop fields (Dessie & Kleman, 2007; Zeleke & Hurni, 2001) and grazing lands (Yimer, Ledin, & Abdelkadir, 2007). Population growth and the ensuing restricted access to arable land and scarcity of economic alternatives often force farmers either to reduce fallow periods or to clear forests for cultivation (Sherefa, 2007).

The impact of livestock grazing on natural forests, woodland, and shrub land is substantial. Ethiopia's cattle population of 40.6 million (in 2001–2002) is the largest in Africa, and its sheep, goat, horse, and camel populations are also sizeable (Central Statistical Authority, 2004). Open-range grazing, predominating in the dry woodlands and savannas, is a major contributor to deforestation and forest degradation when herd populations exceed the carrying capacity of the range.

Commercial agriculture

The development of commercial agriculture in Ethiopia began to accelerate in Ethiopia with revised economic policy under Haile Selassie in the 1970s that emphasized export crops. Modern agriculture can result in deforestation both directly and indirectly. Having ownership of forested land with the intention of converting it to another use directly impacts forests. Coffee and tea plantations in western highland forests reported by Bishaw (2001), Seyoum (2007), Sherefa (2007), and Jotte (2007), and the rapidly expanding *chat* fields (Dessie & Kleman, 2007) have resulted in the logging of sizeable tracks of highland forests. Commercial agriculture can also indirectly result in deforestation by cultivating the most fertile agricultural land and thereby displacing local subsistence farmers who then relocate on less fertile and less productive forested land.

Fuel wood collection and charcoal making

Wood is the primary source of fuel in Ethiopia. It is collected mostly from unregulated commons and, as a consequence, is susceptible to overexploitation as the population increases. Fuel wood cutting and collection, while not completely destroying the forests and woodlands, significantly impoverishes them and alters the habitat by the selective removal of preferred species (Dessie & Kleman, 2007). With intensive collection over a long period of time, which typically includes the gathering of leaves and twigs, the original trees and shrubs lose their ability to reproduce and die out, giving way to new, invasive plant species (Abate, 2004).

Most Ethiopian households depend almost exclusively on wood, charcoal, and animal dung for their energy needs. Average daily per capita consumption of around 1–2 kg has been reported from different regions (Ethiopian Forestry Action Program, 1994; Environmental Protection Agency & Ministry of Economic Development and Cooperation [EPA/MEDC], 1997) but tends to be much higher in the humid, biomass-rich regions (7 kg) than in arid areas (0.8 kg). More than 50% of all primary energy used nationwide is for baking *enjera*, the country's main food staple, but more than 75% in the highlands (Thomas & Bekele, 2003). Inefficient utilization of biomass fuel in the absence of fuel efficient stoves increases wood consumption. The relative importance of various energy sources used in urban and rural homes is shown in table 7.1.

Logging and agricultural expansion

From the late 1950s to the early 1980s, the greatest increase in logging activity occurred in the Afromontane forests of central, southern, and western Ethiopia. Deforestation in the central and southern rift valley was caused mainly by commercial logging (Dessie & Kleman, 2007). Except in a few cases, logging in all forests has not followed well-planned forest management guidelines and is often characterized by a cut-and-get-out mentality. After clear cutting, farmers and fuel wood collectors move into the deforested areas and clear the land for farming. The history of agricultural expansion in Ethiopia is summarized in chapter 8.

TABLE 7.1. Urban and rural household energy consumption in Ethiopia, by fuel type.

Energy Sources	Relative consumption (percent)	
	Urban	Rural
Fuel wood and charcoal	74.7	82.2
Agricultural residues (animal dung)	6.3	8.4
Kerosene	7.6	*
Electricity	3.0	*

Source. Thomas & Bekele (2003).
*No data available.

Indirect Causes

Fiscal and development policies

Government policies outside the forest sector have profound impacts on the forest resource, as do international policies on debt repayment, structural adjustment, and trade. The expansion of agricultural cash crops means that either forests are cleared directly for these crops or subsistence farmers are displaced by them, often forcing farmers to relocate in forested areas, where they exacerbate the deforestation problem (Sherefa, 2007). The massive resettlement and villagization programs by the Derg government, discussed in chapter 8, resulted in large-scale destruction of forests in much of Ethiopia.

Access to land

With the growth of the Ethiopian farming populations and the inheritance of ever smaller plots of land, many farmers either moved to towns to search for employment or relocated along forest margins clearing trees for new farms (Sherefa, 2007). The impact of the 1975 land reform on access to land is described in chapter 8.

Undervaluation of natural forests. In economic terms, there is little understanding in Ethiopia of the value of the goods and services provided by natural forests or of the real costs of forest management being borne by resource users. As a consequence, forests are undervalued and play a less significant role than they should in the decisions affecting resource allocations, development priorities, and land use (Seyoum, 2007). This attitude made the remaining forests more vulnerable to being converted to land uses perceived to be more beneficial, at least in the short term. Research in tropical Africa and other developing regions indicates that the undervaluation of natural forests and woodlands is largely due to the failure to adequately consider the many products and services they provide and the tendency to view them as free commodities. Even where environmental values are recognized, they may not be evaluated and used to assure efficient resource management (Randall, Narendra, & Mohan, 1995; International Institute for Sustainable Development, 1999).

Weak implementing and regulatory institutions. Many institutional failures contributed to deforestation. The most important one is low status of the Forestry Department within the Ministry of Agriculture relative to competing land uses. Typically, forestry departments are handicapped by poorly paid staff, inadequate budgets, and lack of staff training. As a consequence, the departments have been ineffective in successfully putting forth proforestry arguments to the political decision makers and to the public at large. The frequent changes in institutional structures, operations, and policies during the last three decades further weakened forestry mandates and programs. Except during a short period as an independent authority, the Forestry Department has been associated with the dominant Department of Agriculture. It has also often lacked funding to implement regulations and conservation programs or develop a viable trade in forest products. Forestry also suffered from the lack of strong leadership. For example, due to the lack of adequate legal framework of protection of the 38 wildlife conservation areas, only two are "gazetted" (EPA/MEDC, 1997). Human settlements are encroaching national parks and forest reserves due to weak law enforcement and low monitoring capacity of the underfunded regulatory institutions. As a result, critical ecosystems have been damaged beyond repair. Heavily impacted areas include the Abijata-Shala Lakes National Park, one of the most heavily protected areas in Africa, now densely populated by permanent settlers. Similarly, Awash National Park is severely degraded through over-grazing as a result of trespassing pastoralists.

Culture and religion. While forests tend to be looked upon by planners as impediments to agricultural development in Ethiopia, local communities view them as economic and cultural resources. Different ethnic groups and communities have traditionally managed their forests, but recent changes in the political system have weakened or destroyed these customary uses and regulations. For example, several forests in the southwestern highlands have been managed by local institutions (Jotte, 2007). The local communities' knowledge of natural resources and their traditional skills in managing them have ensured their sustainable use. The major religious institutions in Ethiopia—the Orthodox Church and Islam—also contribute to maintaining forests and biodiversity in holy groves and cemeteries. The religious norms promoting the maintenance of these plantings are easier to enforce than government regulations.

IMPACTS OF DEFORESTATION

Hydrological Effects

According to the conventional wisdom of Ethiopians, original forests recharge and filter groundwater and, thus, contribute to maintaining dry season flow, mitigating floods, reducing erosion, and maintaining water quality. However, while their knowledge is generally borne out by several studies in Ethiopia (Tadele & Förch, 2007) and other countries (Parrotta & Agnoletti, 2007), forests are heavy users of water and thus reduce annual runoff (Food and Agriculture Organization, 2005). At the regional and continental levels, forests appear to promote high precipitation and runoff due to their function as atmospheric moisture pumps, mitigating against desertification (Makarieva & Gorshkov, 2007). These and other studies support the generally accepted view that forests reduce low-level flooding by local and regional biological and physical processes, although they cannot prevent heavy floods (Ives, 2006).

Planted forests negatively affect the water budget. The best known example is fast-growing eucalyptus, evapotranspiring more water than rain-fed food crops and, thus, resulting in correspondingly less surface flow. Moreover, poorly managed planted forests may negatively impact water quality and erosion (Ives, 2006; Calder, 2007). On the other hand, well designed, located, and managed planted forests may reduce the volume of sediments, nutrients, and salt transported into river systems, reduce smaller landslides (Van Dijk & Keenan, 2007), provide cleaner surface waters, and reduce flooding in small watersheds, although they appear to have little impact on major floods (Food and Agriculture Organization, 2005; Ives, 2006).

Both groundwater and surface water quality is improved chemically and microbially by the filtering function of forests. Low soil erosion rates under forests assure the efficiency of this filtration system. Trees also provide physical barriers that keep people and domestic animals away from surface waters, protecting their integrity. The impact of deforestation on water quality was demonstrated by research of various forest management systems in different countries (Calder, 2007). In Ethiopia, springs in and around forests are the most sought-after water source by the population because of their good water quality, and several empirical studies have associated higher surface water quality with forests than deforested areas (Embaye, 2000; Gebre-Mariam, 1998).

Land Degradation and Nutrient Loss

In the Ethiopian mountain environment, deforestation increases rates of soil erosion, by increasing the amount of runoff and reducing the protection of the soil provided by tree roots and litter. The downward spiral in soil degradation set in motion by deforestation is examined further in chapter 8, but a few studies are cited here to illustrate the impact of deforestation on soil fertility losses and the beneficial effect of the old Ethiopian practice of retaining scattered trees on farms. Clearing areas of *Acacia* woodland was found to increase moisture deficit and reduce nitrogen content of soil and to reduce crop productivity (Itana, 1998), a relationship reported from other African countries (Young, 1997). Conversion of a forest in the Ethiopian highlands into cropland during a 15-year period significantly reduced the original amount of soil organic carbon by 31%

and total nitrogen by 32% in the top 1 m soil layer (Yimer et al., 2007). Similarly, soil carbon and total nitrogen from the upper 0–10 cm soil layer were depleted following conversion of humid *Podocarpus* forest to farm land (Lemenih & Itana, 2004). Increasing the cultivation period after clearing natural forest led to an overall reduction of soil organic carbon contents (SOC) on farms (Lemenih, Karltun, & Olsson, 2005). In a forest in southwestern Ethiopia, forest clearing followed by continuous cultivation of crops caused the loss of nearly half the magnesium and three quarters of SOC during a 75-year period (Lemma, Kleja, Nilsson, & Olsson, 2006). There is also supporting evidence from studies on the contribution of indigenous tree species to soil fertility through agro-forestry practices (Asfaw, 2003; Gindaba, Rozanov, & Legesse, 2005; Yadessa, Itanna, & Olsson, 2001). Increased soil fertility after deforestation as a result of the decomposition of organic matter in sunlight is temporary before the characteristically sharp deterioration of the soil, especially in Ethiopia, where application of fertilizers on farm land is inadequate.

Soil Erosion

Soil erosion caused by deforestation, degradation of high forest to shrub vegetation, overgrazing, and plow agriculture is particularly severe in Ethiopia. A number of studies from mountain environments worldwide have documented increased erosion rates in modern times as a result of vegetation cover change and their reduction to nearly the extremely low rates of virgin forests after reforestation of steep slopes (Vanacker et al., 2007). The impact of deforestation and farming on soil erosion and the implications of siltation for reservoirs in Ethiopia, Sudan, and Egypt are examined in chapters 2 and 8.

Shading Effect on Water Temperature and Coffee Production

Stream temperature is important in protecting aquatic life and maintaining dissolved oxygen and nutrient levels (Brown, 1976). The water temperature lowering effect of forests has been known for many years (Chapman, 1962). The shade coffee system of Ethiopian small land holders consists of many tree species with established roles: supporting generally high biodiversity mainly due to structurally and floristically diverse canopy of shade trees, which provide habitat for a diverse associated flora and fauna (Perfecto, Vandermeer, Mas, & Soto Pinto, 2005; Wakjira, 2006). Shade trees also protect coffee bushes from harsh winds, excessive light, and soil erosion, and regulate temperature and humidity resulting in better quality coffee products (Muschler & Nair, 2001; Soto-Pinto, Perfecto, & Caballero-Nieto, 2002). Hence, multistoried and multifunctional natural coffee forests are known as ecologically healthy and socially sound type systems (Soto-Pinto et al., 2002). However, the forest structure and the ecology of the montane rainforest are disturbed under continuous and intensified management of coffee systems (Schmitt, 2006). Serious disturbances can eventually limit both the regeneration of coffee plants and the natural forest (Wakjira, 2006).

Biodiversity

Probably the most serious consequence of deforestation is the loss of biodiversity. Although no exact figure on species loss exists in Ethiopia, deforestation is eroding the country's biodiversity, both in terms of flora and fauna, due to its rich variety of environmental conditions. Plant genetic resources in Ethiopia include more than 7,000 recorded flowering plant species, of which at least 12% are endemic (Thomas & Bekele, 2003). Measuring the total rate of habitat conversion (defined as change in forest area plus change in woodland area minus net plantation expansion) for the 1990–2005 period, Ethiopia lost 3.6% of its forest and woodland habitat. Ethiopia has some 1,408 known species of amphibians, birds, mammals, and reptiles according to figures from the World Conservation Monitoring Centre. Of these, 7.0% are endemic and 4.6% are threatened. Ethiopia is home to at least 6,603 species of vascular plants, of which 15.1% are endemic (International Union for Conservation of Nature, 2000). Studies in some natural forests indicate that traditionally managed forests with diverse species provide local communities with diverse goods and services. In Sheka Forest in western Ethiopia, for example, Woldemariam (2007) recorded 190 and 180 plant species in two districts, equivalent to 32 and 27 species per

500 m^2, respectively. These montane forests are known for maintaining threatened species, including *Prunus africana* and *Canarina abyssinica.* They provide diverse nontimber forest products, including *Aframonum corrorima, Piper capense, Coffea arabica,* various tuber plants, honey, and medicinal plants. Coffee is native to Ethiopia, the only country where wild coffee trees of the highland variety (*Coffee arabica*) are found, and constitutes about 80% of the country's exports. However, continuous wild coffee cultivation suppresses natural tree regeneration and reduces tree density and diversity while promoting coffee plants (Senbeta & Denich, 2006). A large portion of honey production is coming from the natural forests. Ethiopia has the largest number of bee hives (10 million) and ranks first in honey production in Africa (Hartman, 2004), and 10th in the world (Fichtl & Admasu, 1994). In addition, various fruits, bamboo, gum arabica, and other products are collected from the natural forests. Deforestation reduces not only habitats and biodiversity, but also traditional knowledge of sustainable management of forest resources (Jotte, 2007).

Habitat loss as a result of deforestation can also influence agricultural diversity, productivity, and, thus, food security. The agricultural biodiversity of Ethiopia is one of the world's 12 Vavilov centers of genetic diversity. Ethiopia is the sole or the most important center of genetic diversity for arabica coffee, *teff (Eragrostis tef)*, *enset* (*Ensete ventricosum*), and anchote (*Coccinia abyssinica*), and for sorghum, finger millet, field pea, chickpea, cow pea, perennial cotton, safflower, castor bean, and sesame. Genetic erosion in other parts of the world has led scientists to Ethiopia and they now recognize Ethiopia as the most important center of genetic diversity for durum wheat, barley, and linseed. The diversity of both natural trees and food and fiber crops have been maintained through accumulated on-farm-level indigenous knowledge in a number of well-managed agro-forestry systems (Abebe, 2005; Asfaw, 2003; Kippie, 2002; Tsegaye, 2002). Loss of indigenous knowledge and inadequate protection of intellectual property rights tend to hamper indigenous conservation measures, as are overharvesting of medicinal plants, many of them rare and endangered species (Giday, 2008).

Climate Change

Forests play a double role in affecting climate change. First, forests influence local climate and weather through evapotranspiration and, thus, create conditions favorable for precipitation (Bruijnzeel, 1990). Second, forest ecosystems can contribute to reducing the rate of climate change. Much evidence suggests that biological sequestration of carbon plays an important role for reducing greenhouse gas emissions (Tavoni, Sohnge, & Bosetti, 2007) because forest ecosystems have the potential to sequester carbon from the atmosphere during the growing period and retain it in the biosphere (biomass and soil), thus functioning as carbon sinks. Similarly, reduction of deforestation and promotion of sustainable forest practices can significantly reduce CO_2 emissions from motor vehicles and industries, a growing problem around Addis Ababa and other cities. Third, forest burning is one of the major sources of greenhouse gas emission. Burning of tropical forests contributes about 20% of the carbon emitted from all sources worldwide, or more than the combined total contributed by the world's transportation sector (Interregional Information Network, 2007).

Forest Conservation History and Practices

Recorded forest conservation activities in Ethiopia go back to the reign of Emperor Zerea Yacob (1434–1468), earlier than elsewhere in Africa. Zerea Yacob ordered the planting of tree seeds on a deforested mountain near present-day Addis Ababa (Pankhurst, 1995). In the 17th century, the *qero* system was implemented in the central highlands, allowing equitable use of fuel wood, thatching grass, and pastures and contributing to maintaining sizeable areas of montane forests. The *qero* system survived the socialist regime under Mengistu, attesting to its time-tested resilience during many political changes (Tefera, 2007). Groves of sacred forest were maintained around many Orthodox churches through the centuries in the highlands (Bekele, 1994). Many other conservation practices did not survive recent political and social changes due to the loss of indigenous knowledge and institutions. In many parts of Ethiopia, forests and woodlands have traditionally been managed by local communities as part of their subsistence economy and within the context of cultural practices and religious beliefs (Jotte, 2007).

The 1975 land reform, the advent of commercial plantations, resettlement, and acculturation, in particular, impacted traditional lifestyle and caused the breakdown of social institutions and the loss of individual and collective rights (Jotte, 2007), further discussed in chapter 8. The Ethiopian government issued the Conservation Strategy of Ethiopia in 1990, the Environmental Policy in 1997, and several cross-sectional policies, including the forest, soil, and water conservation, population, and social policies, all of which are being implemented to varying degrees (Environmental Protection Authority, 1998). Nevertheless, the currently more integrated and participatory approach to natural resource conservation has achieved few of the stated goals and may be unsustainable due to excessive dependence on external policies and experiences, an overly technocratic orientation, pervasive poverty, and lack of incentives for community participation (Kidane-Mariam, 2003).

Recognizing the shortcomings of most government-sponsored forest conservation programs, the WAJIB (in the Oromifa language, the abbreviation for "Forest Dweller Association") approach, based on a partnership between the regional government and individual forest communities, was initiated in 2000 in the Adaba-Dodola Forest in Oromia Region. The program is characterized by exclusive-use rights to the forest dwellers through consensus in each village community. An agreement specifies the rights and responsibilities of both users and the government toward optimum use and sustainability of the forest. Early assessments indicate that communities are able to manage and utilize the forest in a sustainable manner and that their livelihoods have improved (Bishaw, 2001; Segu, 2007; Tadesse, 2007). A review of another eight participatory forest management projects in Oromia and SNNPR reported similar socioeconomic benefits, as well as forest and wildlife rehabilitation (Tadesse, 2008). Other examples of relatively successful traditional forest conservation programs in Ethiopia are cited in chapter 8.

Conclusion and Recommendations

Deforestation in Ethiopia, dating back to ancient times, has been associated with a broad range of environmental, ecosystem, and economic impacts affecting this and all the other countries in the Blue Nile Basin. The relationships between degradation of the natural vegetation and increased runoff and soil erosion during the rainy season and prolonged drought conditions during the dry season are well established. These changing environmental conditions are jeopardizing efforts by the Ethiopian government and local communities to meet food, energy, and domestic water needs and diminishing biological resources, habitats, and ecological services. Most of these resources and services have not been assessed in monetary terms and are threatened by unregulated overexploitation by poor people.

Political commitment to protect natural forests, sustainable water harvesting practices and water retention capacity of soil, and community ownership needs to be strengthened and alternative energy sources must become a priority. Sustainability of forest resources requires that a fair and protective framework be established to ensure that the benefits of resource use accrue to communities (Bishaw, 2001). Comprehensive valuation of hydrological, biodiversity, and ecosystem services of forests and woodlands may establish the true value of forests and woodlands.

Integration of natural resources management programs, in particular development of effective administrative frameworks, is required to improve coordination between departments responsible for forestry planning, development, use, and conservation. Measures and incentives to manage demand for cultivation of new land and fuel wood toward raising the living standard of both rural and urban populations need to be formulated. Particularly, food security, the development of alternative income, and the development of alternative energy sources are necessary preconditions to reducing dependency on the forests for daily needs. Reform of land tenure and access policies can also facilitate the process of rural households achieving water, food, and energy security. An effective forest management policy and laws that are scientifically grounded but also customary arrangements need to be put in place and enforced to properly conserve, develop, and manage the forest resources and maintain biodiversity. Participation of all stakeholders in the management and use of forest resources, including the public's involvement in community-level projects and a commitment of the worldwide coffee industry will be necessary.

This review suggests that in addition to policy and managerial issues, the socioeconomic status of the population and the productivity of agriculture need to be improved to reduce the rate of deforestation, as well as the unsustainable use of forests and woodlands. Existing environmental and antipoverty policies involving land use, energy development, alternative livelihoods, and climate change issues, discussed in chapters 3 and 8, need to be fully implemented toward promoting more responsible stewardship of forests and other woody vegetation.

References

Abate, A. (2004). *Biomass and nutrient studies of selected tree species of natural and plantation forests: Implications for a sustainable management of the Munessa-Shashemene Forest, Ethiopia* (Unpublished doctoral dissertation). University of Bayreuth, Bayreuth, Germany.

Abebe, T. (2005). *Diversity in home garden agro forestry systems of southern Ethiopia* (Unpublished doctoral dissertation). Wageningen University, Netherlands.

Asfaw, Z. (2003). *Tree diversity management in the traditional AF land use of Sidama, southern Ethiopia* (Unpublished doctoral dissertation). Swedish University of Agricultural Sciences, Uppsala.

Bekele, M. (2000). *Review and improvement of data related to wood products* (Unpublished manuscript). Addis Ababa.

Bekele, T. (1994). *Studies on the remnant afromontane forest on the central plateau of Shewa, Ethiopia* (Unpublished doctoral dissertation). Uppsala University, Uppsala.

Bewket, W. (2003). Household level tree planting and its implications for environmental management in the northwestern highlands of Ethiopia: A case study in the Chemoga watershed, Blue Nile Basin. *Land Degradation and Development, 14*, 377–388.

Bishaw, B. (2001). Deforestation and land degradation in the Ethiopian highlands: Strategy for physical recovery. *Journal of Northeast African Studies, 8*, 7–26.

Bonnefille, R., & Hamilton, A. (1986). Quaternary and late tertiary history of Ethiopian vegetation. *Symbolae Botanicae Upsaliensis, 28*, 48–63.

Brown, G. W. (1976). *Forestry and water quality* (Unpublished manuscript). School of Forestry, Oregon State University, Corvallis.

Bruijnzeel, L. A. (1990). *Hydrology of moist tropical forests and effects of conversion: A state of knowledge review.* Paris: UNESCO and Amsterdam: Free University.

Butzer, K. W. (1981). Rise and fall of Axum, Ethiopia: A geoarchaeological interpretation. *American Antiquity, 46*, 471–495.

Calder, I. R. (2007). Forests and water-ensuring forest benefits outweigh water costs. *Forest Ecology and Management, 251*, 110–120.

Central Statistical Authority. (2004). *Statistical* Abstracts 2003. Addis Ababa: Federal Democratic Republic of Ethiopia.

Chapman, D. W. (1962). Effects of logging upon fish resources of the West Coast. *Journal of Forestry, 60*, 533–537.

Dessie, G., & Kleman, J. (2007). Pattern and magnitude of deforestation in the south central Rift Valley region of Ethiopia. *Mountain Research and Development*, *27*, 162–168.

Embaye, K. (2000). The indigenous bamboo forests on Ethiopia: An overview. *Ambio*, *29*, 518–521.

Environmental Protection Authority. (1998). *National action programme to combat desertification.* Addis Ababa: Federal Democratic Republic of Ethiopia.

Environmental Protection Agency & Ministry of Economic Development and Cooperation. (1997). *Conservation strategy of Ethiopia action plan*. Addis Ababa: Author.

Ethiopian Forestry Action Program. (1994). *Final Report*. Addis Ababa: Ministry of Natural Resources Development and Environmental Protection.

Fichtl, R., & Admasu, A. (1994). *Honey bee flora of Ethiopia*. Weikersheim, Germany: DED and Margalef Verlag.

Food and Agriculture Organization. (2000). *Global forest resources assessment, main report* (FAO Forestry Paper 140). Rome: Author.

Food and Agriculture Organization. (2005). *Forests and floods: Drowning in fiction or thriving on facts?* Rome: Center for International Forestry Research.

Food and Agriculture Organization. (2007). *State of the world's forests*. Rome: Author.

Friis, I. (1992). Forests and forest trees of north tropical Africa: Their natural habitats and distribution patterns in Ethiopia, Djibouti and Somalia. *Kew Bulletin* (Additional series), *15*, 1–396.

Gebre-Mariam, Z. (1998, February). *Human interactions and water quality in the Horn of Africa*. Paper presented at the Symposium on Science in Africa: Emerging Water Management Issues, sponsored by the American Association for the Advancement of Science Africa Program, Philadelphia, PA.

Giday, M. (2008). *Medicinal plants of the Bench, Meinit and Sheko ethnic groups in Ethiopia, with emphasis on use diversity and distribution* (Unpublished doctoral dissertation). Addis Ababa University, Addis Ababa.

Gindaba, J., Rozanov, A., & Legesse, N. (2005). Trees on farms and their contribution to soil fertility parameters in Badessa, eastern Ethiopia. *Biology and Fertility of Soils, 42*, 66–71.

Hagos, F., Pender, J., & Nega, G. (2002). *Land degradation and strategies for sustainable land management in the Ethiopian highlands Tigray Region*. Nairobi: International Livestock Research Institute.

Hartman, I. (2004, March). *No tree, no bee-no honey: Management of resources and marginalization in beekeeping societies of southwest Ethiopia*. Paper presented at the Bridging Scales and Epistemologies Conference, Alexandria.

Hein, L., & Gatzweiler, F. (2006). The economic value of coffee (*C. arabica*) genetic resources. *Ecological Economy, 60*, 176–185.

Hoben, A. (1995). Paradigms and politics: The cultural construction of environmental policy in Ethiopia. *World Development*, *23*, 1007–1021.

International Institute for Sustainable Development. (1999). Global Workshop on Addressing the Underlying Causes of Deforestation and Forest Degradation (Summary report). *Sustainable Development*, *1*, 1–10.

International Union for Conservation of Nature. (2000). *Red list of threatened species*. Retrieved from http://www.redlist.org/info/gallery2000

Interregional Information Network. (2007). Africa: Tackling deforestation critical. *IRIN Newsletter*, Office of the Coordination of the Office of Humanitarian Affairs, United Nations.

Itana, F. (1998). Land degradation in Central Ethiopian Rift Valley: Its extent, implications and ameliorative measures. In T. G. Selassie & S. Sertsu (Eds.), *Understanding soils: Key resources to rural development while protecting environment. Proceedings of the 4th Conference of the Ethiopian Soil Science Society* (pp. 91–97). Addis Ababa: Ethiopian Soil Science Society.

Ives, J. D. (2006). *Himalayan perceptions: Environmental change and the well-being of mountain peoples* (2nd ed.). Lalitpur, Nepal: Himalayan Association for the Advancement of Science.

Jagger, P., & Pender, J. (2003). The role of treets for sustainable management of less-favored lands: The case of eucalyptus in Ethiopia. *Forest Policy and Economics, 5*, 83–95.

Jotte, Z. (2007). The impact of the cultural changes on the people of Sheka and their traditional resource management practices: The case four kebeles in Masha Woreda. In M. Fetene (Ed.), *Forests of Sheka: Multidisciplinary case studies on the impacts of land cover changes, southern Ethiopia* (pp. 89–136). Addis Ababa: MELCA Mahiber.

Kidane-Mariam, T. (2003). Environmental and habitat management: The case of Ethiopia and Ghana. *Environmental Management*, *31*, 313–327.

Kippie, T. (2002). *Five thousand years sustainability? A case study of Gedeo land use, southern Ethiopia* (Unpublished doctoral dissertation). Wageningen University, The Netherlands.

Lemenih, M., & Itana, F. (2004). Soil carbon stocks and turnovers in various vegetation types and arable lands along an elevation gradient in southern Ethiopia. *Geoderma, 123*, 177–188.

Lemenih, M., Karltun, E., & Olsson, M. (2005). Soil organic matter dynamics after deforestation along a farm field chronosequence in southern highlands of Ethiopia. *Agriculture Ecosystems and Environment, 105*, 373–386.

Lemma, B., Kleja, D. B., Nilsson, I., & Olsson, M. (2006). Soil carbon sequestration under different exotic tree species in the southwestern highlands of Ethiopia. *Geoderma, 136*, 886–898.

Makarieva, A. M., & Gorshkov, V. G. (2007). Biotic pump of atmospheric moisture as driver of the hydrological cycle on land. *Hydrology and Earth System Sciences, 1*, 1013–1033.

McCann, J. C. (1999). *Green land, brown land, black land: An environmental history of Africa, 1800–1990.* Portsmouth, NH: Heineman.

Ministry of Economic Development and Cooperation. (1999). *Survey of the Ethiopian Economy: Review of post-reform developments (1992/93–1997/98).* Addis Ababa: Author.

Muschler, R. G., & Nair, P. K. R. (2001). Shade improves coffee quality in a sub-optimal coffee-zone of Costa Rica. *Agroforestry Systems, 85*, 131–139.

Nyssen, J., Poesen, J., Moeyerson, J., Deckers, J., Haile, M., & Lang, A. (2004). Human impact on the environment in the Ethiopian and Eritrean highlands—a state of the art. *Earth-Science Review, 64*, 273–320.

Pankhurst, R. (1995). The history of deforestation and afforestation in Ethiopia prior to World War I. *Northeast African Studies, 2*, 119–133.

Parrotta, J. A., & Agnoletti, M. (2007). Traditional forest knowledge: Challenges and opportunities. *Forest Ecology and Management, 249*, 1–4.

Perfecto, I., Vandermeer, J., Mas, A., & Soto Pinto, L. (2005). Analysis: Biodiversity, yield, and shade coffee certification. *Ecological Economics, 54*, 435–446.

Pohjonen, V., & Pukkala, T. (1990). *Eucalyptus globulus* in Ethiopian forestry. *Forest Ecology and Management, 36*, 19–31.

Population Census Commission. (2008). *Summary and statistical report of the 2007 population and housing census: Population size and sex.* Addis Ababa: Federal Democratic Republic of Ethiopia.

Potter, L. (2006). An overview of migration and deforestation in Southeast Asia 1880–2002. In W. de Jong, A. Ken-ichi, & T.-P. Lye (Eds.), *The social ecology and tropical forests: Migration, populations and frontiers* (pp. 174–194). Kyoto: Kyoto University Press.

Rahmato, D. (2001). The landscape of development: A complex reality. In T. Assefa (Ed.), *Food security through sustainable land use. Population, environment and rural development issues for sustainable livelihood in Ethiopia. Proceedings of the second national workshop of NOVIB partners forum on sustainable land use* (pp. 7–17). Addis Ababa: NOVIB Partners.

Randall, A. K., Narendra, S., & Mohan, M. (1995). *Valuing tropical forests*. New York: World Bank.

Reusing, M. (1998). *Monitoring of forest resources in Ethiopia*. Report prepared for the Ministry of Agriculture, Addis Ababa.

Ritler, A. (1997). *Land use, forests and the landscape of Ethiopia 1699–1865: An inquiry into the historical geography of central/northern Ethiopia.*(Soil Conservation Research Project Report 38). Berne: University of Berne, Center for Development and Environment.

Robinson, I., MacKay, R., Afwerki, S., & Yirgalem, A. (1995). *A review of the agricultural rehabilitation programme in Eritrea*. Bangor: Centre for Arid Studies.

Schmitt, C. B. (2006). *Montane rainforest with wild* Coffea arabica *in the Bonga region (SW Ethiopia): Plant diversity, wild coffee management and implications for conservation* (Unpublished doctoral dissertation), University of Bonn.

Segu, K. (2007). *Participatory forest management and livelihood of the Forest Dwellers Association (WAJIB): The case of Adaba Dodola Regional Forest Priority Area* (Unpublished master's thesis). Department of Environmental Science, Addis Ababa University.

Senbeta, F., & Denich, M. (2006). Effects of wild coffee management on species diversity in the Afromontane rainforests of Ethiopia. *Forest Ecology and Management, 232*, 68–74.

Seyoum, A. (2007). Economic value of afromontane natural forest in Sheka, Zone Southwest Ethiopia. In F. Masresha (Ed.), *Forests of Sheka: Multidisciplinary case studies on the impacts of land cover changes, southern Ethiopia* (pp. 183–217), Addis Ababa: MELCA Mahiber.

Sherefa, B. (2007). Landuse/landcover changes in Anderacha and Masha woredas of Sheka Zone, SNNP Regional State. In F. Masresh (Ed.), *Forests of Sheka: Multidisciplinary case studies on the impacts of land cover changes, southern Ethiopia* (pp. 21–56). Addis Ababa: MELCA Mahiber.

Soto-Pinto, L., Perfecto, I., & Caballero-Nieto, J. (2002). Shade over coffee: Its effects on berry borer, leaf rust and spontaneous herbs in Chiapas, Mexico. *Agroforestry Systems, 55*, 37–45.

Tadele, K., & Förch, G. (2007). Impact of land use/cover change on stream flow: The case of Hare River watershed, Ethiopia. *Proceedings of the 2007 Lake Abaya Research Symposium* (pp. 80–84). Arba Minch: Arba Minch University.

Tadesse, T. (2007). An overview of forest ecosystems of Ethiopia: Function, trend and future direction. In S. Mengistu & C. De Stoop (Eds.), *Environment for survival: Taking stock of Ethiopia's environment.*

Proceedings of the first green forum conference (pp. 33–65). Addis Ababa: Green Forum.

Tadesse, T. (2008). Participatory forest management (PFM) in Ethiopia. In D. Cottorone (Ed.), *Proceedings of the workshop on communal forest ownership: An option to address the underlying causes of deforestation and forest degradation in Ethiopia?* (pp. 8–14). Addis Ababa: MELCA Mahiber.

Tavoni, M., Sohnge, B., & Bosetti, V. (2007). Forestry and the carbon market response to stabilize climate. *Energy Policy*, *35*, 5346–5353.

Tefera, Z. (2007). The Ethiopian highland ecosystem. In S. Mengistu & C. De Stoop (Eds.), *Environment for survival: Taking stock of Ethiopia's environment. Proceedings of the First Green Forum Conference* (pp. 75–96). Addis Ababa: Green Forum.

Tekle, K., & Hedlund, L. (2000). Land cover changes between 1958 and 1986 in Kalu District, Southern Wello, Ethiopia. *Mountain Research and Development*, *20*, 42–51.

Thomas, I., & Bekele, M. (2003). *Role of planted forests and trees outside forests in sustainable forest management: Case study, Republic of Ethiopia: Planted forests and trees* (Working paper FP/29E). Rome: Food and Agriculture Organization.

Tsegaye, A. (2002). On indigenous production, genetic diversity and crop ecology of enset *(Ensete ventricosum* (Welw.) Cheesman) (Unpublished doctoral dissertation). Wageningen University, Wageningen, The Netherlands.

Vanacker, V., von Blanckenburg, F., Govers, G., Molina, A., Poesen, J., Deckers, J., & Kubik, P. (2007). Restoring dense vegetation can slow mountain erosion to near benchmark levels. *Geology*, *35*, 303–306.

Van Dijk, A. I. J. M., & Kennan, R. J. (2007). Planted forests and water in perspective. *Forest Ecology and Development, 251*, 1–9.

Von Breitenbach, F. (1962). National forestry development planning: A feasibility and priority study on the example of Ethiopia. *Ethiopian Forestry Review*, *3/4*, 41–68.

Von Breitenbach, F. (1963). *The indigenous trees of Ethiopia* (2nd ed.). Addis Ababa: Ethiopian Forestry Association.

Wakjira, F. S. (2006). *Biodiversity and ecology of Afromontane rainforest with wild Coffea arabica L. populations in Ethiopia.* Göttingen: Cuvillier Verlag.

Woldemariam, T. (2007). The impact of landuse/landcover changes on biodiversity in Masha and Anderacha woredas of Sheka Zone, SNNP Regional State. In M. Fetene (Ed.), *Forests of Sheka: Multidisciplinary case studies on the impacts of land cover changes, southern Ethiopia* (pp. 57–88). Addis Ababa: MELCA Mahiber.

Yadessa, A., Itanna, F., & Olsson, M. (2001). Contribution of indigenous trees to soil properties: The case of scattered trees of *Cordia africana* Lam: Croplands of western Oromia. *Ethiopian Journal of Natural Resources*, *3*, 245–270.

Yimer, F., Ledin, S., & Abdelkadir, A. (2007). Changes in soil organic carbon and total nitrogen contents in three adjacent land use types in the Bale Mountains, south-eastern highlands of Ethiopia. *Forest Ecology and Management, 242*, 337–342.

Young, A. (1997). *Agroforestry for soil management*. Wallingford, U.K.: CAB International and Nairobi: ICRAF.

Zeleke, G., & Hurni, H. (2001). Implications of land use and land cover dynamics for mountain resource degradation in the northwestern Ethiopian highlands. *Mountain Research and Development*, *21*, 184–191.

Chapter 8

Land Degradation and Water Problems in the Highlands

Causes, Impacts, and Interventions

Helmut Kloos, Yohannes Gebre Michael, and Alula Pankhurst

Introduction

Continuing environmental degradation as a result of overpopulation and unsustainable land use practices in Ethiopia's mountainous terrain has been associated with a downward spiral of changes in natural resources and socioeconomic wellbeing. Definitions of land degradation recognize the involvement of two complex systems: the natural ecosystem and the human social system, with the implications for production potential, sustainability, and stability (Yang & Conacher, 2009). The paramount land degradation processes, particularly soil erosion, and consequent loss of soil depth, biomass, fertility, and water holding capacity of the soil, are both the cause and result of ever more exploitive, nonsustainable cultivation and livestock grazing by the rapidly growing poor, rural Ethiopian population. Soil degradation worldwide often leads to agricultural drought caused not by lack of rainfall but by poor soil management. These manmade droughts can be prevented or ameliorated by adopting soil, crop, and water management strategies that secure adequate soil water/green water availability and maximize plant water uptake capacity (Rockström, et al., 2009). The objective of this chapter is to examine the nature and determinants of land degradation in the Ethiopian Highlands within the context of different land use practices and systems, soil, and water conservation programs (including government interventions), socioeconomic, cultural, and political parameters, and various developments since the 1974 Ethiopian revolution. We argue that land degradation and associated water scarcity faced by Ethiopia during the socialist period from 1974 to 1991 and during the subsequent transition from a socialist state to a market economy was primarily a socioeconomic and political problem, physico-biotic vulnerability playing a secondary role.

In Ethiopia, declining agricultural productivity tends to inhibit investment in land management and encourages the clearing of new land for agriculture rather than intensification of production. This practice often leads to further land degradation, completing a vicious cycle that leaves the landscape increasingly vulnerable to drought, rainfall-related erosion, and floods. Ethiopia's mountainous topography, with 79% of its land surface having slopes greater than 16 degrees and one third of this area with slopes of 30 degrees or more, is particularly prone to erosion (Campbell, 1991). An estimated 2.0–3.5 billion tons of soil are being eroded annually in Ethiopia, mostly in the highlands (Admassie, 2000; Food and Agriculture Organization, 1993), but amounts range from about 10 tons per ha in the southwestern part of the country to 1,500 tons ha in the northern and eastern parts (World Bank, 2006). A number of studies estimate that 30,000 ha of agricultural land are lost annually due to water erosion, with over 2 million ha already severely damaged (Berry, 2003).

Custodianship of the imperial government of Haile Selassie and the Derg socialist regime over forest, land, and water resources has been described as "a dismal failure" that "in many cases led to mismanagement and loss of natural resources" (Rahmato, 2001). This poor record is highlighted by a number of soil and water conservation efforts spearheaded by the Derg regime in the 1980s, which failed, largely due to a lack of understanding the complexities of the relationships between resource conservation and agriculture and of farmers' survival needs and attitudes toward sustainable land management (Zeleke, Kassie, Pender, & Yesuf, 2006) and because of deficient implementation of the programs.

This chapter uses the UN Convention to Combat Desertification definition of land degradation as the reduction or loss of the biological or economic productivity of the terrestrial bio-productive system. Although this definition overlaps with the term desertification, defined as land degradation in the arid, semiarid, and subhumid areas (Food and Agriculture Organization, 2005), we use it here sparingly because of the predominantly subhumid and humid climate in Ethiopia's highlands.

Ethiopia's Soil Resources and Degradation

Land degradation in Ethiopia is centered on the highlands due to the seasonally heavy, intense rainfall, large deforested areas, the mountainous terrain, and prevailing land use practices. Sociopolitical changes and misguided government policies and development programs have contributed to land degradation in recent decades.

Soil Types

No detailed soil survey has been undertaken in Ethiopia, owing to the great diversity of soil types in the geologically, topographically, and climatically complex country. The soil map prepared by the Ministry of Agriculture shows 17 major soil groups with a wide range in fertility, depth, water retention capacity, and erodibility. Thin, stony, infertile lithosols are the most common soils, covering 17% of the country's area, mostly in the lowlands and eastern highland escarpments. These stony soils are mostly less than 10 cm deep, have developed over hard rock, and have no agricultural potential. Dystic and neutric nitosols cover 12.5% of the land, mainly in the high rainfall areas of western and southern Ethiopia. The nitosols are some of the best soils as long as they remain under forest or grass cover, but rapidly lose their fertility under continuous plow cultivation. Black vertisols (10% of the land), which are relatively fertile but difficult to work and have poor drainage and water infiltration properties, are found on flat lands in the grain growing areas of the northern, central, and southern highlands. These clay soils form a crusty, deeply cracked surface during the dry season, when they are too hard to plow, and take up water through their deep cracks during the first rains, when they can be plowed. They swell and self-seal during continuing rain and become impervious to water, drowning germinating seeds and causing much runoff, requiring that farmers plow and plant during the window of opportunity when soil conditions are suitable (Guiral, 2009). In the northern and much of the central and eastern highlands, thin cambisols, renzinas, and phaeozems with limited agricultural potential prevail, together with lithosols on steeper slopes. The remaining common soils in the highlands, chromic luvisols, acrisols, and cambisols, characterized by different degrees of agricultural suitability, are found in small areas throughout the highlands (Ethiopian Mapping Authority, 1988).

Water-Related Soil Erosion Rates

According to the Ethiopian Highland Reclamation Study, 80% of the total population, 60% of the livestock, and 90% of land suitable for rainfed agriculture are found at elevations above 1,500 m, rendering the highlands highly susceptible to anthropogenic impacts (Hurni, 1988a). The highest soil erosion rates and the most degraded areas in Ethiopia have been found in the drought-prone Low Potential Cereal Zone identified by the Ethiopian Highland Reclamation Study in the northern highlands of Amhara (including the former Wello and Gondar administrative regions) and Tigray Regions. About 4% of this area has become totally unproductive; severe degradation was found in most areas, and only 20% of the total area of the highlands was classified as slightly to moderately eroded.

Soil losses measured on test plots in different parts of the Ethiopian highlands were reported to range from 3.4 to 84.5 tons per ha per year (Berry, 2003) but reached 122–425 ha/ton/year on steep slopes in the Blue Nile and upper Baro basins (Netherlands Engineering Consultants, 1997). Comparison of these rates with the much lower soil formation rates shown in table 8.1 indicates high rates of net soil loss. In the most highly degraded areas, which extended from the central part of the highlands northward into Tigray Region (Hurni, 1988b), annual soil production declines in the absence of conservation measures were estimated to be as high as 2–3% (Constable, 1985; Hurni, 1988a). Former crop land, degraded and lacking adequate vegetation cover, was most intensively eroded, followed by land planted to annual crops. Swamps and forests had the lowest erosion rates (table 8.2). Soils tend to be shallowest in the high-altitude *dega* and *wurch* zones, where rainfall is relatively high, soil formation rates low, and slopes among the steepest (table 8.1). While this progressive land degradation process argues for soil and water conservation in the Ethiopian Highlands and against the notion that surface runoff in Africa may be more beneficial than harmful if stored for dry-season use (Falkenmark & Rockström, 2004), the conservation programs by the Ethiopian government did not achieve their targets and even accelerated resource depletion. On the other hand, the negative reaction to forced labor the colonial regimes used to build anti-erosion physical structures, which was associated with the failure of conservation programs after independence in Kenya, Tanzania, and several other countries (Falkenmark & Rockström, 2004) was not an issue in Ethiopia, which was spared from these programs during its short colonial history.

TABLE 8.1. Average soil formation rates and mean annual rainfall in Ethiopia, by altitude/climatic zone.

Altitudinal/ climatic zone	Altitude (m)	Mean annual rainfall (mm)	Soil formation rates*
High *wurch*	>3,700	No data	2
Wet *wurch*	3,200–3,700	>1,400	4
Moist *wurch*	3,200–3,700	900–1,400	3
Wet *dega*	2,300–3,200	>1,400	10
Moist *dega*	2,300–3,200	900–1,400	8
Wet *woina dega*	1,500–3,200	>1,400	16
Moist *woina dega*	1,500–2,300	900–1,400	12
Dry *woina dega*	1,500–2,300	800–1,100	6
Moist *kolla*	500–1,500	600–900**	6
Dry *kolla*	500–1,500	<300–600**	3
Berha, desert	<500	<300**	1

*Tons per hectare per year.
Sources. **Modified from Hurni (1983a, 1983b), based on unpublished data of the Ethiopian National Meteorological Services Agency.

In the lowlands, freshwater resources have declined and sand and dust storms have increased due to desertification (Environmental Protection Authority, 1999). The decline of surface waters and rapid population growth on the densely populated highland plateaus is contributing to severe water shortages in larger cities, including Addis Ababa, where water rationing has become more frequent in recent years.

Factors and Processes in Soil Degradation

Population pressure, land fragmentation, scarcity of arable and grazing land, repeated drought and famine, and poverty have been instrumental in decreasing the productivity of the land, increasing the number of landless peasants over the years in Ethiopia. This situation prompted several land

TABLE 8.2. Estimated soil loss by sheet and rill erosion for various land uses in Ethiopia.

Land cover	Area (%)	Soil loss*
Grazing land	47	5
Uncultivable	19	5
Crop land	13	42
Wood and bush land	8	5
Former crop land	4	70
Wetlands	4	0
Forests	4	1
Perennial crops	2	8

*Tons per ha per year
Source. "Sources" Hurni (1983b; 1985)

redistributions after the 1975 land reform, resulting in ever smaller land holdings. As a survival strategy, many households sent family members as laborers to large commercial farms or to towns, or moved as family units and settled in more promising parts of the country. Spontaneous settler movements predated and continued during and after the government resettlement program described below. These movements took millions of farmers to higher elevations and steeper slopes in the highlands and to lowlands, especially after the commencement of malaria control programs around 1960 (Dejene, 1990; Ezra, 2001; Kloos, 1982; Pankhurst & Piguet, 2009; Rahmato, 2001). A survey of 2,000 households in 40 peasant associations in degraded areas in the former northern Shewa, Wello, and Tigray regions reported that the poorest people living in severely degraded areas were most likely to out-migrate during the 1984–1994 period (Ezra, 1997). These spontaneous migrations were in contrast to the government resettlement programs, which largely failed (Pankhurst & Piguet, 2009).

The soil degradation process consists of three components of soil degradation: biological, chemical, and physical degradation, which operate in combination, although their individual contributions vary in different areas due to differences in climate, soil type, slope, erosion, vegetation cover, land management practices, and the degree of degradation (Dubale, 2001). Although still poorly understood, the complexity of the climate, characterized by changes within short distances from arid tropical to warm subhumid, cool humid, and alpine types between lowlands and high mountains, generates widely varying amounts of runoff and erosion under different types and densities of vegetation cover and cultivation systems. Rainfall distribution, vegetation cover, slope, land use, soil texture, organic matter content, and amount of stones constitute the major driving forces in soil erosion in the Ethiopian highlands, with land use practices the underlying culprit in the degradation process. Mean annual runoff was up to 40 times higher on largely nonvegetated than well-vegetated experimental plots on steep slopes receiving high rainfall (Nyssen et al., 2006). Increasing amount and intensity of precipitation with altitude and on windward (mostly western and southern) slopes accounts for much of the high runoff in the Nile and Baro-Akobo basins (see chapter 3). A longitudinal study of hydrometric records of 168 meteorological stations revealed that the erosive potential of rainfall in the Ethiopian Highlands increases from south to north and from west to east, with the highest values in the traditional famine areas (Osman, 2001). In addition to the mountainous topography and large areas of clay soils, some of the highest rain erosivity rates worldwide (determined by density and size of drops), further increase the vulnerability of these highly degraded areas (Nyssen et al., 2005). Catchments lose about half of all precipitation after the initial seasonal rains saturate the soil (Liu et al., 2008).

Physical soil degradation

As indicated earlier in this section, physical soil degradation due to soil movement, compaction, reduction in aeration, reduced permeability, and decreased soil depth cause much biological and chemical erosion in Ethiopia, especially in the cereal agro-ecological zone. Although both water and wind erosion are widespread in Ethiopia, the former in the highlands and the latter in the arid southern, eastern, and northern lowlands, our discussion is confined to water erosion due to lack of information on the impact of wind erosion on water availability and the hydrological cycle in Ethiopia. According to Constable and Belchaw (1989), an estimated four fifths of the erosion in the highlands is due to the overexploitation of croplands, and the remainder is the result of overgrazing and deforestation.

Land use patterns play a central role in physical degradation in Ethiopia. An experimental study of soil erosion reported changes in the texture and structure of soil in eroded soils that were associated with declining organic matter and nitrogen levels as well as soil productivity (Tegene, 1992). Another experimental study, in the Chercher Highlands in eastern Oromo Region, found runoff and soil loss to be 6–7 times greater on plowed plots devoid of conservation structures than on slopes with level bunds along field contours or grass strips planted along contours (Tolcha, 1998). Cultivation of the staple crop *teff (Eragrostis tef)*, which involves 5 to 8 plowings, including cross-plowings for seedbed preparation at the onset of the rains, renders the soil highly susceptible to erosion. Tillage systems involving the digging stick or hoe and long-fallow, traditionally prevailing in the western and southern highlands, are the least vulnerable to erosion (Leye, 2007). In a study of the sediment trapping capacity of exclosures installed on steep slopes in the highlands of Tigray Region, a widely used conservation strategy in that region, slope gradient and length of slope, and characteristics of the sediment source area were important factors (Descheemaeker et al., 2006).

Biological soil degradation

Biological soil degradation is a process characterized by a decline in the soil humus content of the soil through mineralization, reduction in organic matter content and carbon from biomass, and decreases in the soil fauna. The potential for replenishing and rebalancing the biological components of soil under cultivation is small in Ethiopia, especially in the highland grain growing areas, where crop residues are commonly used for livestock feed or roof thatching, the remaining crop stalks are grazed on by livestock, and dung is mostly used as fuel. The mineralization of nutrients is further reduced by plowing, which is increasingly carried out under accelerating population pressure. Deterioration of the biological soil environment contributes to reducing the water infiltration and retention capacity of the soil under permanent seed/plow cultivation, in contrast to ecologically compatible long-fallow systems in western Ethiopia (Dubale, 2001). Several agronomic conservation measures continue to be widely used throughout much of the grain growing highlands, although population pressure is making inroads. They include crop rotation involving grains and leguminous crops (which increase the crop canopy), water infiltration and retention (which reduce runoff and erosion and add nitrogen to the soil), organic and green manuring and the application of ashes along the contour, leaving crop stalks in fields to promote soil fertility and control erosion, fallowing on steep slopes, and agro-forestry. Poor farmers applied the fewest conservation measures, largely due to lack of bovine livestock and shortage of labor (Gebre Michael, 1999; Tegene, 1998).

Chemical soil degradation

Similar to biological and physical soil degradation, rates of nutrient depletion are particularly high in densely populated and erosion-prone mountainous countries, including Ethiopia, Kenya, Malawi, and Rwanda (Scherr, 1999). In the Ethiopian highland agro-ecosystem, chemical soil degradation due to intensive cultivation in the absence of adequate soil management has contributed to declining soil fertility, especially through further acidification of the naturally acidic highland soils and loss of phosphorous and nitrogen. In some lowland areas, the accumulation of salt has impacted crop production in some irrigated areas (Dubale, 2001; Food and Agriculture Organization, 2005). A spatial analysis of national data on crop production, fertilizer use, and land management practices by the Central Statistical Authority found that soil nutrients were decreasing in all regional states except in areas under permanent crops (coffee and *enset, Enset ventricosum*) and vegetable crops. Whereas nutrient losses under annual cereal crops were caused mostly by erosion, losses under permanent and vegetable crops were mainly due to the removal of residues and harvested products, as well as leaching, indicating the protective nature of permanent vegetation (Haileselassie, Priess, Veldkamp, Teketay, & Lesschen, 2005). Leaching is common only in the humid southwestern part of the country, where the prevailing nitosol soils rapidly lose their fertility after deforestation and under continuous plow agriculture. In one of the first quantitative assessments of the impact of exclosures (areas off limits to livestock grazing) on reversing the process of land degradation in Ethiopia, Mekuria et al. (2007) found that grazing areas closed for 5 and 10 years had significantly higher levels of soil organic matter, total nitrogen, and available phosphorous (the two nutrients most affected by soil erosion) than free-grazing lands, demonstrating that exclosures are effective in restoring pastures, improving soil nutrient status, and reducing erosion.

Although duration of cereal/plow cultivation after deforestation has been associated with long-term soil exhaustion, recent studies have demonstrated that significant soil regeneration can take place during periods as short as several decades. In a southern highlands cereal growing area, both soil organic carbon and total nitrogen values were inversely correlated with the duration different fields had been cultivated since forest clearance (for periods ranging between 7 and 53 years); the highest nutrient levels were recorded for soils in a nearby forest (Lemenih, Karltun, & Olsson, 2005). These and other studies indicated that unsustainable land use patterns are largely responsible for chemical soil degradation. Ethiopian government planning and implementing agencies have tended to attribute soil and water conservation problems to physical-environmental and technical factors without considering the role of the socioeconomic and cultural environment (Berry, 2003; Tolcha, 1998). In the absence of farmer-centered programs guided by farmer experiences, concerns, and preferences, farmers rarely adopted conservation measures introduced from the outside (Tolcha, 1998). Thus, the sustainability of interventions has been difficult and occasionally compromised by lack of relevant information or models developed under different conditions and assumptions elsewhere (Nyssen, et al., 2004; Tegene, 1992).

Indigenous Land Use and Conservation

A brief review of major indigenous land use practices can provide background information on soil and water conservation efforts and facilitate the examination of the impacts of the 1975 land reform, the Ethiopian Highland Reclamation Project, and the villagization program in the mid-1980s and the 1990 economic reform.

It is generally recognized that full assessment of the processes of soil degradation under natural conditions and within conservation schemes requires not only an understanding of changes in the stock of physical and nutrient capital but also of the underlying socioeconomic and cultural factors (Beyene, Gibbon, & Haile, 2006). Seed/plow cultivation in Ethiopia has been a major factor historically in land degradation and continues to make major inroads into the older, ecologically more compatible hoe/digging stick agricultural systems in the western and southern parts of the country (McCann, 1995). It has been known for decades that conventionally plowed fields worldwide produce erosion rates between tens and hundreds of times higher than rates under native vegetation and long-term geological erosion. Erosion rates from no-till agriculture come close to rates of soil production and, thus, form the foundation for sustainable agriculture (Montgomery, 2007). Additional benefits of no-tillage were increased availability of water to crops through increased water infiltration and reduced evaporation, sequestering carbon on soil, reducing energy needs by eliminating draft animals, and increased yields. Several field experiments in Ethiopia revealed significantly lower runoff and higher crop yields using no-till cultivation (Welderufael, 2006). However, the fact that no-till technologies are effective only with the use of crop residue as mulch (Lai et al., 2007) and the central role of oxen in the highland plow culture (McCann, 1995), reduces their potential use in Ethiopia, where crop residues are a major source of fodder for oxen. In addition to using crop residues for animal feed, highland farmers have traditionally maintained hillside pastures, a contentious issue with government agencies, who consider communal pasture lands as wasted or undeveloped and push for their afforestation as part of rehabilitation efforts (Pankhurst, 2001). Recent field trials in conservation tillage using modified forms of the simple *maresha* wooden plow have been promising as they resulted in soil erosion rates and *teff* and maize yields similar to those achieved with the traditional digging stick and hoe in conjunction with mulching (Leye, 2007; Temesgen, Rockström, Sanenije, & Hoogmoed, 2007). However, the broad-bed maker technology has not been widely adopted by farmers, who feared that this would obligate them to participate in political activities (Pankhurst, 2000).

Technical experts and policy makers are increasingly considering indigenous agricultural systems, knowledge, and practices in tropical countries as the most effective and sustainable systems in soil and water conservation and most externally imposed projects as inappropriate for African environments (Critchley, Reij, & Willcocks, 2006). In the central and northern highlands of Ethiopia, farmers have traditionally constructed drainage ditches in fields, stone walls and simple terraces on steep slopes, and practiced strip cultivation. In the highlands of South Wello and Northern Shewa zones, discontinuous (staggered or interlocked) bunds and terraces continue to be used as one of the main soil conservation methods. The advantage of this indigenous technology over newly introduced conservation measures is that it is compatible with the local environment, land use, and farming system and is easy to implement and maintain, ensuring that it is socially accepted, maintained, and protected (Gebre Michael, 1999; Tegene, 1998). Conservation tillage systems developed for *teff* and maize have been shown to reduce surface runoff, increase soil moisture, and increase yields (Leye, 2007; Welderufael, 2006). Together with small-scale irrigation, agro-forestry, and properly managed rangelands, conservation tillage can, according to the UN Food and Agriculture Organization, make agricultural systems more resilient to climate change and also increase carbon sequestration (Interregional Information Network, 2009).

The most elaborate and effective soil and water conservation techniques in Ethiopia were developed by the Konso and other peoples in southwestern Ethiopia, who constructed high-density stone terrace systems, stone-lined canals and reservoirs, and used the hoe, manure, and a large variety of drought and pest-resistant crop varieties in their semi-arid, infertile homelands (Beshaw, 2003; Hallpike, 1970; Kruger, Fantaw, Gebre Michael, & Kajela, 1996). The arrival of the plow and the imposition of new labor systems by the northern conquest at the beginning of the 20th century (Amborn, 1998) and the formation of state-controlled peasant associations shortly after the 1975 land reform contributed to the decline of the terracing structures and traditional institutions in those areas. These changes led to reduced resilience of local agro-ecological systems, including increasing shocks from drought (Beshaw, 2003; see also chapter 4). Similar effects were reported from other developing countries developing new production systems that failed to consider the complexity of human and natural systems that calls for adaptive rather than exploitive practices (Walker & Salt, 2006).

Cognizant of the fact that the recent government soil and water conservation projects in Ethiopia have not achieved their objectives and the long tradition of Ethiopian farmers to deal with soil degradation problems (Beyene et al., 2006), the current government is promoting greater participation of farmers in land management decisions as stipulated in the National Water Resources Management Policy in an effort to engage local land users in decision making (Ministry of Finance and Economic Development, 2002). This constitutes a departure from the government's social and rural development programs during the revolution, starting with the 1975 land reform. Nevertheless, in only a few pocket areas could this participatory approach using new technologies be successfully launched.

In most areas, the agricultural extension service continues to operate in a top-down fashion in introducing to farmers preselected technologies, with little capacity to provide the most appropriate solutions to local soil and water problems (Zeleke et al., 2006). Ethiopian policy makers are trying to strike a balance between state-operated and community-based jurisdictions and programs, as indicated by continuing efforts in policy revisions in the area of agriculture (Teshome, 2007) and poverty alleviation (Federal Democratic Republic of Ethiopia, 2006). There are some indications that a combination of local initiative and the strengthening of community management capacity, followed by community and external inputs and the formation of institutional linkages to empower local people to be agents of their own development may be feasible (Bergdall & Powell, 1998; Liu et al., 2008). But such initiatives must include a broader set of partners, including private firms and water users themselves, and also be sufficiently inclusive to integrate viable elements of local cultures (Namara, Awulachew, Namara, Babel, & Gupta, 2007).

The environmental impacts of the 1975 land reform, the Highland Reclamation Project, and the resettlement, villagization, and food aid programs are summarized below.

Government Policies and Programs

The 1975 Land Reform

The Ethiopian land reform of 1975, the first national measure undertaken by the socialist government after the overthrow of the absolute monarchy of Haile Selassie, had profound immediate and long-term economic, institutional, and environmental impacts. By abolishing landlordism, tenancy, and other forms of dependency on land owners, and by establishing peasant associations and other rural mass organizations, the reform allocated, for the first time in recent Ethiopian history, land to the rural population, and allowed for greater decision making by peasants in regard to the management and use of natural resources, including the development of land conservation projects (Rahmato, 1984). A major drawback of the land reform was that the state remained the real owner of the land, with peasants having only usufruct rights over the land, which precluded private ownership and land sales. The distribution of land previously owned by large land holders to the large and rapidly growing rural population resulted in ever smaller peasant plots, sharecropping and leasing/renting of land (even though this was considered illegal), and the hiring of farm labor. These practices arose out of the inability of many farmers to farm their land due to lack of oxen, lack of labor, or insufficiency of land holdings (Gamachu, 1990). Moreover, the grassroots institutions were gradually highly politicized and many leaders became political appointees. Land fragmentation and diminution of per capita production thus continued as before the revolution. Plans by the government to incorporate nonviable, fragmented plots into larger producer cooperatives as part of its plan to further socialize production failed due to the disinterest of peasants, who feared further restrictions on their family economy (Gamachu, 1990) and traditionally had plots in different areas as a strategy of risk minimization from crop failure and to exploit different microclimates and soil conditions.

The land reform failed to raise the socioeconomic status of the rural population, which continued by en large to exploit the environment in a survival mode as had happened during the prerevolutionary period, especially in the northern highlands, where communal land tenure and associated land fragmentation and plot diminution had existed prior to the revolution. Many investigators have emphasized that this pattern was reinforced by insecure tenure resulting from the 1975 land reform and advocated private land ownership to solve this and other problems associated with the existing public land ownership (Admassie, 2000; Rahmato, 1990). The overall effect of the land reform from the perspective of state powers and intervention was that it made the state the sole proprietor of the land, making it possible for the government to effectively move forward with the rural social transformation and environmental conservation initiatives and programs. There have been additional land distributions in recent years, for example in Amhara Region in 1997 and more recently in Tigray and Amhara, giving mostly pasture and forest land to young people, which has, of course, environmental consequences. There has also been land certification in several regions that could affect tenure.

The Highlands Reclamation Project

The socialist Derg government, through the Ministry of Agriculture and the peasant associations, and with the support of the Food and Agriculture Organization and the World Bank, carried out the Ethiopian Highlands Reclamation between 1979 and 1989. This massive program reflected state intervention advocated by the Eastern Block and prevailing views among donors about linkages between drought and deforestation. Strong bilateral, multilateral, and nongovernmental support involved the United Nations, more than a dozen European governments, and numerous organizations. The University of Berne managed the Soil Conservation

Project in collaboration with the Ethiopian Ministry of Agriculture. The results of the project's soil erosion experiments in three small watersheds showed high erosion rates, justifying the urgency of proponents of soil and water conservation to press ahead with their plans. Donors, recognizing the social cause/environmental effect linkage and the improved possibilities to intervene in this process, heavily supported the project financially and materially. Immense efforts and resources were devoted to soil and water conservation on the premise that a rural infrastructure consisting of bunds, terraces, and community forests managed by local peasant associations could be put in place through the hierarchy of the state. The underlying rationale for this approach was based on the assumption that farmers would best play only a passive role in conservation and could only gain from this long-term investment (Campbell, 1991).

The social constraints that had caused the failure of earlier afforestation programs, particularly the complex land tenure system involving absentee landlords, local administrators, the church, and farmers holding land titles under private and freehold tenure (Ezra, 2001; Campbell, 1991) were removed through the 1974 revolution, and conservation work accelerated rapidly after the 1975 land reform. The creation of peasant associations made it possible to organize rural development work, including food-for-work, on a large scale. This process was facilitated by a range of government policies aimed at the collectivization of agriculture and the 1984–1985 famine through the huge international support it generated.

The government followed an integrated catchment treatment approach, consisting of on-farm and off-farm soil conservation structures, hillside afforestation and enclosures, and road construction. Between 1979 and 1985, an estimated 600,000 km of soil and stone bunds and 470,000 km of hillside terraces were built and 80,000 ha of steep slopes enclosed for regeneration (People's Democratic Republic of Ethiopia, 1990). After finding a measure of support during the 1984 famine, the campaigns were soon recognized as a means to reinforce state power by exerting control over large pasture and farming areas (Admassie, 2000). Nevertheless, farmers continued to participate as they received grain and cooking oil for their labor. The World Food Program of the United Nations, the largest food-for-work program worldwide, supported the Highland Reclamation Project operating in 117 river catchments. By 1987, a total of 640,000 ha had been afforested or treated with on-farm soil and water conservation structures (Admassie, 2000).

In spite of the impressive temporary achievements of both programs in physical terms, they failed to become community-supported and sustainable for a number of administrative, technical, and policy-related problems. In particular, issues surrounding ownership of rehabilitated areas were not resolved, resulting in lack of incentives for maintaining the structures that were built. The allocation of scarce hillside land by peasant associations (under the directives of the Ministry of Agriculture) for community forestry without considering the priorities and needs of individual households was resented by the rural population (Pankhurst, 2001, 2003). In most areas, peasants participated in the program only because of the food aid they received and their preference for assistance for their private trees or access to community forests (Admassie, 2000; Scones & McCracken, 1989). Campbell's (1991) review of the literature revealed that the hillside closures reduced access to land and increased rural impoverishment and regional labor migration. By 1992, most tree seedlings planted on more than 300,000 ha of hillside closures had perished due to livestock damage, illegal cutting, or unsuitability of some tree species for local environments (Admassie, 2000). According to two surveys, only 30% of soil bunds, 25% of stone bunds, 60% of hillside terraces, 22% of planted trees, and 7% of enclosed areas implemented between 1976 and 1990 survived by the early 1990s (Zeleke et al., 2006). Similarly, most check dams and cut-off drains were not maintained or were destroyed by farmers (Admassie, 2000).

This and other evidence suggests that major reasons for the failure of the government-sponsored rehabilitation program were 1) the top-down approach, which marginalized farmers in decision making, 2) use of a technology consisting of physical structures with a wider range of risks and benefits than indigenous structures, 3) a general lack of interest by poor farmers in land conservation measures imposed on them, 4) the uniform application of similar soil and water conservation measures in catchments within different agro-ecological zones and socioeconomic characteristics, rendering some measures inappropriate for certain land use systems, thwarting their adaptation by local farmers, and 5) tenure insecurity (Admassie, 2000; Gebre Michael, 1999; Tolcha, 1998).

Although soil erosion decreased and crop yields increased on some bunded fields (Gebre Michael, 1999), the environmental effects of the Highland Reclamation Project are more profound and must be evaluated within the context of prevailing socioeconomic conditions and policy framework, as noted by Campbell (1991). He concluded that a number of conditions compounded the effects of reduced access to land, including 1) the elimination by the government of alternative income opportunities by restricting small-scale trade, grain marketing, and the planting of certain cash crops (which had been important in offsetting food deficits in the past), 2) intensification of agriculture in overpopulated areas without increased technological inputs, and 3) technological conservatism and various peasant survival strategies. These conditions resulted in cropping of erosion-prone slopes and climatically risky lowlands and river valleys. One of the recommendations of the Highland Reclamation Study was to resettle peasants from the most severely degraded highland areas in agriculturally more promising areas of the country (Ministry of Agriculture, 1986).

Although comprehensive cost-benefit analyses of conservation interventions which consider not only economic but also social and ecological parameters have not been undertaken in Ethiopia, the few existing studies are instructive. According to the model Sonneveld (2002) developed for the assessment of current and future food production in Ethiopia, the loss of agricultural value due to land degradation between 2000 and 2010 would be US$ 7 billion and net food production would be about 50% higher with than without soil conservation. A number of small-scale studies have shown variable benefits of soil conservation. Whereas increasing crop yields were reportedly achieved in two study areas over several years of soil conservation (Ketteman, 1992; Tesfaye, 1988), no net benefits were achieved in another area studied cross-sectionally for the cost benefits of introduced bunds and grass strips (Tolcha, 1998). The estimated monetary benefits achieved during this pilot experimental study, including increased water infiltration, reduced erosion, and increased crop yields, were smaller to the loss of land to conservation structures and infestation of the grass strips with field mice and runner grasses. Lack of immediate significant economic returns, which received high priority by farmers, became a source of negative attitudes toward soil conservation measures as local farmers did not value anticipated long-term ecological benefits (Tolcha, 1998). A study of the determinants of farmers' decisions to adopt and continue to use introduced stone terraces showed that mostly socioeconomic factors and perception of soil fertility, but not perceptions of erosion or land tenure security, were the main decision-making determinants (Amsalu, Stroosnijder, & de Graaff, 2007).

The inability of farmers to increase yields in the absence of fertilizer and technology often led to changes in land use practices that increased water and food insecurity. In one highland catchment where crop yields did not increase during a 43-year study period in spite of a quadrupling of the population, farmers tended to change from annual grain cultivation to tree planting and livestock production to cope with soil degradation, water scarcity, and ever smaller farm plots. However, by turning to selling wood and cattle dung as the major livelihood strategy, without investing in soil and water conservation, they perpetuated ecologically unsustainable practices (Amsalu et al., 2007). Thus, the eucalyptus revolution, which used to be an elite symbol under the Haile Sellassie regime, spread under the Derg in part as a result of the so-called community forests but also because peasants saw the opportunities of a cash crop (Guiral, 2009). The adverse environmental consequences of these wood lots in terms of erosion, acidification of the soil, and drying up of moist areas have not been assessed.

Voluntary Resettlement Versus Government-Sponsored Resettlement Programs

Much of the settlement history of Ethiopia has been characterized by voluntary migration. The Tigrean and Amhara migrations to the south and west over the centuries and the Oromo migrations to the north and west between the 16th and 18th centuries in response to population pressures represent the largest population movements. Between 1957 and 1972, nearly 2 million ha of land were distributed by the government of Haile Selassie to individuals and groups in the southern and western parts of Ethiopia (Cliffe, 2004), contributing to the spread of the northern seed-plow agricultural system and attendant environmental problems, an issue which has not received due attention by researchers.

Government planned and sponsored resettlement dates back only to the late 1950s and 1960s, when malaria was being controlled in the lowlands and the first five-year development plans emphasized commercial agriculture and industrialization during the latter part of Haile Selassie's reign. The larger resettlement program of the Mengistu regime began in 1976 and resulted in the resettlement of 1.7 million persons between 1980 and 1990. The 1984 national census showed that the major sending regions—Wello, Tigray, Shewa, and Gondar—had experienced a net loss in population during the previous 10 years due to migration and that the major receiving regions (Illubabor, Bale, and Arsi) reported net gains (Ezra, 1997).

The resettlement program implemented by the government in 1985 was burdened by some of the same underlying problems of hastily planned and implemented top-down interventions, lack of information about the local biophysical environment, and discontinuity between central and indigenous decision-making structures and the Highland Reclamation Project (de Wet, 2004; Pankhurst & Piguet, 2009). The relocation of around 600,000 people from the most degraded highland areas in newly constructed villages and land preparation at the local level (villagization) and national level (resettlement) caused extensive deforestation in all regions. In Metekel (Amhara Region), one of the largest resettlement schemes, the construction of 42 villages and clearance of more than 100,000 ha of land for mechanized agriculture resulted in massive destruction of forests, woodlands, and tree savannas, erosion, and fertility decline. Land clearance using heavy mechanized equipment in this and other resettlement schemes removed the biomass and much topsoil. Disputes and clashes ensued between settlers and indigenous people over the indiscriminate destruction of natural resources and land rights

(Abbute, 2002; Rahmato, 1988; Yntiso, 2009). The settlers, having been transformed overnight from farmers into daily laborers in a vastly different agro-ecological zone than in their home areas and, thus, being unable to grow their accustomed staples of *teff* and wheat, were unfamiliar with the modern agricultural methods used on the schemes and experienced poor and declining crop yields. They, thus, pursued similar destructive activities in all resettlement schemes as a survival strategy (Dejene, 1990). By cutting down and burning forests, the settlers not only caused significant land degradation and loss of biodiversity but also destroyed the economic basis and cultural heritage of the Berta, Gumuz, and other indigenous people, who depended on the forests for food (plants, honey, and wild animals) and medicines and had maintained sacred trees and forests consistent with their ritual and cultural values. Coffee shade trees also became victims of settler expansion, impacting commercial coffee production (Piguet & Lamessa, 2004). The indiscriminate destruction of ecological resources occurred in the absence of government regulations and oversight and stands in sharp contrast to the customary tenure rules of the indigenous communities which assured efficient and ecologically compatible use of resources. The Tsamako people in the Woito Valley, for example, value trees highly and harmonize private and communal interests, thus assuring the equitable and sustainable use of their tree savanna resources, but their environment has been increasingly impacted by outside private investment and settlers in recent years (Getu, 2009). In the wetlands of the western Illubabor Region, large-scale deforestation and resettlement caused soil fertility decline and hydrological changes resulting in increased river flooding (Woube, 1999). The 1984–1985 resettlement program was terminated as a failure in 1991, and many settlers returned to their home areas as government support dried up for the settlements, most of which had not become self-supporting (Pankhurst & Piguet, 2009). Because this and earlier resettlement programs were implemented on the basis of political motives and short-sighted economic gains and were not integrated into other development programs, they aggravated land-use and ethnic conflicts, resources degradation, and poverty (Woube, 2005).

The Villagization Program

The villagization program was implemented by the central government during the drought and famine in 1985 to allegedly provide services to the rural population and to stimulate voluntary self-help among villages. However, the underlying aim in forcibly moving 33 million, or two thirds of the country's rural population, into villages by 1994 was to transform rural society toward agricultural collectivization. By 1988, 35% of the rural population had been moved into 2.7 million hastily constructed dwelling units in spite of much resistance by the peasants (Lirenso, 1989). The arbitrary, generally poor site selection, often on prime agricultural land, and the construction of houses adversely affected the natural vegetation (especially forests), agricultural production, and the social and physical wellbeing of the population. Villagization deprived peasants of independent organization and action as well as promised social services. Perhaps environmentally most damaging was the movement of the population from their widely dispersed farmsteads into nucleated villages, which resulted in the intense use of nearby water, soil, grazing, and forest resources in the absence of land-intensive technologies and of land rotation to prevent overgrazing and land exhaustion (Cohen & Isaksson, 1987; Lirenso, 1989). The move into villages also increased the distance between places of residence and farm plots, pastures, and water sources. The reduction in water consumption was associated by peasants with the spread of various diseases (Arnesen, 1991). The decline of social and religious institutions which had traditionally played a key role in protecting natural resources contributed to accelerated destruction of natural resources during this period (Tafesse, 1995). Many newly created villages were disbanded in the early 1990s for these and other reasons. In some places, particularly where irrigation was practiced and markets were accessible, settlers remained in the new villages. In all fairness, we want to point out that the Ministry of Agriculture, the organization best suited to implement both the resettlement and villagization programs, was in effect rarely consulted by top-level government officials and often disputed the soundness of the decisions made (Dejene, 1990).

The Current Government Resettlement Program

Following the demise of the highland conservation, villagization, and resettlement programs, the government developed a more participatory approach to natural resource conservation (Ministry of Finance and Economic Development, 2002), in line with the 1991 economic policy of the transitional government and the decentralization program. However, the government launched another large resettlement program in 2003 that aimed at resettling 440,000 chronically food-insecure families (2.2 million people) from degraded highlands in central, northern, and eastern Ethiopia in the four regions of Amhara, Tigray, Oromia, and SNNPR. Although officially described as a voluntary program that receives adequate government support, it suffers from some of the same deficiencies as the earlier program, especially inability of the administration to provide sufficient on-farm and infrastructural services for this large settler population, resulting in considerable return migration. The negative economic effects of the influx of settlers in regard to increased population pressure on the receiving areas and their need for assistance has not been adequately considered (Tadele, 2004; Hammond & Dessalegn, 2003). Fears of inequitable and unsustainable development have led to confrontations between settlers and indigenous people who resented unregulated deforestation by poor settlers trying to generate income.

The Economic Reform of 1993 and Food Aid

After the downfall of the Derg regime in 1991, the incoming Ethiopian Peoples' Democratic Revolutionary Front (EPDRF) government issued a new economic policy involving greater market liberalization, removal of restrictions previously imposed on the private sector, and desocialization of the agricultural sector. The management of natural resources other than the state forests was transferred to the *kebeles* (the smallest administrative unit). Many forests and planted trees were illegally cut and burned during the turbulent interim period before the new government exerted control. The conflict between farmers and the administration over the issues of communal versus private (household) management of hillside pastures and forests and priorities between land distribution and rehabilitation persisted (Harrison et al., 2001).

The food-for-work program for land rehabilitation activities was continued by the EPDRF government after 1991 with generous international donor support after the demise of the Highland Reclamation Project, resulting in an ever worsening dependency syndrome. Ethiopia was the most food aid dependent country in the 1990s, receiving 795,000 metric tons of food aid annually between 1990 and 1999 (about 10% of total domestic grain production), increasing to 997,000 tons between 2002 and 2003 (11.5% of total domestic grain production) (Gebreselassie, 2006). Ethiopia continued to receive more food than any other country except Sudan in 2006 (World Food Program, 2008). The net benefits of the food aid program are probably small and affect farmers' water and soil conservation efforts. While food aid probably benefited the environment in the short term by meeting immediate livelihood needs, a major reason for farmers to pursue unsustainable exploitation of natural resources, bought their labor for environmental rehabilitation programs, and certainly improved acute food security of many households, it exacerbated the dependency on outside agencies at the expense of essential indigenous knowledge and initiatives. The plan by the World Food Program to change course and purchase grain directly from Ethiopian farmers rather than importing it from the U.S. may help to stabilize uncertainties in the marketplace, which caused farmers to grow excessive food beyond their survival needs at the expense of land and water conservation (Guth & Thurow, 2008). Important from a longer term perspective of water, soil, and vegetation conservation and rehabilitation is the fact that the food aid program became a major disincentive for farmers to use their well-honed coping mechanisms and resilience to withstand temporary shocks during droughts, including migration (Coulter, 2008). In addition, the program was beset with inefficiencies and poor implementation of the rehabilitation program, and failed to provide the means to effectively stem exacerbation of the poverty problem, including the erosion of farmers' productive assets.

Another disincentive was the tendency of food aid to depress local grain prices to a point where farmers could not earn enough from their crops to cover production costs, the result of the World Food Program purchasing food from large suppliers rather than small local farmers and the U.S. food aid policy requiring that U.S.-grown food aid be sent to feed the hungry (Guth & Thurow, 2008). The opposition of the U.S. farm lobby to use cash as a more versatile aid instrument, as done for more than a decade by European donors, further reduced the benefits of food aid (Thurow, 2008). The unreliability of food aid as a means to meet food needs and support environmental programs on a sustainable basis is further demonstrated by its use toward political ends (Gebreselassie, 2006). The U.S. repeatedly used food as a tool of foreign policy (Shepherd, 1985), the Derg regime threatened to withhold food aid from communities linked with the opposition (Pankhurst, 2001), and the current government blockaded emergency food aid in Somali Region where opposition groups are active (Gettleman, 2007). Critics also argue that instead of external infusion of aid, more emphasis needs to be given to strengthening local institutions, innovative initiatives, and ways to reduce the decline in farmer productivity and environmental degradation that have impacted the food security of both rural and urban populations (Gebreselassie, 2006). The formation and strengthening of community watershed management organizations carrying out integrated water management activities reduced the dependence on food aid through rehabilitation of eroded gulleys, the establishment of microenterprises, and increased confidence and capabilities of communities in one pilot project in northern Ethiopia (Liu et al., 2008).

Owing to growing concern about increasing numbers of chronically poor households, the Ethiopian government developed in 2005 a new approach with multidonor support in 2005 known as the Productive Safety Net Programme (PSNP). The main objective is to protect poor households from selling their key assets during times of drought. A dual approach was developed: On the one hand, selected households with labor are involved in public works for which they receive payments in cash or kind. On the other hand, households without labor or other support are eligible for direct support without labor contributions. The PSNP public works were designed with a participatory watershed management approach in mind.

The PSNP was launched by the government in 2005 with the support of development partners. Initially, 4.8 million chronically food insecure people were targeted in 192 food-insecure *woredas* in Amhara, Oromia, SNNP, Tigray, Dire Dawa, and Harar regions. This was increased in 2006 to 7.19 million people selected from 234 *woredas* in seven regions. In 2007, the operation targeted 7.19 million people in 274 *woredas*. Since January 2008, PSNP has also involved a pilot pastoral program in Afar, Oromia, and SNNPR and Somali Regions (Gilligan et al., 2008).

The increasing numbers involved in the PSNP raised the questions of if and how households could 'graduate' from dependence on the safety-nets program. It was realized that poor households may not be able to graduate easily unless they are able not just to protect their assets from depletion but can increase them through involvement in productive activities. Linking the PSNP with the Other Food Security Programme (OFSP) came to be seen as an important way of addressing the question of 'graduation,' and households that were involved in both PSNP and OFSP were believed to be more likely to increase their assets and improve their livelihoods than those that were only involved in the public works (Devereux et al., 2008). However, in practice, the overlap between the two programs has not been systematic and varied among regions and vulnerable households may face shocks that prevent them from using loans appropriately (Slater et al., 2006). At a community level in the public works, there was a change in design to enable watershed management to be carried out on private plots and thereby encourage conservation. However, it remains to be seen whether issues to do with user rights, management, and maintenance of rehabilitated watersheds, and the consequences on pasture of significant increase in livestock herds due to food security and other loans, can be resolved in ways that ensure equitable participation and sustainability so that the failures of the public works programs of the 1980s are not repeated.

CONCLUSION

Land degradation in the Ethiopian Highlands is severe and caused by a combination of an ecologically fragile mountain environment, prevailing land use patterns, a rapid population increase, major social upheavals in recent decades, and the weakness of indigenous institutions at the community level, ill-conceived government programs, land tenure problems, and external influences. These forces have directly or indirectly contributed to soil degradation, through various biophysical processes, exacerbating the historical decline in the productivity and sustainability of the agricultural base and mountain ecosystem and creating more arid conditions. The land is likely to continue to further degrade in the near future unless policy changes can effectively create an enabling environment that permits the population to emerge from poverty and a survival mentality. This requires going beyond the current emphasis on economic diversification toward industrialization, which may provide farmers with nonagricultural employment and technology permitting more intensive and sustainable agriculture and water security. These various systemic impediments will have to be eliminated by innovative policies and incentives for better stewardship of land and water resources.

Although decentralization is expected to facilitate the development of community-based water management, it is unlikely to change the top-down legacy in the short term, considering Ethiopia's long history of centralized, hierarchical systems of control (Teshome, 2007). In Gambela Region, for example, decentralization strengthened the position of the government and interfered with the development of local institutional arrangements in wetland use (Maconachie, Dixon, & Wood, 2009). Opposition to decentralization, fueled by "centrist attitudes" of many government officials and their suspicion of rural people's intentions and capabilities goes back to the emergence of the new East African states (Rontinelli, 1981). With continuing decentralization, the role of the government at all levels will need to change from a top-down administrator to a facilitator of regional and community water management programs that are based on proven indigenous knowledge and practices.

References

Abbute, W.-S. (2002). *Gumuz and highland settlers: Differing strategies of livelihood and ethnic relations in Metekel, northwestern Ethiopia* (Unpublished doctoral dissertation). University of Göttingen, Germany.

Admassie, Y. (2000). *Twenty years to nowhere: Property rights, land management and conservation in Ethiopia.* Lawrenceville, NJ: Red Sea Press.

Amborn, H. (1998). Agricultural intensification in the Burji-Konso cluster of south-western Ethiopia. *Anzania, 4*, 71–83.

Amsalu, A., Stroosnijder, L., & de Graaff, J. (2007). Long-term dynamics in land resources use and the driving forces in the Beressa watershed, highlands of Ethiopia. *Journal of Environmental Management, 83*, 448–459.

Arnesen, O. (1991). Any time for space? Farming systems in northern Ethiopia. In P. T. W. Baxter (Ed.), *When the grass is gone: Development intervention ion African arid lands* (pp. 90–14). Uppsala: Scandinavian Institute of Africa Studies.

Bergdall, T., & Powell, F. (1998). Grassroots empowerment in Ethiopian villages. In J. Burbidge (Ed.), *Beyond prince and merchant: Citizen participation and the rise of civil society* (pp. 83–96). New York: Pact Publications.

Berry, L. (2003). *Land degradation in Ethiopia: Its extent and impact.* Report prepared for FAO, Rome.

Beshaw, T. (2003). *Understanding farms: Explaining soil and water conservation in Konso, Woleita and Wello, Ethiopia* (Unpublished doctoral dissertation). Wageningen University, Wageningen, The Netherlands.

Bewket , W., & Sterk, G. (2005). Dynamics in land cover and its effects on stream flow in the Chemoga Watershed, Blue Nile Basin, Ethiopia. *Hydrological Processes, 19*, 445–458.

Beyene, A., Gibbon, D., & Haile, M. (2006). Heterogeneity in land resources and diversity in farming practices in Tigray, Ethiopia. *Agricultural Systems, 88*, 61–74.

Campbell, J. (1991). Land or peasants?: The dilemma confronting Ethiopian resource conservation. *African Affairs, 90*, 5–21.

Cliffe, L. (2004). International and Ethiopian resettlement experiences: Lessons for planning in Oromia Region. In A. Pankhurst & F. Piguet (Eds.), *People, space and the state: Migration, resettlement and displacement in Ethiopia* (pp. 191–210). Addis Ababa: Society of Ethiopian Sociologists, Social Workers, and Anthropologists.

Cohen, J. M., & Isaksson, N. I. (1987). *Villagization in the Arsi Region of Ethiopia*. Uppsala: Swedish University of Agricultural Sciences.

Constable, M. (1985). *Resources for rural development in Ethiopia. Ethiopian Highlands Reclamation Study* (Working paper No. 24). Addis Ababa: Ministry of Agriculture.

Constable, M., & Belchaw, D. (1989). The Ethiopian Highlands Reclamation Study: Major findings and recommendations. In *Proceedings of the National Workshop on Food Stategies for Ethiopia* (pp. 142–179). Addis Ababa: Office of the National Committee for Central Planning.

Coulter, L. (2008). *Integrating water economy analysis and HEA*. Report prepared for the Livelihood Integration Unit, USAID, Addis Ababa.

Critchley, W. R. S., Reij, C., & Willcocks, T. J. (2006). Indigenous soil and water conservation: A review of the state of knowledge and prospects for building on traditions. *Land Degradation and Development, 5*, 293–314.

Dejene, A. (1990). *Environment, famine, and politics in Ethiopia: A view from the village*. Boulder and London: Lynne Rienners.

Descheemaeker, K., Nyssen, J., Rossi, J., Poesen, J., Haile, M., Raes, D., … Deckers, G. (2006). Sediment deposition and pedogenesis in exclosures in the Tigray highlands, Ethiopia. *Geoderma, 132*, 291–314.

de Wet, C. (2004). Conclusion: Comments on the Workshop. In A. Pankhurst & F. Piguet (Eds.), *People, space and the state: Migration, resettlement and displacement in Ethiopia* (pp. 655–656). Addis Ababa: The Ethiopian Society of Sociologists, Social Workers, and Anthropologists.

Dubale, P. (2001). Soil and water resources and degradation factors affecting productivity in Ethiopian Highland-ecosystems: A strategy for physical recovery. *Northeast African Studies, 8*, 27–51.

Environmental Protection Authority. (1999). *National action programme to combat desertification*. Addis Ababa: Federal Democratic Republic of Ethiopia.

Ethiopian Mapping Authority. (1988). *National atlas of Ethiopia*. Addis Ababa: Author.

Ezra, M. (1997). Demographic responses to ecological degradation and food insecurity: Drought-prone areas in northern Ethiopia. Unpublished doctoral dissertation, University of Groningen, Groningen, The Netherlands.

Ezra, M. (2001). Ecological degradation, rural poverty, and migration in Ethiopia: A contextual analysis (Department of Health and Senior Services, State of New Jersey, Policy Research Division Research Working Paper No. 149). Trenton, NJ.

Falkenmark, M., & Rockström, F. (Eds.). (2004). *Balancing water for humans and nature: The new approach in ecohydrology*. Tall Gables, U.K.: CPL Scientific Publishing.

Federal Democratic Republic of Ethiopia, Ministry of Finance and Economic Development. (2006). *Ethiopia: Building on progress, a plan for accelerated and sustained development to end poverty (PASDEP) (2005/06-2009/10)*. Addis Ababa: Author.

Food and Agriculture Organization of the United Nations. (1993). *Extent and causes of land degradation*. Land and Plant Nutrition Management Service, MADS-SEA Network. Rome: Author.

Food and Agriculture Organization of the United Nations. (2005). *Properties and management of drylands*. Rome: Author. Retrieved from http://www.fao.org/ag/ag11/drylands/definitions.htm

Gamachu, D. (1990). Environment and mass poverty. In S. Pausewang, F. Cheru, S. Brüne, & E. Chole (Eds.), *Ethiopia: Rural development options* (pp. 164–173). London: Zed Books.

Gebre Michael, Y. (1999). *The use, maintenance and development of soil and water conservation measures by small-scale farming households in different agro-climatic zones of northern Shewa and southern Wello, Ethiopia* (Soil conservation research report 44). Berne: University of Berne, Center for Development and Environment.

Gebreselassie, S. (2006, March). *Food aid and small-holder agriculture in Ethiopia: Options and scenarios*. Paper presented at the Future Agriculture Consortium Workshop, Institute of Development Studies, Addis Ababa University.

Gettleman, J. (2007, July 22). Ethiopia is said to block food aid to rebel region. *The New York Times*. Retrieved from http://www.nytimes.com/2007/07/22world/africa/22ethiopia.html?_r=1&oref=slogin

Getu, M. (2009). The effects of investment on the livelihoods of the Tsamako in the Weyto Valley. In A. Pankhurst & F. Piguet (Eds.), *Moving people in Ethiopia: Development, displacement & the state* (pp. 81–92). London, James Currey.

Gilligan, D. O., Hoddinott, J., & Tafessse, A. S. (2008). An analysis of Ethiopia's Productive Safety Net Program and its linkages (Report prepared for the International Food Policy Research Institute). Washington, DC.

Guiral, C. (2009). *Diversité des jardins et strategies paysannes de long d'un gradient altitudinal en Éthiopie Centrale* (Unpublished doctoral dissertation). Muséum National d'Historié Naturelle, Paris.

Guth, R. A., & Thurow, R. (2008, September 25). Weaving Africa's bread basket. *Wall Street Journal*, p. A18.

Haileselassie, A., Priess, J., Veldkamp, E., Teketay, D., & Lesschen, J. P. (2005). Assessment of soil nutrient depletion and its spatial variability on smallholders' mixed farming systems in Ethiopia using partial versus full nutrient balances. *Agriculture, Ecosystems & Environment*, *108*, 1–16.

Hallpike, C. R. (1970). Konso agriculture. *Journal of Ethiopian Studies*, *8*, 31–41.

Hammond, L., & Dessalegn, B. (2003). *Safara: An assessment of current and future voluntary resettlement in Ethiopia.* Report prepared for USAID-Ethiopia, Addis Ababa.

Harrison, E., Pankhurst, A., Serra, A., Yibabie, T., Watson, E., & Admassie, Y. (Eds.). (2001). *Management of natural resources: Report of workshop in Addis Ababa.* Brighton, Sussex: School of African and Asian Studies, University of Sussex and Addis Ababa: Forum for Social Studies.

Hurni, H. (1983a). *Soil formation rates in Ethiopia.* FAO/Ministry of Agriculture Joint Project, Ethiopian Highlands Reclamation Study, Working Paper No. 2. Addis Ababa: Ministry of Agriculture.

Hurni, H. (1983b). Soil erosion and soil formation in agricultural ecosystems in Ethiopia and northern Thailand. *Mountain Research and Development*, *32*, 131–142.

Hurni, H. (1988a). Degradation and conservation of the resources in the Ethiopian Highlands. *Mountain Research and Development*, *8*, 123–130.

Hurni, H. (1988b, December). *Ecological issues in the creation of famines in Ethiopia.* Paper presented at the National Conference on a Disaster Prevention and Preparedness Strategy for Ethiopia, sponsored by the Office of the National Committee for Central Planning, Addis Ababa.

Interregional Information Network. (2009, April 2). Global: Put small-scale farmers on the climate talks agenda. *IRIN Newsletter.* Office of the Coordination of Humanitarian Affairs, United Nations.

Kloos, H. (1982). Farm labor migrations in the Awash Valley of Ethiopia. *International Migration Review, 1*, 133–168.

Kruger, H.-J., Fantaw, B., Gebre Michael, Y., & Kajela, K. (1996). Creating an inventory of indigenous soil and water conservation measures on selected sites in Ethiopia. In I. S. C. Reij, I. Scoones, & K. Kajela (Eds.), *Sustaining the soil: Indigenous soil and water conservation in Africa* (pp. 170–180). London: International Institute for Environment and Development.

Lai, R., Reicosky, D. C., & Hanson, J. D. (2007). Evolution of the plow over 10,000 years and the rationale for no-till farming. *Soil and Tillage Research, 93*, 1–12.

Lemenih, M., Karltun, E., & Olsson, M. (2005). Soil organic matter dynamics after deforestation along a farm field chronosequence in southern highlands of Ethiopia. *Agriculture, Ecosystems & Environment, 109*, 9–19.

Leye, M. T. (2007). *Conservation tillage systems and water productivity: Implications for smallholder farmers in semi-arid Ethiopia.* London: Taylor and Francis.

Lirenso, A. (1989). *Villagization and agricultural production in Ethiopia: A case study of two regions* (IDR Research Report No. 37). Addis Ababa: Addis Ababa University, Institute of Development Research.

Liu, B. M., Collick, A. S., Zeleke, G., Adgo, E., Easton, Z. M., & Steenhuis, T. S. (2008). Rainfall-discharge relationships for a monsoonal climate in the Ethiopian highlands. *Hydrological Processes, 22*, 1059–1067.

Maconachie, R., Dixon, A. B., & Wood, A. (2009). Decentralization and local institutional arrangements for wetland management in Ethiopia and Sierra Leone. *Applied Geography, 29*, 269–279.

McCann, J. C. (1995). *People of the plow: An agricultural history of Ethiopia, 1800–1990.* Madison: University of Wisconsin Press.

Mekuria, W., Veldkamp, E., Haile, M., Nyssen, J., Muys, B., & Gebrehiwot, K. (2007). Effectiveness of exclosures to restore degraded soils as a result of overgrazing in Tigray, Ethiopia. *Journal of Arid Environments, 69*, 270–284.

Ministry of Agriculture. (1986). *Ethiopian Highlands Reclamation Study: An assessment of the causes, severity, extent and probable consequences of degradation*. Addis Ababa: Author.

Ministry of Finance and Economic Development. (2002). *Ethiopia: Sustainable development and poverty reduction. Strategic paper for promoting development and poverty reduction*. Addis Ababa: Federal Democratic Republic of Ethiopia.

Montgomery, D. R. (2007). Soil erosion and agricultural sustainability. *Proceedings of the National Academy of Sciences of the United States of America, 104*, 13268–13272.

Namara, D. C., Awulachew, S. B., Namara, R. E., Babel, M. S., & Gupta, A. D. (2007). Indigenous systems of conflict resolution in Oromia, Ethiopia. In B. C. P. Koppen, M. Giordano, & J. Butterworth (Eds.), *Community-based water law and water resources management reform in developing countries* (pp. 145–157). Wallingford: CABI.

Netherlands Engineering Consultants. (1997). *Tekezze River Basin integrated development master plan project.* Report prepared for the Ministry of Water Resources, Addis Ababa.

Nyssen, J., Vandenreyken, H., Poesen, J., Moeyersons, J., Deckers, J., Haile, M., & Govers, G. (2005). Rainfall erosivity and variability in the Northern Ethiopian Higjlands. *Journal of Hydrology, 311*, 172–187.

Nyssen, J., Harageweyn, N., Descheemaker, K., Gebremichael, D., Vancampenhout, K., Poesen, J., & Govers, G. (2006). Comment on "Modeling the effect of soil and water conservation practices in Tigray, Ethiopia." (Agriculture, Ecosystems & Environment, 105, 29–40). *Agriculture, Ecosystems & Environment, 106*, 407–411.

Nyssen, J., Poesen J., Moeyersons, J., Deckers, J., Haile, M., & Lang, A. (2004). Human impacts on the environment in the Ethiopian and Eritrean highlands: A state of the art. *Earth-Science Reviews*, *64*, 273–320.

Osman, M. (2001). *Rainfall and its erosivity in Ethiopia with special consideration of the central Highlands* (Bonner Bodenkundliche Abhandlungen No. 37). Bonn: University of Bonn.

Pankhurst, A. (2000). *Conditions relating to the adoption of broad-bed makers by farmers in Ethiopia.* Unpublished report prepared for the International Livestock Research Institute, Addis Ababa.

Pankhurst, A. (2001). Resource management institutions in post-conflict situations: Lessons from Yegof State Forest, South Wello Zone, Amhara Region, Ethiopia. In A. Pankhurst (Ed.), *Natural resource management in Ethiopia. Proceedings of the Workshop Organized by the Forum for Social Studies in Collaboration with the University of Sussex* (pp. 85–104). Addis Ababa: Forum for Social Studies.

Pankhurst, A. (2003). Conflict management over contested natural resources: A case study of pasture, forest and irrigation in South Wello, Ethiopia. In P. Castro & E. Nielsen (Eds.), *Natural resource management conflict case studies: An analysis of power, participation and protected areas* (pp. 59–80). Rome: FAO, Community Forest Unit.

Pankhurst, A., & Piguet, F. (2009). Migration, resettlement & displacement in Ethiopia: A historical and spatial overview. In A. Pankhurst & F. Piguet (Eds.), *Moving people in Ethiopia: Development, displacement and the state* (pp. 1–22). London: James Currey.

People's Democratic Republic of Ethiopia. (1990). Ethiopia, national conservation strategy. Phase I Report. Addis Ababa: Author.

Piguet, F., & Lamessa, D. (2004). Review of voluntary migration and resettlement programmes up to the end of 2001. In A. Pankhurst & F. Piguet (Eds.), *People, space and the state: Migration, resettlement and displacement in Ethiopia* (pp. 133–162). Addis Ababa: Ethiopian Society of Sociologists, Social Workers, and Anthropologists.

Rahmato, D. (1984). *Agrarian reform in Ethiopia.* Trenton, NJ: Red Sea Press.

Rahmato, D. (1988). Resettlement and resettlement in Metekel, western Ethiopia. *Africa* (Rome), *63*, 14–43.

Rahmato, D. (2001). Opening address. In E. Harrison, A. Pankhurst, A. Serra, T. Yibabie, E. Watson, & Y. Admassi, (Eds.), *Management of natural resources: Report of a workshop* (p. 4). Addis Ababa: Forum for Social Studies.

Rockström, J., Falkennmark, M., Karlberg, L., Hoff, H., Rost, S., & Gerten, D. (2009). Future water availability for global food production: The potential of green water for increasing resilience to global change. *Water Resources Research*, 45, doi 10.1029/2007WR006767.

Rontinelli, D. A. (1981). Government decentralization in comparative perspective: Theory and practice in developing countries. *International Review of Administrative Sciences*, *47*, 22–42.

Scherr, S. J. (1999). *Soil degradation: A threat to developing-country food security by 2020?* Washington, DC: International Food Policy Institute.

Scones, I., & McCracken, J. (Eds.). (1989). *Participatory rapid rural appraisal in Wello: Peasant association planning for natural resource management*. London: ERCS/IIED.

Shepherd, J. (1985). Ethiopia: The use of food as an instrument of U.S. foreign policy. *A Journal of Opinion*, *14*, 4–9.

Slater, R., Ashley, S., Tefera, M., Buta, M., & Esubalew, D. (2006). *PSNP policy, programme and institutional linkages*. ODI, the IDL Group and Indak. Retrieved from www.odi.org.uk/resources/download/3034.pdf

Sonneveld, B. G. J. S. (2002). *Land under pressure: The impact of water erosion on food production in Ethiopia.* Maastricht, The Netherlands: Shaker Publishing.

Tadele, F. (2004). The new resettlement programme in Ethiopia: Reflections on the design and implementation approach. In A. Pankhurst & F. Piguet (Eds.), *People, space and the state: Migration, resettlement and displacement in Ethiopia* (pp. 211–218). Addis Ababa: Ethiopian Society of Sociologists, Social Workers, and Anthropologists.

Tafesse, T. (1995). *Villagization in northern Shewa, Ethiopia: Impact assessment.* Münster: Lit Verlag.

Tegene, B. (1992). *Erosion, its effects on properties and productivity of eutric nitosols in Gunamo area, southern Ethiopia, and some techniques of its control* (African Studies Series, No. 49). Berne: University of Berne, Center for Development and Environment.

Tegene, B. (1998). Potential and limitations of an indigenous structural soil conservation technology of Welo, Ethiopia. *East African Social Science Research Review, 14*, 1–18.

Temesgen, M., Rockström, J., Sanenije, H. H. G., & Hoogmoed, W. B. (2007). Assessment of strip tillage systems for maize production in semi-arid Ethiopia: Effects on grain yield and water balance. *Hydrology of Earth Systems and Science Discussions, 4*, 2229–2271.

Tesfaye, M. (1988). *Soil conservation experiments on cultivated land in Maybar area, Wello Region, Ethiopia* (Unpublished master's thesis). Addis Ababa University, Addis Ababa.

Teshome, A. (2007, June 23). Bottom up policy process: An agenda for future agricultures in Ethiopia. *Ethiopian Herald*, p. 8.

Thurow, R. (2008, May 2). Bush pushes shift in food aid. *Wall Street Journal.* Retrieved from wsj-investment.blogspot.com/2008_05_01_archive.html

Tolcha, T. (1998). Results of soil conservation experiments and the scope for implementation: The case of Chercher Highlands. *Ethiopian Journal of Development Research, 20*, 25–63.

Walker, B., & Salt, D. (2006). *Resilience thinking: Sustaining ecosystems and people in a changing world.* Washington, DC: Island Press.

Welderufael, W. A. (2006). *Quantifying rainfall-runoff relationships on selected benchmark ecotypes in Ethiopia: A primary step in water harvesting research* (Unpublished doctoral dissertation). University of Free State, Bloemfontheim, South Africa.

World Bank. (2006). *Ethiopia: Managing water resources to maximize sustainable growth*. Washington, DC: Author.

World Food Program. (2008). *WFP in 2006: A quick glance*. Rome: Author.

Woube, M. (1999). Floods ad lack of sustainable land-water management in the lower Baro-Akobo river basin. *Applied Geography, 9*, 235–251.

Woube, M. (2005). *Effects of resettlement schemes on the biophysical and human environments: The case of Gambela Region, Ethiopia*. Boca Raton, FL: Universal Publishers.

Yang, X., & Conacher, A. (2009). Guest editors' introduction, land degradation and rehabilitation in ecologically fragile areas: Research status and perspectives. *Geographical Research, 47*, 1–3.

Yntiso, G. (2009). Why did resettlement fail? Lessons from Metekel. In A. Pankhurst & A. Piguet (Eds.), *Moving people in Ethiopia: Development, displacement & the state* (pp. 119–137). London: James Currey.

Zeleke, G., Kassie, M., Pender, J., & Yesuf, M. (2006, April). *Stakeholder analysis for sustainable land management (SLM) in Ethiopia: Assessment of opportunities, strategic constraints, information needs, and knowledge gaps*. Paper prepared for the Economics Policy Forum for Ethiopia (EPFE), Addis Ababa.

FURTHER READING

Food and Agriculture Organization of the United Nations (1984). *Ethiopian Highlands Reclamation Study (EHRS). 2 vols.* Rome: Author.

CHAPTER 9

PROBLEMS FOR PASTORALISTS IN THE LOWLANDS

RIVER BASIN DEVELOPMENT IN THE AWASH AND OMO VALLEYS

Helmut Kloos, Worku Legesse, Stuart McFeeters, and David Turton

INTRODUCTION

Pastoralists represent about 12% of the Ethiopian population (Helland, 2006), or 9.5 million people belonging to more than 29 ethnic groups (Central Statistical Authority, 1998; Gebre Michael, 2004). They raise about 40% of all cattle in Ethiopia, 75% of goats, 20% of equines, and 100% of camels and occupy about 40% of the land, practically all in the lowlands (Kassa, 2002). They are, like all pastoralists in Africa, facing increasing challenges to survive in an environment marked by dwindling water and pasture resources and outside forces impacting their way of life. Ethiopia's stepped-up river basin development, in particular, is of prime concern to pastoralists, who largely remain outside the national decision-making processes in water development and land use planning.

Ethiopia's pastoralists pursue livestock and environmental management strategies in the lowlands that have evolved over several millennia. This system of land use was, until recently, seen as environmentally exploitative and economically irrational. According to the equilibrium theories, which were adhered to by most development agencies in the past, traditional communal land tenure and rangeland management practices inevitably lead to the overutilization of natural resources in an unsustainable manner (Devereux, 2004). But in spite of the negative perception of pastoralism by many administrators, the public, and some earlier rangeland and development experts, it is increasingly recognized to be the most sustainable form of land use in semiarid and arid environments (Helland, 2006). According to the "new range ecologists," pastoralism is a system in dynamic disequilibrium, and pastoralists are expert at living with risks during the cycles of accumulation, collapse, and rebuilding that are inherent in this livelihood system (Devereux, 2004).

Pastoralists maximize the use of rangelands by seasonally migrating with their herds to take temporary advantage of areas with high rainfall, high nutrient forage, and suitable watering sites. Their use of this strategy makes meaningless the idea that temporarily unused rangeland constitutes a no-man's land (Helland, 2006). Pastoralists are not the aimlessly wandering "nomads" often described by officials and depicted in popular accounts. Their movements reflect quick and expert decision making concerning both land tenure and natural resource management, to ensure continued and sustainable access to water and forage. Some pastoralist systems in Africa, including that of the Borana Oromo in Ethiopia, achieve significantly higher returns of both energy and protein per hectare than industrialized ranching systems in Australia (United Nations OCHA, 2007).

But pastoralists in Ethiopia and other countries in Northeast and East Africa face political and economic marginalization, inappropriate development policies, increasing resource competition and climate change (Oxfam, 2008). These once self-reliant people have increasingly become poverty-stricken famine relief clients, eking out an existence at the margins of society (Helland, 1998). Although droughts recurred for centuries in lowland Ethiopia, there is evidence that those of the 1970s and 1980s were amongst the most severe in terms of human and livestock losses (Kloos & Lindtjørn, 1993). This was due to the combined effect of drought, the development of commercial farms, hydroelectric dams, and national game reserves, and the resettlement of large numbers of people from the highlands, all of which decreased access to traditional rangelands and watering points. The control of malaria in the lowlands, starting in the 1960s, and the development of drought-resistant crops, particularly maize and sorghum, also encouraged the settlement and commercial exploitation of large lowland areas by outside agencies and populations. As a result of these dramatic changes in population distribution and land use, increasing numbers of pastoralists have been displaced from their dry season pastures and permanently settled on farms and in towns, where most of them make a living under precarious conditions (Kassa, 2001; Kloos, 1982).

An overriding factor in this deteriorating situation has been the focus of the state on harnessing the resources of pastoral lands for the national economy, without compensating the pastoralists. Under Haile Selassie, a new development course was charted, based on commercial agriculture and the exportation of agricultural products. The Awash Valley was selected as an ideal area in which to implement this policy. Under the Derg, communal pastoral tenure was abolished as part of the nationalization of all Ethiopian lands, and pastoral land was redefined as tribal grazing areas or tribal grazing reserves. The new tenure system was enforced through heavy-handed administrative control, including a strong military presence, restricting pastoralist mobility across regional and international boundaries and encouraging or forcing herders to abandon their pastoral way of life (Afar Human Rights Organization, 2007; Helland,

2006; Kassa, 2002). By 1993, 466,640 ha of land, 353,730 ha in the Awash Valley alone, had been converted to national parks and controlled hunting areas to which pastoralists had no access (Yemane, 2003).

Although large tracts of land have been returned to pastoralists and traditional rights to land and water were restored under the 1994 constitution, there is no public policy that secures access for pastoralists to water or pastures in contested areas. Meanwhile, unregulated land transactions between clan leaders and commercial enterprises create new inequities in access and perpetuate the eviction of herdsmen. Helland (2006) has noted that, among the Afar, the weak capacity of the new Afar Administrative Region has made it impossible to prevent an upsurge in such clandestine lease agreements. This has led to inequitable access to resources among different clans (Hundie, 2006), a situation which had its antecedents in the 1970s (Harbeson, 1975). The relative neglect of pastoralist populations by NGOs in the past (Christian Relief and Development Association, 2001)—beyond the distribution of food aid—exacerbated their socioeconomic plight. There has been a recent trend in Ethiopia away from food relief programs toward more fundamental development assistance, aimed at combating chronic food insecurity (Hammond, 2006). This holds the promise of more sustainable development, but only if other pressing issues, such as land tenure and the promotion of the democratization process (Interregional Information Network, 2003b), are successfully addressed.

The developments we have described have had the following impacts, amongst others. First, pastoralists have been displaced from their home ranges and have settled on commercial farms and in towns. Second, there has been increasing intergroup competition for dwindling water and grazing resources. And third, pastoralists have experienced increasing food insecurity as a result of their inability to rebuild their herds following famine, as well as increasing poverty, environmental risk, and economic and social marginalization (Hammond, 2006; Helland, 2006). In this chapter, we describe the actual and potential impacts of river basin development on pastoralists in the Awash and the lower Omo valleys. The Awash Valley, located mostly in Afar Regional State, has undergone perhaps the most dramatic transformation of any Ethiopian lowland area, while the lower Omo Valley, in the Southern Nations, Nationalities, and Peoples' Regional State (SNNPR), seems likely to take over this dubious distinction in the years to come.

The Afar and Kereyu Oromo of the Awash Valley

Pastoralists in the Awash Valley have been severely affected by development projects, population increase, and diminishing resources within both the valley and the surrounding highlands. Large-scale agricultural and hydroelectric development started under the Haile Selassie regime in the early 1950s with the development of Wonji sugar plantation and the construction of Koka High Dam. The four pastoralist groups occupying the Awash Valley, the Afar, Kereyu, Arsi, and Jile Oromo, have been similarly affected by recent development activities and are, thus, discussed together in this section.

Crucial preconditions for the events that occurred in the valley during the past five decades include, first, the irrigation potential of the Awash floodplains, the prime dry-season pastures of all four pastoralist groups; second, the proximity of these potentially irrigable areas to the national and commercial capital of Addis Ababa; third, the relatively well-developed infrastructure of the Awash Valley, including Ethiopia's first large dam and rail and road links between Addis Ababa and the Red Sea ports of Asseb and Djibouti; and fourth, the large hydroelectric potential of the Awash Basin. Koka Dam became operational in 1960 for hydroelectric power generation, flood control, and irrigation. Two smaller dams and power stations were constructed downstream of Koka Dam in 1966 and 1971. By 1973, on the eve of the Ethiopian revolution, 52,370 ha of prime dry-season pastures were irrigated by 23 commercial schemes, resulting in the displacement of local pastoralists. The controlled water releases from Koka Dam sharply reduced flooding downstream and, combined with the construction of access roads and flood protection dikes along the Awash River, facilitated the development of Ethiopia's largest irrigation complex. But this also caused the displacement of several pastoralist groups and the deterioration of their dry-season grazing areas and livestock watering points on the fertile Awash flood plains (Kassa, 2001; Kloos, 1982). Similar impacts have been felt by agro-pastoralists of the lower Omo Valley as a result of national park development since the 1960s (Turton, 1995, 2002) and, as we explain below, the construction of the Gibe III hydroelectric dam in the Omo-Gibe Basin is likely to have even more drastic impacts on both the environment and the peoples of the Lower Omo in the next few years.

Water resource development in the Awash Valley continues. In 2009, two additional large dams, the Kessem and Tendaho dams, were nearing completion. The objectives are to irrigate 90,000 ha of sugar cane mainly for ethanol production and sugar export, generate 100 MW of hydropower, and facilitate socioeconomic development (Beyene, 2006). Realization of the last objective, particularly the creation of alternative livelihoods for the Afar, is difficult to envisage, given the impact on them of the earlier dam and irrigation projects. The Afar Forum (2008) in Brussels pointed out the absence of socioeconomic and environmental impact studies for the Tendaho sugar project and expressed concern over ongoing repercussions of this project on the Afar. Having lost most of their prime pastures to large irrigation schemes starting in the 1960s, the Afar and other local pastoralists fear that the dams will further reduce their water and grazing resources. They have not been involved in the planning process and are therefore opposing these projects (Afar Human Rights Organization, 2007). Several government projects aimed at settling local Afar on mechanized irrigation schemes, where they were to grow maize and practice animal husbandry and dairying on irrigated pastures, failed for a number of administrative and health reasons (Gebre & Kassa, 2009), further aggravating their grievances.

By pursuing herd diversification and seasonal migration strategies which optimize the use of available rangeland and water resources, Ethiopian pastoralists use more refined techniques to utilize water and protect the environment than highland farmers (Edwards, Classen, & Schrote, 1983). Different Afar and Kereyu clans kept variable proportions of camels, cattle, and goats in different localities depending on the grazing and browsing resources of their respective areas. Both groups, as well as the Arsi, seasonally moved between their dry-season pastures on the flood plains of the Awash River and the plains and foothills of the Rift Valley escarpment (Voelkner, 1974). The Afar traditionally dug effluents across the natural levees along the Awash River and into the floodplains to sustain the highly productive swamps and grazing areas on the Awash flood plains during the dry season. The swamps produced forage for large livestock herds as well as water lilies (*Nymphea coerulea*), the rhizomes of which constituted a highly prized Afar food, and bulrush (*Scirpus* sp.), used for weaving mats for huts and sleeping areas. The lush flood plains supported large livestock and human populations, amounting to about 10,000 persons on the plain near the town of Gewani alone (Kloos, Polderman, DeSole, & Lemma, 1977). During the annual floods, the Afar moved with their livestock to their wet-season pastures away from the Awash River. The dikes along the Awash River, which had been built by irrigation schemes to prevent the flooding of their land, disrupted the annual floods that replenished the floodplain pastures. This resulted in the drying up of nearby swamps, the deterioration of floodplain pastures, and the reduction of groundwater levels in the shallow hand-dug wells away from the river. The annual floods continue to damage the farms and irrigation systems in most years, requiring the strengthening and extension of the dikes (Interregional Information Network, 2003a). Meanwhile, the pastoralists frequently break the dykes to induce the flooding of their pastures (Achamyeleh, 2003).

Exclusion from, or limited access to, their former dry-season areas and watering points, together with changing environmental conditions, severe droughts, and intertribal warfare exacerbated the water and pasture problems. Some Afar clans and the Kereyu Oromo now have to migrate more frequently and travel greater distances to find sufficient feed and water for their cattle and camels at certain times of the year. The longer wet-season migrations of the Afar away from the Awash River, their main source of water, although constrained by lack of water due to declining rainfall in many surrounding areas and fierce competition with the Isa Somali, have become more extensive in some areas (Hammond &

Dessalegn, 2003; Kassa , 2002; Voelkner, 1974). Wells are often abandoned in the areas contested by the Afar and Isa, further exacerbating water stress experienced by both groups (United Nations, 2003). Less rainfall increases pressure on the riverine dry-season pastures in at least two ways. First, it results in the deterioration of wet-season pastures that sustain fewer livestock for a shorter period. Second, the later departures from, and earlier returns to, the dry-season pastures make them more susceptible to overgrazing, a phenomenon reported from other African pastoral nomadic areas (Edwards et al., 1983).

Interethnic conflicts between the Afar, Kereyu, Isa Somali, and other neighboring groups over decreasing pasture and water resources and the control of major illicit trade routes have increased in recent decades (Kassa, 2001; Kloos, 1982). The most severe confrontations take place between the Afar and Isa Somali, who have gradually encroached onto Afar dry-season pastures along the Djibouti-Addis Ababa road. A combination of decreasing rainfall, increasing population densities, and environmental degradation has prompted some Afar groups to seek out pastures at higher elevations along the more humid Rift Valley escarpment. Springs and pastures here are more productive later in the dry season than in the lowlands, where they increasingly compete with farmers from the highlands who are encroaching on their customary grazing lands. Population pressure in the highlands, exacerbated by land degradation, is resulting in the expansion of rain-fed agriculture toward the lowland Afar pastoralist areas, a movement which is expected to accelerate under the current government's poverty reduction and resettlement programs (Hammond & Dessalegn, 2003).

Conflicts between pastoralists and highland farmers are particularly intense during droughts, as observed in the overstocked Borkena River Valley in eastern Ethiopia in 2000 (Guinand, 2000), and during the main cropping season. In an attempt to stop the incursion by highland farmers, some Afar have taken up farming near the escarpment. The effects of conflict include the loss of material assets and human lives, especially men in their most productive years, the interruption of the livestock trade between pastoralists and highland farmers, and the underutilization of pastures and watering points in contested areas (Hundie, 2006). In spite of their need to migrate further to utilize dwindling resources, the movements of many Afar, Somali, and Kereyu pastoralists have been increasingly curtailed by fear of livestock raiding and conflicts over land and water which have become more deadly with the wide circulation of automatic weapons (Belay, Beyene, & Manig, 2005). Although endemic low-level struggles over resources, particularly between herders and farmers, are common in Sahelian Africa (Kevane & Gray, 2008), climate change is expected to intensify these conflicts (see chapter 9).

The Kereyu Oromo developed coping responses similar to those of the Afar. They lost large tracts of grazing and browsing land to Metahara and Abadir irrigation farms, to the Awash National Park, and, more recently, to the rising Lake Beseka. While some clans continued hazardous transhumance movements on the nearby Mount Fentale, where they often clash with the Afar, others began to venture into the upper Awash Valley and beyond Lake Zway. The Kereyu have also begun to construct water harvesting systems, with the help of a local NGO, to meet their domestic water needs (Alamerew & Natea, 2004). The inundation of pastures by the alkaline/saline Lake Beseka has been due mainly to the dumping of drainage water from Abadir Farm (Alemayehu, Ayenew, & Kebede, 2006; Gebre, 2002).

Water pollution is further reducing the water resources of the Awash Valley. The discharge of untreated residential, commercial, and industrial wastewater in Addis Ababa, a city of more than 3 million inhabitants, is increasingly affecting the Awash River. Downstream of the capital, nitrogen, BOD (biochemical oxygen demand), and lead levels have been found, in and below Lake Koka, which are up to three times as high as the World Health Organization's (WHO) maximum acceptable level (Redda Teklehaimanot, 2003). These waters are used for drinking and cooking by the local Jile Oromo, who have been displaced by Lake Koka and Wonji sugar plantation (Kloos, 1982). Downstream of Koka High Dam, high pesticide, herbicide, and fertilizer use in two dozen commercial farms and agro-industries increasingly imperil the water resources of the Kereyu and Afar pastoralists, who use the irrigation systems for domestic use and livestock watering. The indiscriminate use of these agro-chemicals and the health risks of high levels of naturally occurring fluoride in the Awash Valley and other parts of the Ethiopian Rift Valley are described in chapter 13.

After the creation of the Afar National Regional State in the mid-1990s, large tracts of grazing land and a number of former state farms were returned to the Afar, as guaranteed in the 1994 constitution. However, the administrative weakness of this newly formed state meant that they could not secure access to their traditional pastures and water points because the irrigated farms had been redeveloped by commercial investors. The entrepreneurs had made these mostly clandestine lease agreements with representatives of the few clans which had secured land rights, with most Afar having little say in the matter (Helland, 2006). This is reminiscent of the development of irrigated farms under private tenure by the former Afar leader Sultan Ali Mira and his family in the early 1970s (Harbeson, 1975). Unequal access to resources, together with the other constraints on the pastoral way of life we have described, is putting the survival of the pastoral Afar in question. Recent studies of the perceptions of the Afar and Kereyu in regards to their grazing and water resources show that although these groups still prefer communal land tenure, they no longer consider that the sustainable utilization of their deteriorating rangeland ecosystem is possible under the traditional tenure system (Abule, Snyman, & Smit, 2005).

AGRO-PASTORALISTS OF THE LOWER OMO VALLEY AND AROUND LAKE TURKANA

The account we have just given of the impact of dams and irrigation schemes on the people and environment of the Awash Valley holds important lessons for the development of the Omo-Gibe River Basin in the southwest of the country. The hydropower potential of the Omo-Gibe is second only, amongst Ethiopian rivers, to that of the Blue Nile (table 2.5, chapter 2). As yet, however, only one dam has been completed along its course. Known as Gilgil ('Lesser') Gibe I, this began operating in 2004 and is currently Ethiopia's single largest supplier of electricity. A power plant known as Gilgel Gibe II was completed in late 2009. This does not have its own dam but will draw water through a 25 km tunnel from the Gilgil Gibe I reservoir. Neither the existing dam nor the Gilgel Gibe II power plant has, or will have, an appreciable impact on the hydrology of the lower Omo. Another dam is now under construction, however, which most certainly will.

This dam is known as Gibe III (although it is actually on the Omo River). Construction began in 2006 and is scheduled to be completed in 2011 or 2012. At 240 m high, or six times the height of Gilgil Gibe I, and with a reservoir 150 km in length (making it the second largest in Africa), it is expected to double the country's electricity-generating capacity. This will far exceed the predicted domestic demand, and it is planned to export up to 50% of the electricity generated to neighboring countries. Two more dams are planned further downstream (Gibe IV and V), the second of which will be located at the point where the entrenched valley of the Omo opens into its lower basin (figure 9.1). These developments represent a huge financial investment and an impressive feat of civil engineering which it is hoped will bring great economic benefits to Ethiopia. But, of course, big dams everywhere do not come without their drawbacks, most of which are felt at the local level. The impact of Gibe III on the people and environment of the lower Omo has only recently begun to be addressed by planners, prompted no doubt by the need to satisfy the loan conditions of multilateral development banks.

Possible Impacts of Gibe III Dam in the Lower Omo Valley

The flood plain of the lower Omo Valley is occupied by eight different groups with populations ranging roughly from 2,000 to 40,000, a total of about 200,000. They speak six different languages, belonging to two of the four major African language families—Nilo-Saharan (Bodi, Mursi, Suri (Chai), Kwegu/Muguji, and Nyangatom) and Afro-Asiatic (Hamar, Kara, and Dassanech). All, apart from the Kara, are pastoralists, but all

FIGURE 9.1. Dams and hydropower plants in the Omo Basin and the peoples of the lower Omo Valley.

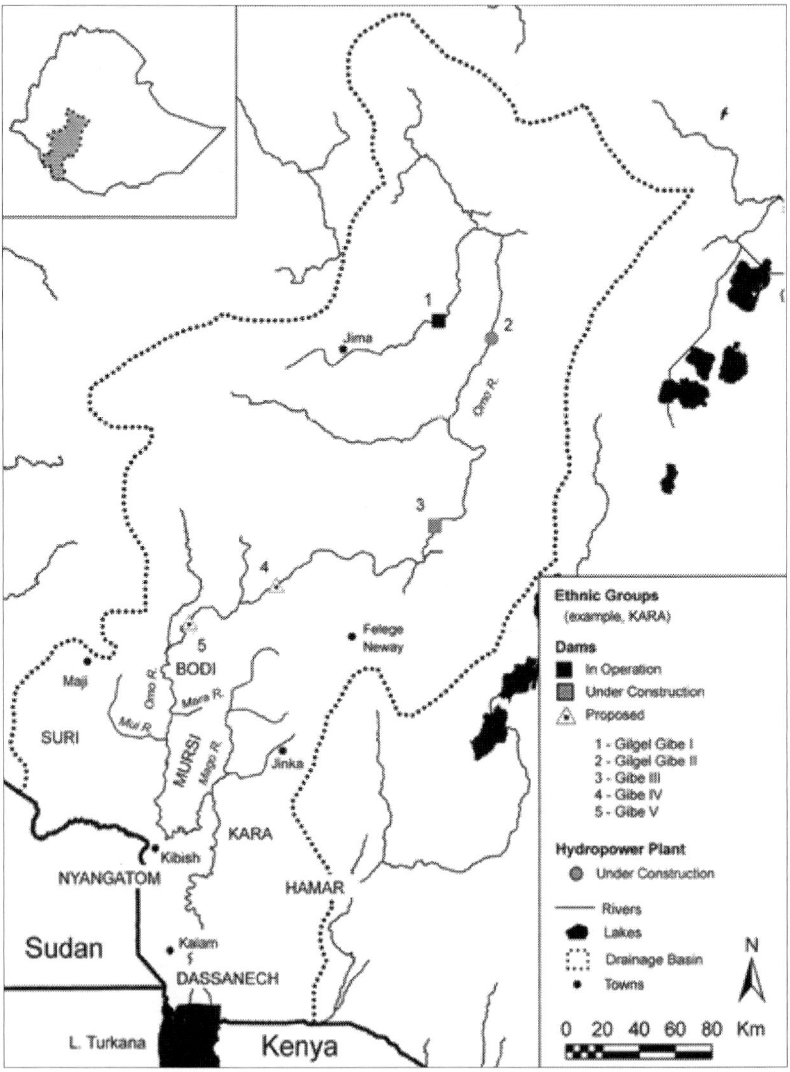

Sources. Food and Agriculture Organization (n.d.) and Centro Ellectrotecnico Sperimentale Italiana & Mid-Day International Consulting Engineers [CESI & MDICE] (2008).

depend heavily on cultivation. The main crop is sorghum, but maize, beans, and chick peas are also grown. For those living closest to the Omo itself (Bodi, Mursi, Kara, Kwegu/Muguji, Nyangatom, and Dassanech), land liable to be inundated by the annual flood is the most valuable agricultural resource they possess. This is for two main reasons. First, flood-retreat cultivation depends on the heavy rain that falls over the highland catchment area of the Omo, rather than on the erratic flood plain rainfall. And second, the annual deposition of fertile flood silts maintains the productivity of the same areas without the use of fertilizer. Since flooded areas are more extensive on the inner bends of meanders, the amount of land available for planting is determined by the sinuosity of the river, which increases from north to south. The particular contribution made by the Omo flood to the economy of a group depends, therefore, on where that group is located along the Omo.

For the Mursi and Bodi, who live in the northern part of the lower basin, a relatively limited amount of land is inundated by the flood, and flood-retreat cultivation must, therefore, be combined, not only with pastoralism, but also with the less reliable shifting cultivation. This does not mean that the contribution of flood cultivation to the economies of these groups is any less important than the contribution of either shifting cultivation or pastoralism. Each would be insufficient on its own, or even in combination with one of the other two, but taken together they form the basis of the long-term viability of the economy. As the Mursi put it, "You need three stones to support a pot over the fire."

Further downstream, not only are the meanders more pronounced, but the flood also submerges larger areas away from the immediate banks of the river, including oxbow lakes. This enables the Kara and Dassanech to produce all or most of the grain they need from flood-retreat cultivation. For the Dassanech, who occupy the Omo Delta on the northern shore of Lake Turkana, the flood is also vital for their pastoral activities. During the driest months of the year, between November and March, the recently flooded 'flats,' which are more extensive than can be cultivated by the available labor, provide excellent grazing when no alternative pasture is available (Almagor, 1978; Carr, 1977).

Given the reliance of all these groups on the Omo flood, it is clear that the most immediate and direct impact of Gibe III Dam on the people of the lower Omo will come from a reduction in river flow. We cannot say with any certainty, based on the available evidence, what the results of this reduction will be, in terms of lost crops and pasturage. But we can say that, if the floods were eliminated or greatly reduced, this would have a devastating impact on the livelihoods of up to 200,000 people in the lower Omo Valley and an even larger number around Lake Turkana. A second major impact of the dam, although an indirect one, will come from the commercial irrigation schemes which are likely to follow in its wake. Given the history of dam building and irrigation development in the Awash Valley, one cannot but be pessimistic about the impact of such schemes on the people and environment of the lower Omo. We consider first the question of river flow and flood level.

The original environmental impact assessment, completed in 2006 by the Italian consulting firm Centro Elettrotecnico Sperimentale Italiana (CESI), said nothing about the impact of the dam on the Omo flood plain and its peoples. The only "downstream" populations referred to were those living in the immediate vicinity of the dam and its reservoir. Two years later, and two years into the construction of the dam, a revised and expanded version of this document was produced, now entitled "Environmental and Social Impact Assessment" (CESI & MDICE, 2008). This includes summaries of relevant passages from a third report, subtitled "Additional Study of Downstream Impact," dealing specifically with the lower Omo (Agriconsulting Spa & Mid-Day International Consulting Engineers, 2008). Two main measures are proposed in this report to mitigate the impact of reduced river flow: the maintenance of a "minimum environmental flow" and the annual release of a "controlled flood," over a 10-day period in late August or September. It is claimed that these measures will "mitigate and fully compensate all adverse effects" of reduced river flow (CESI & MDICE, 2008, p. 223; Gibe III Dam faces challenges, 2009), and bring more benefits besides. Unfortunately, it is difficult to share this degree of optimism.

In the first place, the argument made in the Environmental and Social Impact Assessment (CESI & MDICE, 2008) that one of the major benefits of Gibe III will be flood control in the lower Omo Valley is erroneous. Most people living on flood plains, whether along the Awash (Kloos, 1982), the Nile in Sudan (Sileet, 2007), or in the lower Omo Valley, have a profound knowledge of flooding which they see as life-giving rather than life-threatening. Years of extensive flooding are celebrated as years of plenty. This indicates the need for greater emphasis on the development of flood early warning systems, preparedness, and management capacity, further discussed in chapters 10 and 11.

In the second place, the size of both the minimum flow and the controlled flood appears to have been worked out purely from mathematical models of the river flow, coupled with the need to minimize water loss from the reservoir. There is no evidence that these calculations were informed by a detailed understanding of the environmental knowledge and agricultural and pastoral practice of local people. One is left wondering how an artificial flood, taking place over 10 days and with much of its nutrient-rich sediment load having been trapped behind the dam wall, could compensate for the productive potential of the current flood regime. This lasts from March–April, when the river begins to rise, until September when it begins to fall, after having reached its peak level in August. It is also not clear how local people—meaning not local administrators, but those actually dependent on the controlled flood for their livelihoods—will be able to influence how it is managed and how its effectiveness is monitored.

Managed releases from dams are a relatively new and untried method of sustaining floodplain ecosystems, and they involve a high degree of technical and social complexity (Acreman, 2000). One would have expected, therefore, that the proposed managed releases from Gibe III would have merited rather more discussion than the brief and superficial treatment given to them in the "additional study" (CESI & MDICE, 2008, pp. 165–168). Although some examples of managed releases from dams elsewhere are mentioned, and although there is a list of 'critical steps' required for the successful implementation of a controlled flood, there is no consideration given to the specific lessons that could be learnt by the Gibe III project from experience elsewhere. One very important lesson would have been that controlled floods must be designed and implemented on the basis of a detailed understanding of local knowledge and practices. This requires that those whose livelihoods will be most affected by the reduced flow should be informed and consulted from the very start of the planning process, a condition that has manifestly not been met by the Gibe III project.

Turning now to the second major downstream impact of Gibe III, and with the Awash Valley again in mind, we need to consider how large-scale commercial irrigation schemes might affect the population of the lower Omo Valley. The fact that the regulation of the river flow will encourage investment in irrigation by 'farmers and companies' is mentioned only in passing by the authors of the 'additional study' (CESI & MDICE, 2008, p. 136). They do note, however, elsewhere in the report, that the site of what was an irrigated state cotton farm, in Dassanech territory on the east bank of the Omo, "has recently been acquired by some investors and is apparently being developed to grow oil palm" (CESI & MDICE, 2008, p. 96). In fact, these investors, including Fri el Green Power of Italy which describes itself as an "alternative energy company," have acquired a concession amounting to 30,000 ha. This includes the original farm but is three times its size (Editorial, 2007). The plan is to grow oil palm, and possibly also *Jatropha,* as feedstock for the production of biofuel. Other "investors" are rumored to have acquired, or to be in the process of acquiring, concessions of equal size for the production of biofuel feedstock in Kara and Nyangatom territory.

It seems reasonable to suppose that, once the Gibe III Dam is fully operational, there will be a significant expansion of large-scale commercial irrigated agriculture in the lower basin. At present, and although we have little information to go on, the signs are that this will take the form, predominantly, of biofuel feedstock plantations. Two arguments are frequently made in favor of investment in the production of biofuel feedstock in the developing world. One is that crops such as oil palm and *Jatropha* can be grown on 'marginal' land, unfit for the production of food crops, and the other is that the commercial production of biofuel feedstock can provide much needed employment for local people. The obverse of these arguments, from the point of view of the investor, is that land and labor can be acquired at minimal cost.

The first argument fits perfectly a stereotypical view of the lower Omo, often found amongst neighboring highlanders, as a hot, dangerous, and disease-ridden 'desert,' occupied by anarchic and violence prone 'nomads.' In fact, the grasslands of the lower Omo are a valuable productive resource for thousands of agro-pastoralists, whose economies depend on the careful integration of cattle herding with flood-retreat and rain-fed cultivation. The second argument rests upon another fallacy, namely that local people will be those most likely to benefit from employment opportunities created by plantation agriculture. Having no previous experience of wage labor, and with little or no knowledge of the national language, they will be in a very poor position to compete, even with unskilled labor migrants from the nearby highlands. Unless effective mitigating measures are taken, therefore, it is likely that large-scale irrigation development in the lower Omo will have much the same social impact as we described above for the Awash Valley (see also Kloos, 1982): an influx of migrant laborers from the highlands working on the irrigation schemes, while thousands of displaced and destitute pastoralists are fed at nearby famine relief stations.

Public health impacts of Gibe III are similarly underestimated. The assumption that the proposed 90 m annual drawdown of the Gibe III Reservoir will prevent vectors of malaria and schistosomiasis to proliferate (Ethiopian Electric Power Corporation, 2009) is not borne out by dam impacts in a number of African countries, where schistosomiasis and malaria became endemic or more prevalent at dam sites or downstream as a result of ecological and demographic changes (see chapter 13).

Possible Dam Impacts on the Ecology and People of Lake Turkana

The Omo River supplies 90% of the freshwater inflow into Lake Turkana. The ecology of the lake is therefore likely to be seriously affected by a greatly reduced water flow during the estimated five-year period of reservoir filling, by subsequent changes in the flow regime, and by the lower silt load of the lower Omo River. It has been estimated that decreased river flow will lower the lake level by 7–10 m (Africa Resource Working Group, 2008). Three small rivers at the southern end of the lake, particularly the Turkwell River, are increasingly used for irrigation and were, together with recent prolonged droughts, associated with the decline in the lake's level between 1973 and 1995 (Campbell, 1998). Acceleration of this decline must be expected for this terminal lake in a desert environment after the completion of Gibe III and the start of irrigation hydropower and irrigation in the Omo Valley, similar to the decline of Lake Chad (United Nations Environmental Program, 2008).

Lake Turkana is characterized by high biodiversity, maintained in part because of its location in Kenya's remote northwestern desert region and because of the untapped water of the Omo River until a few years ago. In addition to about 48 species of fish, Lake Turkana provides habitat for hippos, crocodiles, and terrestrial vertebrates, as well as numerous birds, attesting to the lake's biological integrity. Lake Turkana is a moderately saline and alkaline lake. Salinity is estimated to increase from 2,330 mg/L to about 2,800 mg/L as a result of the falling lake level (Africa Resource Working Group, 2008; Olago & Odada, 1996). Detailed environmental information is limited (Cohen, 2004), however, making it difficult to monitor ecological changes related to river regulation and abstraction. Of more immediate concern are the depletion of nutrients and major changes in Lake Turkana ecology through the Omo dams. Changes in Omo flow after completion of Gibe III and the other two large dams are anticipated to include increasing salinity/alkalinity levels, decreases in the size and biodiversity of the lake, and resulting impacts on local economies. Unfortunately, none of these potential effects were assessed in the dams' environmental and social impact assessments.

In addition to the Dassanech, who live in the vicinity of the Omo Delta, the Turkana, Borana, Rendille, Samburu, Ariaal, Gabbra, and Elmolo peoples in Kenya depend on Lake Turkana for their livelihoods. All these groups pursue to varying degrees pastoralism, agro-pastoralism, fishing, gathering, and hunting, and their reliance on any one of these activities depends on environmental factors and changing social relations (Mirzeler, 2006). The dependence of the Turkana on the lake, for example, has increased in recent years because of their adoption of fishing. Increasing consumption of fish as human food by this and most other pastoralist groups in Northeast and East Africa, whose food habits excluded fish in the past, indicates the desperation and adaptability of local groups in a desert environment (Africa Resource Working Group, 2008; Mirzeler, 2006; Simoons, 1974). However, fish catches and the size of some species decreased in the 1980s due to habitat changes related to lake level decreases that decimated spawning grounds and overfishing (Odada et al., 2006). Altogether, an estimated 300,000 people rely on Lake Turkana for these and other subsistence activities. Accelerated overgrazing, tree cutting, and hunting due to the growing population are depleting resources, putting additional pressure on the Lake Turkana Basin, and constitute a major threat to the lake's biodiversity. While environmental pressures on terrestrial and aquatic ecosystems were relatively insignificant in the past due to the isolation of Lake Turkana, they may, when compounded with the impending riverine changes, result in livelihood crises if existing environmental policies in the Lake Turkana Basin are not enforced (Gichuk, Oyieke, & Tere, 2005).

Transboundary Issues

Water resources development in both the Awash and Omo valleys constitute transboundary issues that have not been officially agreed on by Ethiopia, Djibouti, and Kenya. Lake Abe decreased in size by more than one third between the late-1940s and 1984 due to increasing abstraction of Awash River water for irrigation (from 400 km^2 to 250 km^2) and apparently continues to decrease. Although this lake is used little by the local Afar, it provides, together with the four nearby lakes upstream on the Awash (lakes Bario, Afambo, Gamari, and Gargori), habitat for endemic and migratory birds. The biology of Lake Abe and changes in the lake's chemistry have not been studied (BirdLife International, 2009;

Hughes & Hughes, 1992). We believe that acquiescence of Djibouti to the extensive irrigation development in the Awash Valley without insistence on a bilateral agreement is part of a long-standing economic symbiosis involving Ethiopia's use of the port of Djibouti for international trade.

Similarly, no bilateral agreement between Ethiopia and Kenya on the use of the Omo-Turkana waterway is known to exist. It was an activist group in Kenya which first raised the transboundary water issue in late 2008, but we had not been able, by the time of this writing, to obtain information on any official documentation (Gibe III Dam faces challenges, 2009).

PROSPECTS

Hopefully it is not too late, even at the equivalent of the eleventh hour, for the potentially adverse outcomes described in this chapter to be prevented in the lower Omo and around Lake Turkana. If they are not prevented, this will not be because of some unfortunate oversight or lack of diligence on the part of the planners. It will be because Gilgel Gibe III is an example, albeit an extreme one, of a certain way of looking at the world, which James C. Scott, in *Seeing Like a State* (1998) has called, "the hegemonic planning mentality of high modernist state-craft." According to Scott, the reason why so many large-scale, state-sponsored schemes to improve human welfare fail is that those who design and implement them, however well intentioned, have little or no confidence in, respect for, or understanding of, the knowledge, skills, and experience of ordinary people.

Like other river basin development schemes in Ethiopia and across Africa, the Gibe III Dam was planned, exclusively and from the start, according to the priorities and interests of central government (Duvail & Hamerlynck, 2007; Kloos, 1982). There was no attempt even to inform downstream people of the project until construction of the dam was well underway, let alone to include them meaningfully in the planning process. The "public consultations" described in the "additional study" (CESI & MDICE, 2008, pp. 159–163 and appendix 9) have been understandably criticized on the grounds that they fell far short of what a meaningful consultation process should be (Hathaway, 2009). Their main weakness, however, lay not in the way they were carried out but in the fact that they did not take place at all until two years after construction of the dam had begun.

This illustrates a general and widespread weakness of environmental impact assessments: they tend to be written to support and help forward an existing project, rather than to provide a neutral assessment of the environmental and social costs of one that is still being considered (Acreman, 2000). The impact assessments carried out for Gibe III look like a classic case of this phenomenon. To paraphrase Acreman, their purpose appears to have been more to support the developer than to protect the environment. More big dams are likely to be constructed in Ethiopia in the next few years, not least in the Omo-Gibe Basin itself. Let us hope that their social and environmental impacts are comprehensively and independently assessed at the beginning of the planning process, "when options for development are first considered, not as an afterthought when the project is near completion" (Acreman, 2000, p. 55). The usual dislocation between the capacity of dams to assure the sustainability of downstream economies and societies may be overcome if the complexities of downstream impacts are fully understood and downstream users have a stake in dam releases (Adams, 2000).

Recent declines in water and rangeland resources amongst Ethiopian pastoralists, and the coping mechanisms they have adopted, confirm that pastoral production systems are in dynamic disequilibrium and are optimally managed in spite of regular shocks, such as drought (Devereux, 2004; Helland, 2006). This view, which has rapidly become the current orthodoxy, is borne out by the adaptive water and land use practices of the Afar and Kereyu and the ability of all pastoralists to develop alternative forms of rangeland use and land tenure arrangements. Nevertheless, and as others have also recommended (Helland, 2006; United Nations OCHA, 2007), our account indicates that more attention will have to be given to the productivity and sustainability of natural resources and to the major enabling factors in their equitable use, such as mobility, land and water rights, and effective administration. This is nowhere more evident than in the case of those pastoralists who have become (in the Awash Valley), and who may in the future become (in the lower Omo Valley), the "victims" of river basin development in the form of hydroelectric dams and irrigation schemes.

In 2007, representatives of 15 African countries, including Ethiopia, met at Isiolo in Kenya, to lay the groundwork for the formulation of a continental pastoral policy. This was an important step forward. Further progress will not be made, however, unless individual governments are prepared to implement land tenure systems that make possible the ecologically sound management of land and water resources in the typically fragile grassland and shrub/woodland ecosystems occupied by pastoralists. In Ethiopia, various organizations have recently been formed, at the federal and state levels, to advocate for the rights and socioeconomic progress of all Ethiopian pastoralists (Gebre Michael, 2004; Guleid, 2003). These include the Parliamentary Pastoral Affairs Standing Committee, the Pastoralist Forum Ethiopia, and the Ministry of Agriculture's Pastoralists Extension Team.

The inclusion of pastoralists in the national planning effort will hopefully lead to the design of effective drought management systems, mitigate against future damage to water and rangeland resources, and encourage the development of economic alternatives for destitute pastoralists (Guleid, 2003). These efforts must also include attention to flexible resource tenure systems, the establishment of strong pastoralist organizations at the local and regional levels (Teka, 2003), safeguards against the loss of land and water resources to outside interests, strategies to deal with vulnerability to drought and other shocks, and the control of interethnic warfare. In particularly hard-hit areas, as in parts of Somali Region where food aid is undermining local production systems, or in Afar Region where farming and pastoralism cannot provide long-term food and water security, one long-term solution to assure sustainable livelihoods may be investment in alternative, nonpastoral employment (Devereux, 2004). Ultimately, solutions to the existing water and rangeland crisis experienced by Ethiopian pastoralists will require the commitment of government agencies, nongovernmental organizations, and agricultural and industrial enterprises involved in resource utilization, management, and conservation. But, above all, it will require the active participation, from the start of any planning process, of pastoralists themselves.

REFERENCES

Abule, E., Snyman, H. A., & Smit, G. N. (2005). Comparisons of pastoralists perceptions about rangeland resource utilization in the Middle Awash Valley of Ethiopia. *Journal of Environmental Management, 75*, 21–35.

Achamyeleh, K. (2003). *Integrated flood management, the associate program on flood management*. Report prepared for the World Meteorological Organization, Associated Programme on Flood Management, Geneva.

Acreman, M. (2000). Managed flood releases from reservoirs: Issues and guidance. Prepared for *Thematic review II.1: Dams, ecosystem functions and environmental restoration*. Cape Town: World Commission on Dams.

Adams, W. (2000). *Downstream impacts of dams*. Paper prepared for the World Commission on Dams. Retrieved from www.dams.org/docs/kbase/contrib/soc195.pdf

Afar Forum. (2008, October 17–19). The Afar Forum Conference (Press release). Brussels. Retrieved from www.afarfriends.org/Dok%20t%20websida/Belgien/Afar%20Press%20Release.pdf

Afar Human Rights Organization. (2007, July 4). Half a million Afar pastoralists risk displacement and environmental disaster in the Awash Valley of Ethiopia. *Ethiopian Review News*. Retrieved from http://www.ethiopianreview.com/news/?p=253

Africa Resource Working Group. (2008, May). *Environmental and social impacts of the proposed Gibe III Hydroelectric Project in Ethiopia's lower Omo River Basin* (Unpublished manuscript).

Agriconsulting Spa & Mid-Day International Consulting Engineers. (2008). *Gibe III Hydroelectric Project: Environmental impact assessment additional study of downstream impact 2008*. Retrieved from www.eepco.gov.et/GibeIII-2/Chapter_2.pdf

Alamerew, E., & Natea, S. (2004). *Participatory assessment of rainwater harvesting systems providing domestic water supply: Case studies in the Rift Valley area, central Ethiopia*. Reports prepared for WaterAid Ethiopia, Addis Ababa.

Alemayehu, T., Ayenew, T., & Kebede, S. (2006). Hydrogeochemical and lake level changes in the Ethiopian Rift. *Journal of Hydrology, 316*, 290–300.

Almagor, U. (1978). *Pastoral partners: Affinity and bond partnership among Dassanetch of southwestern Ethiopia*. Manchester: Manchester University Press.

Belay, K., Beyene, F., & Manig, M. (2005). Coping with drought among pastoral and agro-pastoral communities in Eastern Ethiopia. *Journal of Rural Development, 28*, 185–210.

Beyene, T. (2006). *Kessem-Tendaho multipurpose water resources development, Ethiopia*. Paper presented at the Fourth World Water Forum, Mexico City.

BirdLife International. (2009). *Important bird area factsheet: Lake Abe wetland system, Ethiopia*. Retrieved from the Data Zone at http://www.birdlife.org

Campbell, R. W. (1998). Lake Turkana, Ethiopia/Kenya: 1973, 1989, 1995. In *Earthshots: Satellite images of environmental change*. U.S. Geological Survey. Retrieved from http://earthshots.usgs.gov

Carr, C. (1977). *Pastoralism in crisis: The Dassanetch and their Ethiopian lands* (Research paper no. 180). Chicago: University of Chicago, Department of Geography.

Centro Elletrotecnico Sperimentale Italiana & Mid-Day International Consulting Engineers. (2008). *Gibe III Hydroelectric project: Environmental and social impact assessment*. Milano, Italy: Author.

Central Statistical Authority. (1998). *The 1994 Population and Housing Census of Ethiopia: Results at country level* (Vol. 1, statistical report). Addis Ababa: Federal Democratic Republic of Ethiopia.

Christian Relief and Development Association. (2001). *Pastoralism Panel. Proceedings of a workshop on non-governmental organizations and sectoral development*. Addis Ababa.

Cohen, A. S. (2004). Distribution of faunal associations of benthic invertebrates at Lake Turkana. *Hydrobiologia, 141*, 179–197.

Devereux, S. (2004, February). *Food securities issues in Ethiopia: Comparison and contrasts between lowland and highland areas*. Paper presented at the Seminar on Food Security Issues in Ethiopia, sponsored by UN-OCHA, Addis Ababa.

Duvail, S., & Hamerlynck, O. (2007). The Rufiji River flood: Plague or blessing? *International Journal of Biometeorology, 52*, 33–42.

Editorial (2007, April 15). Ethio-Korean farm privatized, Sweet 16. *Addis Fortune*. Retrieved from http://www.addisfortune.com/Vol%207%20No%20363%20%20Archive/Index.htm

Edwards, K. A., Classen, G. A., & Schrote, E. H. J. (1983). *Organization and management of water supplies in tropical Africa* (ILRI Research Report No. 8). Addis Ababa: International Livestock Research Institute.

Ethiopian Electric Power Corporation. (2009). *Gibe III Hydroelectric Project. Executive summary and environmental and social impact assessment*. Addis Ababa: Author.

Food and Agriculture Organization of the United Nations. (n.d.). *GeoNetwork*. Retrieved from http://www.fao.org/geonetwork/srv/en/main.home

Gebre, A. (2002). Change and continuity in pastoral adaptations: The case of the Karrayu in the Upper Awash Valley of Ethiopia. In B. Yimam, R. Pankhurst, D. Chapple, Y. Admassu, A. Pankhurst, & B. Teffera (Eds.), *Proceedings of the 14th International Conference of Ethiopian Studies* (Vol. 2, pp. 661–705). Addis Ababa: Institute of Ethiopian Studies, Addis Ababa University.

Gebre, A., & Kassa, G. (2009). The effects of development projects on the mid-Awash Valley. In A. Pankhurst & F. Piguet (Eds.), *Moving people in Ethiopia: Development, displacement & the state* (pp. 66–80). London: James Currey.

Gebre Michael, Y. (2004a). *Impact assessment of the Ethiopian Pastoralist Day (EPD) commemorations (1999–2004).* Addis Ababa: Pastoralist Forum Ethiopia.

Gebre Michael, Y. (2004b). *Economic diversification in the pastoral areas of Ethiopia: Opportunities and challenges.* Report prepared for Oxfam International, Addis Ababa.

Gibe III Dam faces challenges from Lake Turkana community. (2009, March 7). *The Reporter.* Retrieved from http://en.ethiopianreporter.com/content/view/781/9/

Gichuk, N. N., Oyieke, H. A., & Tere, T. (2005). Status and root causes of biodiversity loss in the eastern Rift Valley lakes, Kenya. In E. O. Odada, D. O. Olago, W. Ochola, M. Ntiba, & S. Wandiga (Eds.), *Proceedings of the 11th World Lakes Conference* (Vol. 2, pp. 511–516). Nairobi: Kenya Ministry of Water and Irrigation and International Lake Environmental Committee.

Guinand, Y. (2000). *Afar pastoralists face consequences of poor rains: Rapid assessment mission 19–24 April 2000.* Report prepared for the UN-Emergencies Unit for Ethiopia. Addis Ababa.

Guleid, A. A. A. (2003). The way forward: Plenary discussion. In G. Gebru, S. Desta, & D. L. Coppock (Eds.), *Summary of the proceedings of the meeting pastoralism in Ethiopia and the policy environment: Linking research, development actors, and decision-makers* (pp. 14–17). Addis Ababa: International Livestock Research Institute.

Hammond, L. (2006). Famine and food security in Ethiopia: Causes, implications and response. In Y. Berhane, D. Haile Mariam, & H. Kloos (Eds.). *The epidemiology and ecology of health and disease in Ethiopia* (pp. 110–128). Addis Ababa: Shama Books.

Hammond, L., & Dessalegn, B. (2003). *Safara: An assessment of currant and future voluntary resettlement in Ethiopia*. Report prepared for USAID-Ethiopia. Addis Ababa.

Harbeson, J. W. (1975). Afar pastoralists and Ethiopian rural development. *Rural Africana, 28*, 71–86.

Hathaway, T. (2009). *Facing Gibe 3 Dam: Indigenous communities of Ethiopia's lower Omo valley*. International Rivers. Retrieved from http://www.internationalrivers.org/en/node/2794

Helland, J. (1998). Institutional erosion in the drylands: The case of the Borana pastoralists. *Eastern Africa Social Science Research Review, 14*, 49–72.

Helland, J. (2006). *Pastoral land tenure in Ethiopia*. Paper presented at the Colloque International (Unpublished manuscript). Christian Michelson Institute, Bergen, Norway.

Hughes, R. H., & Hughes, J. S. (1992). *A directory of African wetlands*. Nairobi: United Nations Environmental Program.

Hundie, B. (2006). *Property rights among Afar pastoralists of northeastern Ethiopia: Forms, changes and conflicts* (Unpublished manuscript). Humboldt University, Berlin.

Interregional Information Network. (2003a, July 23). Ethiopia: ADF supports study on harnessing Awash River. *IRIN Newsletter*. Office of the Coordination of Humanitarian Affairs, United Nations.

Interregional Information Network. (2003b, October 15). Ethiopia: Government admits pastoralists excluded from democratic process. *IRIN Newsletter*. Office of the Coordination of Humanitarian Affairs, United Nations.

Kassa, G. (2001). *Among the pastoral Afar in Ethiopia: Tradition, continuity and socio-economic change*. Utrecht: International Books and Addis Ababa: OSSREA.

Kassa, G. (2002). Overview of interventions in pastoral areas, resource scarcity, competition and conflicts and prospects for social development.

In B. Yimam, R. Pankhurst, D. Chapples, Y. Admassu, A. Pankhurst, & B. Teferra (Eds.), *Proceedings of the 14th International Conference of Ethiopian Studies* (pp. 785–811). Addis Ababa: Addis Ababa University, Institute of Ethiopian Studies.

Kevane, M., & Gray, L. (2008). Darfur: Rainfall and conflict. *Environmental Research Letter, 3*.

Kloos, H. (1982). Development, drought, and famine in the Awash Valley of Ethiopia. *African Studies Review, 25*, 21–48.

Kloos, H., & Lindtjørn, B. (1993). Famine and malnutrition. In H. Kloos & Z. A. Zein (Eds.), *The ecology of health and disease in Ethiopia* (pp. 103–120). Boulder, CO: Westview Press.

Kloos, H., Polderman, A. M., DeSole, G., & Lemma, A. (1977). Haematobium schistosomiasis among seminomadic and agricultural Afar in Ethiopia. *Tropical and Geographical Medicine, 29*, 399–406.

Mirzeler, M. K. (2006). Guest editor's prelude. *Ethnohistory, 53*, 3–11.

Odada, E. O., Olago, D., Kulindwa, K. A. A., Bugenyi, F., & West, K. (2006). *East African Rift Valley lakes* (Global International Water Assessment No. 47). Kalmar, Sweden: University of Kalmar on behalf of the United Nations Environmental Program.

Olago, D. O., & Odada, E .O. (1996). Some aspects of the physical and chemical dynamics of a large rift lake: The Lake Turkana North Basin, northwest Kenya. In T. C. Johnson & E. O. Odada (Eds.), *The limnology, climatology and paleoclimatology of East African lakes* (pp. 413–441). Amsterdam: Gordon and Breach.

Oxfam. (2008). *Survival of the fittest: Pastoralism and climate change in East Africa* (Oxfam Briefing Paper No. 116). Oxford: Oxfam International.

Redda Teklehaimanot, R. (2003). A literature based study into the expected pollution of the Awash River, Ethiopia (TOX-80316), preliminary toxic risk evaluation of the Awash River (TX-80328) (Unpublished master's thesis). Wageningen University, The Netherlands.

Scott, J. C. (1998). *Seeing like a state: How certain schemes to improve the human condition have failed.* New Haven and London: Yale University Press.

Sileet, T. M. (2007). *The past and future of flood management in the Eastern Nile Basin*. Retrieved from http://www.inbo-news.org/ag2007/comms/Nile_Flood_Management_abstract.pdf

Simoons, F. J. (1974). Rejection of fish as human food in Africa: A problem in history and ecology. *Ecology of Food and Nutrition, 3*, 89–109.

Teka, H. (2003). Pastoral development strategy in the Oromia Region. In G. Gebru, S. Desta, & D. L. Coppock (Eds.), *Summary of the Proceedings of the Meeting Pastoralism in Ethiopia and the Policy Environment: Linking research, development actors, and decision makers* (pp. 13–14). Addis Ababa: International Livestock Research Center.

Turton, D. (1995). *Pastoral livelihoods in danger; Cattle disease, drought, and wildlife conservation in Mursiland, south-western Ethiopia*. Oxford: Oxfam.

Turton, D. (2002). The Mursi and the elephant question. In D. Chatty & M. Colchester (Eds.), *Conservation and mobile indigenous peoples: Displacement, forced settlement and sustainable development* (pp. 97–118). Oxford and New York: Berghahn Books.

United Nations. (2003). *Update on the humanitarian situation in Ethiopia* (Field report prepared by the UN Country Team). Addis Ababa: United Nations Office Ethiopia.

United Nations Environmental Program. (2008). *Africa's lakes atlas of our changing environment*. Nairobi, Author.

United Nations OCHA. (2007). *The future of pastoralism in Ethiopia: 4 possible scenarios*. Addis Ababa: Pastoralist Communication Initiative.

Voelkner, H. E. (1974). *The social feasibility of settling seminomadic Afar on irrigated agriculture in the Awash Valley, Ethiopia* (Informal Technical Report UNDP/FAO/ETH 72/006 No. 23). Prepared on behalf of the State Rivers and Water Supply Commission for the Food and Agriculture Organization, Rome.

Yemane, B. (2003). *Food security situation in the pastoral area of Ethiopia*. Unpublished report prepared for Oxfam, Addis Ababa.

Chapter 10

Flooding in Ethiopia

Recent History and the 2006 Flood

Semu Moges, Yirga Alemu, Stuart McFeeters, and Worku Legesse

Introduction

Historically, floods have been intertwined with the livelihood of human beings in at least two ways. First, they have been cherished and have facilitated the emergence of early civilizations along riverbanks, as in Egypt and Mesopotamia; and second, they have been considered as evil acts or punishment from spiritual forces. Today, flooding is a broad term that includes flash floods, river floods, and coastal floods. Their negative effects are most severe in the poorest countries, where populations and governments are least able to deal with this scourge. In 2006, floods were reported from several countries in northeast and east Africa, the U.S., and east and southeast Asia. In Kenya, 723,000 people were affected by floods (Interregional Information Network, 2006a), and in Somalia, where shallow wells became contaminated and food stocks were destroyed by flood waters, an estimated 900,000 people were affected (Interregional Information Network, 2006b). The 2006 floods in the eastern U.S. were the worst since 1979, and those in central and eastern Europe in 2000–2001 and the ones in southeast Asia broke all previous records. This worldwide upsurge in floods has been associated with climate change (Intergovernmental Panel on Climate Change, 2007), although other factors appear to be involved as well (Van Aalst, 2006). In Ethiopia and many other African countries, chronic incipient environmental changes promoting flooding have been associated with a lack of land use planning, poverty, and nonsustainable land use (Woube, 1999; see also chapters 7 and 8).

Floods in many parts of Ethiopia traditionally assured the livelihood of riverine farmers and pastoralists in arid and semiarid lowland areas. The ecological integrity of flood plains and many other wetlands can be maintained only through seasonal floods (see chapter 9). In recent years, due to stepped up efforts to use these typically fertile and well-watered ecosystems for irrigated agriculture as a means of boosting food production, floods are increasingly seen as destructive (Achamyeleh, 2003; Ministry of Finance and Economic Development, 2002). Widespread poverty in Ethiopia renders many local populations vulnerable to the destructive effects of floods (Helmer & Hilhorst, 2006).

The objective of this chapter is to briefly trace the recent history of flooding in Ethiopia and examine the distribution, main causes, and consequences of the 2006 floods in Ethiopia, with a focus on spatial and temporal patterns in rainfall and flooding and possible interventions. The study uses a hydrological perspective in analyzing the 2006 and earlier climatological and river runoff and lake level records of the National Meteorological Service. Special emphasis is given to the major floods in Dire Dawa Town and its small river catchment in the eastern highlands, Lake Tana Sub-Basin, an intermediate sized catchment, and the large Omo Basin. These three basins were selected because the Dire Dawa and lower Omo areas were most severely affected by the 2006 flood in terms of human deaths, and the hydrology of the Lake Tana Sub-Basin is of considerable interest because of its role as the major source of the Greater Nile and its significant role in hydropower production in the Blue Nile Basin.

Recent Floods in Ethiopia

Flooding has occurred for many years in nearly all river basins in Ethiopia. The Awash River, providing water for the largest irrigation complex in the country starting in the early 1960s, has the longest flood record. Severe flooding of the Awash flood plains occurred in 1973, 1987, and 1995, damaging large cotton irrigation schemes, buildings, and properties. Consequently, during the last 40 years, river protection works such as construction and maintenance of long embankments, construction of breakaway channels, dykes, earth dams, and clearing of siltation and other obstructions from the river bed had become routine, expensive activities. The Awash changed its main course in its meandering lower reaches before 1974 and again in 1988. As a result, the Dit Bahari cotton scheme was unable to irrigate and the downstream settlements were facing shortage of water, requiring them to shift from gravity use to more expensive motor pumps. Moreover, some of the traditional flood plain grazing areas were permanently inundated. Additional floods inundated the Awash flood plains in 1993, 1996, 1997, 1998, 1999, 2005, and 2006, some of which caused extensive damage to state and private farms and claimed the lives of human beings and livestock (Abebe, Taddese, & Genet, 2006). The 2006 floods in Ethiopia were exceptionally severe due to their intensity, water volume, geographical coverage, and the magnitude of the damage. Floods occurred in eight regional states, where more than half a million people were vulnerable, about 200,000 people were affected, and a total of 639 people were reported to have lost their lives (Disaster Prevention and Preparedness Agency, 2006), probably underestimates owing to poor communication in rural areas.

Seasonal flooding occurred again in 2007 in northeast Africa. According to the UN Office for Coordination of Humanitarian Affairs (Interregional

Information Network, 2007a), due to the increase of the level of rivers, lakes, and dams, an estimated 103,000 people have been affected and about 20,000 people have been displaced in the different regions of Ethiopia. This underlines that in 2007 floods occurred also in non-flood-prone areas and a trend of increasing floodwater area inundation was noticed. Floods in Sudan were the worst in living memory, affecting 400,000 people and putting 3.5 million people at risk of disease (Interregional Information Network, 2007b). In Somalia's Middle Wabe Shebele Region, 4,000 ha of farmland were destroyed, affecting 12,000 people, but the damage elsewhere was even higher (Interregional Information Network, 2007c).

THE EXTENT AND IMPACTS OF THE 2006 FLOODS

By the end of August 2006, at least 75 *woredas* in eight regions, a seventh of all *woredas*, had been affected by the floods (figure 10.1). According to the reports by the Ethiopian Government Disaster Prevention and Preparedness Agency (DPPA, 2006), more than 500,000 people were vulnerable and about 200,000 people had been affected, with 639 deaths (368 people in South Omo, 256 people in Dire Dawa, and 15 people in other parts of the country). Thousands of livestock were killed, 228 tons of harvested crops were washed away, 147 tons of export coffee beans were lost (and machinery ruined), and 42,229 ha of crop land were inundated.

THE 2006 FLOODS IN THE LAKE TANA SUB-BASIN, DIRE DAWA TOWN, AND OMO BASIN

This section is based on monthly rainfall data of the National Meteorological Service Authority (NMSA) for 12 stations in the Lake Tana Sub-Basin (Bahir Dar, Gondar, Debre Tabor, Enjbara, Dangila, Aykel, Woreta, Addis Zemen, Tsege, Maksegnit, Delgi, and Gorgora) for the period 1976–2006 and hydrological data of the Hydrology Department of the Ministry of Water Resources (MWR) for the same 31 years for

FIGURE 10.1. *Woredas* reporting flooding in Ethiopia in 2006.

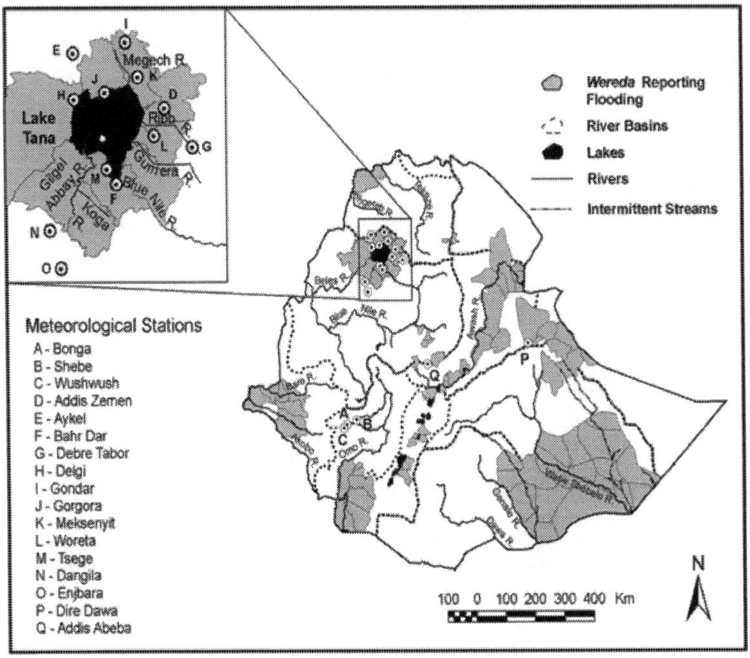

Source. Based on Disaster Prevention and Preparedness Agency (2006).

the four major headwater tributaries (Gumera, Ribb, Megech, and Koga/ Gilgel Abbay) and Lake Tana water surface level fluctuations. Using descriptive statistics, including mean, standard deviation, and coefficient of variation, the temporal and spatial variations of rainfall and stream flow data associated with Lake Tana were analyzed. Rainfall and stream flow variability were analyzed to investigate the cause of flooding. Similarly, Lake Tana level fluctuations were analyzed based on longitudinal and 2006 flood data. Rainfall and hydrological data for 2006 were compared to long-term means for comparative analysis.

For each major perennial river, the largest flood in each year of record was retrieved from the database. Flood characteristics were analyzed by using basic flood statistical parameters such as mean annual flood volume in cubic meters per second, specific mean annual flood volume in $m^3/sec/km^2$, and coefficient of variation for each data series. Specific mean annual flood is defined here as the spatial intensity of the mean annual maximum flood, while the coefficient of variation is a measure of flood variability from year. These two measures can depict the essential characteristics of a catchment's flood series. Information on the impacts of 2006 flooding in Ethiopia was collected from different reports, mainly from the major report issued by the federal Disaster Preparedness and Prevention Agency (2006).

THE FLOOD IN LAKE TANA SUB-BASIN

Rainfall variability analysis

Monthly rainfall for 2006 from May to September (the main rainy period in the northwestern highlands) was compared with the long-term mean values (1976–2005) of four stations (Bahir Dar, Woreta, Addis Zemen, and Tsege) in the Lake Tana Sub-Basin to investigate the temporal rainfall distribution during these four months. The precipitation in July and August 2006 was higher than the mean values at each of these four stations, as shown by the Woreta climatological data, which are similar to those of the other three stations (figure 10.2). Long-term mean annual precipitation at these four stations for the period 1988–2006 was 1,388 mm (figure 10.3). Thus, both seasonal and annual rainfall precipitation in 2006 were above normal.

Stream flow variability

River flow was analyzed for the five major rivers flowing into Lake Tana (Gilgel Abbay, Koga, Gumera, Ribb, and Megech) to compare flows during 2006 with long-term means. Annual flow of the Gilgel Abbay deviated more sharply from mean values in 2006 than during any other year since 1973 (figure 10.4). Similarly, during the period July–September 2006, all but 2 of the 15 comparison data pairs had higher mean values than for those months in 1976–2005. The flow increase ranged

Figure 10.2. Mean rainfall (2001–2006) and 2006 rainfall, by month, at Woreta Station.

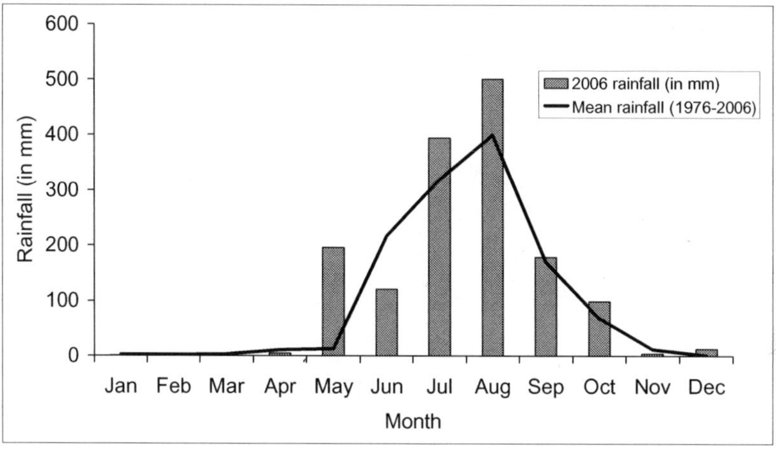

Source. Based on unpublished data of Woreta climatological station.

Figure 10.3. Mean annual rainfall at four stations in the Lake Tana Sub-Basin (Woreta, Bahir Dar, Tsege, and Addis Zemen), 1988–2006.

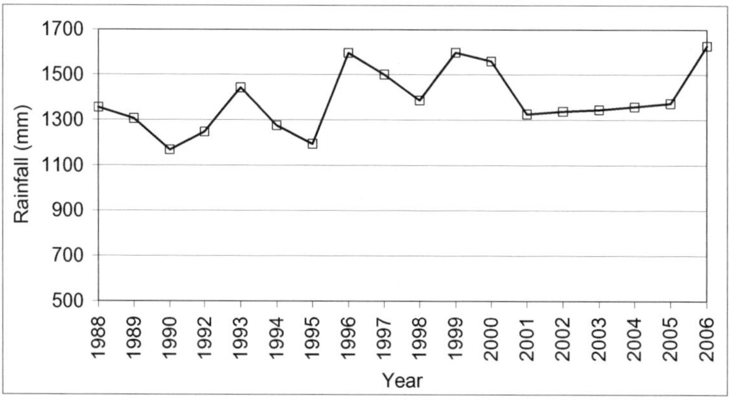

Source. Based on unpublished climatological data of Woreta, Bahir Dar, Tsege, and Addis Zemen climatological stations.

FIGURE 10.4. Mean annual flood deviation in the Gilgel Abbay River for 1973–2006 (104 m^3 per second).

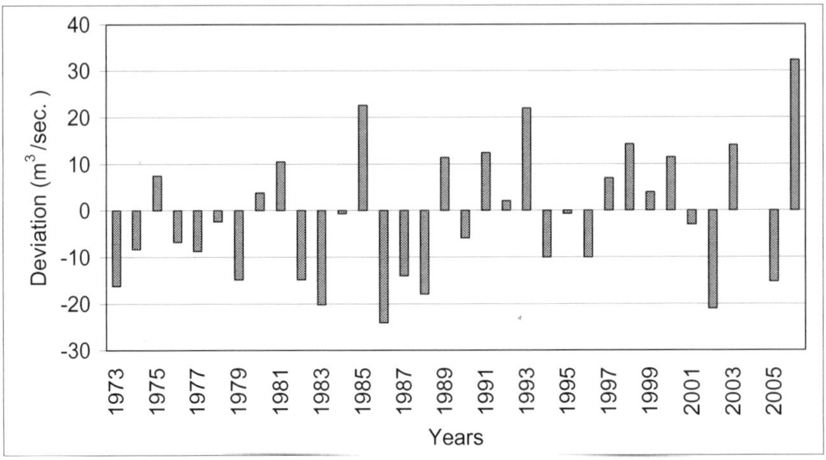

Source. Based on unpublished hydrological data of Gilgel Abbay hydrological station.

from 22% for the month of July to 34% in August and 47% in October (table 10.1). The increase in flow did not, however, result in an increase in Lake Tana to flood level.

Water level variability of Lake Tana

Historical hydrological data shows that annual maximum water levels of Lake Tana fluctuated since 1985 between a maximum of 1,787.82 m above sea level (masl) recorded on September 8, 1998, and a minimum level of 1,784.26 masl, recorded on June 30, 2003. The highest level of the lake in 2006 was 1,787.16 masl, only 16 cm above the normal water surface level and below the maximum levels for the period 1996–2001. Runoff during the 2006 flood could only partly compensate for the sharp drop in lake level during the 2001 drought (figure 10.5).

TABLE 10.1. Comparison of 2006 maximum runoff with long-term mean maximum runoff (1976–2005) between July and October in five rivers in the Lake Tana Sub-Basin, by month.

River	July flood (m^3/s)			August flood (m^3/s)			September flood (m^3/s)		
	Mean max. (1976–2005)	2006 max.	Increase (%)	Mean max. (1976–2005)	2006 max.	Increase (%)	Mean max. (1976–2005)	2006 max.	Increase (%)
G/Abbay	300	293	–2	319	355	11	255	273	7
Koga	29	39	37	46	104	124	25	44	77
Gumera	215	260	21	258	274	6	179	221	23
Ribb	90	93	3	107	99	–8	80	97	22
Megech	69	106	53	139	186	34	44	65	49
Mean	141	158	22	174	203	34	117	140	35

Source. Based on unpublished data of the Gilgel Abbay, Koga, Gumera, Ribb, and Megech hydrological stations.

FIGURE 10.5. Annual maximum flood levels of Lake Tana, 1966–2006.

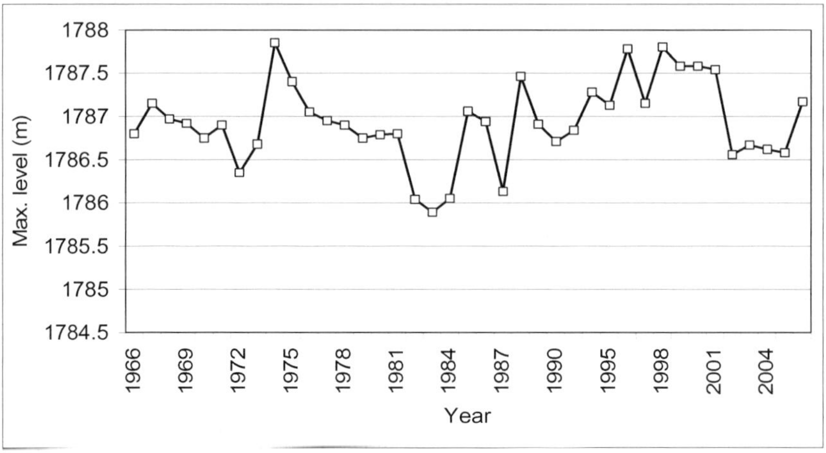

Source. Based on unpublished data of Bahir Dar hydrological station.

Dire Dawa

The Dire Dawa area is the smallest and most arid of the three catchments but the most vulnerable to flash floods, consistent with the pattern in semiarid, degraded areas.

Rainfall variability

The main rainy season in the Dire Dawa catchment extends from June to August. Seasonal as well as daily precipitation data were compared for these months. Between June and August 2006, rainfall totaled 384 mm, compared to 300 mm for 1997–2006. Mean monthly rainfall for August was particularly heavy in 2006, about 60% of the seasonal total, but June-August totals were similar to those in 1999, 2001 and 2003 (figure 10.6). During those three years, only one flood of low magnitude was reported in Dire Dawa—in 2001—further indicating that the August 2006 flood was the result of unusually heavy rainfall.

Runoff in the small Dire Dawa catchment is dominated more by short duration rainfall than seasonal totals. Therefore, data from the six

FIGURE 10.6. Mean seasonal rainfall during June–August (JJA Mean) and per year (JJA RF), 1997–2006, at three stations in the Dire Dawa catchment area (Dire Dawa, Alemaya, and Dengego).

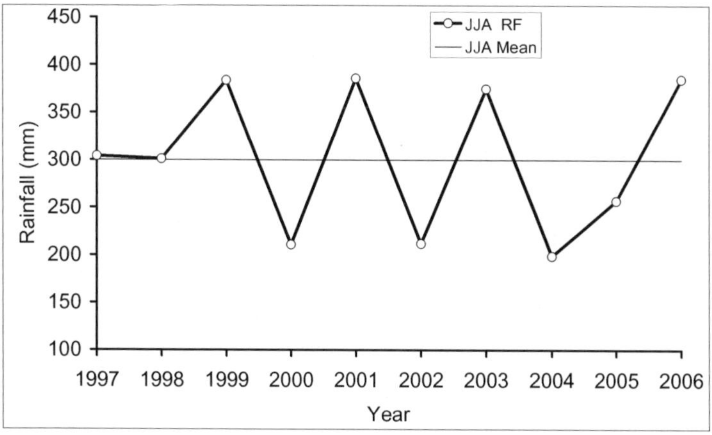

Sources. Unpublished data of the Dire Dawa, Alemaya, and Dengego meteorological stations.

rainfall stations located in the catchment of the four small rivers causing the flood in Dire Dawa and having daily meteorological time series were analyzed. We examined three possible explanations for the severe flooding in this town. First, flash floods may be caused when isolated, unusually heavy rain storms occur in catchments with poor vegetation cover. Second, flash floods may occur when moderate rainfall continues over all or the larger part of a catchment. A third possibility may be created by a combination of the above two scenarios, where four factors may govern the formation and intensity of flash floods: 1) the duration of storm rainfall, 2) the intensity of the storm, 3) the areal extent of the storm, and 4) the time of concentration.

Dire Dawa Town and the three upstream stations of Alemaya, Kersa, and Kulubi reported sharp increases in precipitation on August 5, 2006, the day before the flood hit Dire Dawa (figure 10.7), with totals between

FIGURE 10.7. Daily rainfall at four stations in the Dire Dawa catchment area, August 2006.

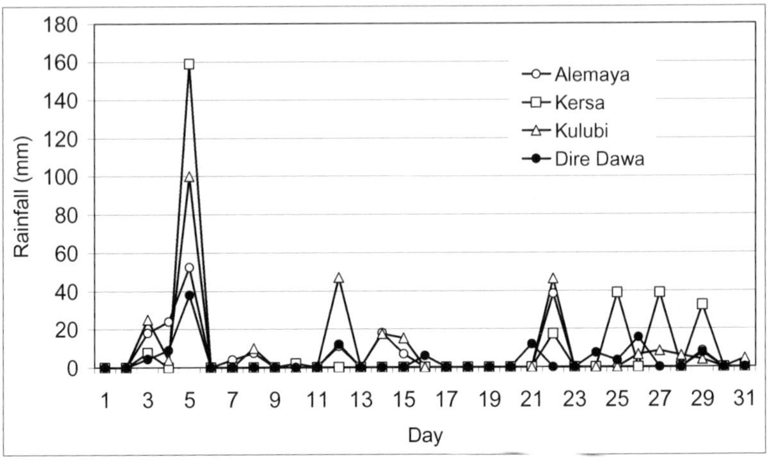

Sources. Unpublished data of the Alemaya, Kersa, Kulubi, and Dire Dawa meteorological stations.

52 mm and 159 mm at the upstream stations. These amounts were several times larger than during most other rainy days in August 2006, indicating that the entire catchment received heavy rain on August 5, 2006. In addition, all three rainfall stations are within 30 km of Dire Dawa Town, resulting in a spatial and temporal concentration of storm and runoff that caused severe flooding where the four rivers of this small catchment converge in this town (see chapter 11). The heavier than normal rains in July and early August 2006 resulted in the saturation of the soil, further contributing to this major flood.

Stream flow variability

In the absence of a river gauging station in Dire Dawa Town to estimate the magnitude of the flood, we estimated the discharge using a rational formula and the SCS method of runoff estimation techniques described by Adeba (2007). On the basis of these methods, the flood generated from the Dire Dawa catchment is estimated to have varied between 642.3 m^3/s and 852.4 m^3/s during the critical three-hour period on August 5. These estimated volumes are equivalent to the major river's normal runoff during a 200-year period (Adeba, 2007).

Omo Basin

Seasonal and monthly precipitation variability of the wet season months June, July, and August were also analyzed for several Omo Basin stations. Most stations in the upper Omo Basin did not receive significantly more rain in 2006 than in previous years, although Shebe Station recorded the highest precipitation since 1998 (figures 10.8 and 10.9).

During the peak rainy period of June to August 2006, many stations reported significantly higher amounts in 2006 than in other years. Comparison of the daily maximum flow at different stations shows that flooding in the large Omo Basin (78,200 km^3) may have been triggered by

FIGURE 10.8. Mean annual rainfall and total for June to August at Shebe Station in the Omo Basin, 1996–2006.

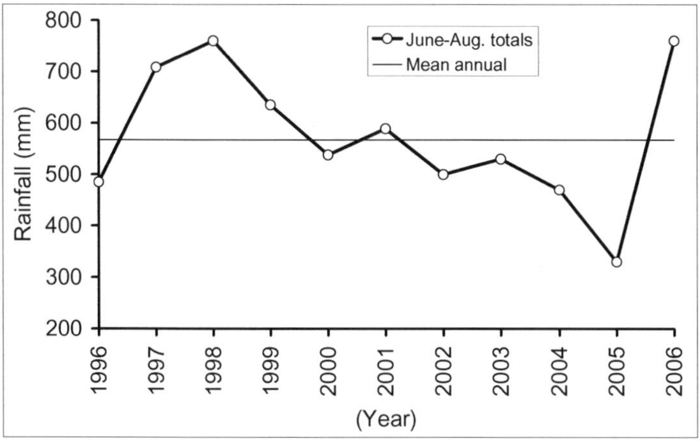

Source. Unpublished data of Shebe meteorological station.

FIGURE 10.9. Mean annual maximum daily rainfall and mean maximum daily rainfall in July and August (1996–2006) at Wushwush (A) and Bonga (B) stations.

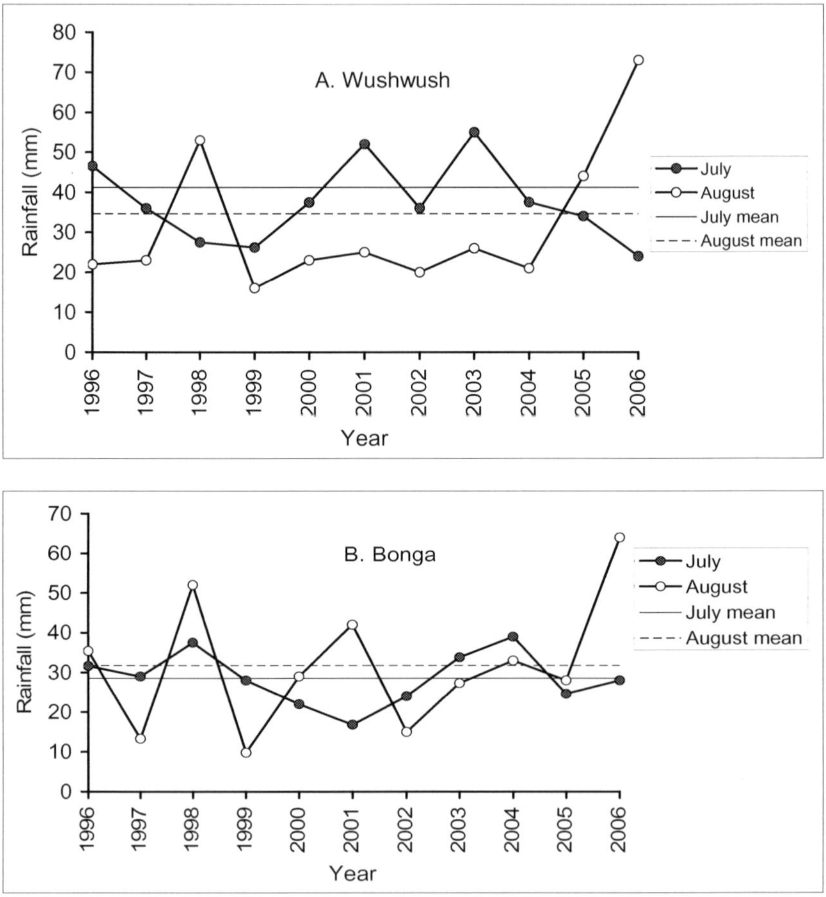

Sources. Based on unpublished data of Wushwush and Bonga climatological stations.

heavier daily storm rainfall over most of the basin. Typical daily rainfall for Bonga and Wushwush stations (figures 10.9A and 10.9B) tend to confirm this possibility. Thus, the flooding of the Omo River in 2006 appears to have been due not only to the general increase in annual and seasonal rainfall amount but mainly due to unusually high daily rainfall throughout the catchment.

Comparison of Rainfall and Flooding Patterns in the Three Catchments

Examination of the precipitation and runoff data from stations in different parts of Ethiopia indicates a pattern of increases in 2006 in both the seasonal and areal rainfall above normal values. The areal and intensity increases in rainfall persisted in spite of the shorter duration of rainfalls in the country. This has been confirmed from satellite-derived real rainfall imagery for the three study areas, showing that there is a generally greater concentration in seasonal rainfall over shorter periods (data accessed from the IRI/LDEO Climate Data Library at http://iridl.ldeo.columbia.edu/).

In the Lake Tana area, the cause of the August 2006 flood appears to be due to one or more of the following factors: 1) the backwater effect of the lake, 2) overbank flow from the major tributaries, and 3) high storm rainfall over the flood plains. As shown in the data analysis section, the lake level was only 16 cm above the normal level during the flood. Inflow from the tributaries was not large enough to recover the low level of the lake in 2005 and much water might have been released through the regulating weir. Hence, it is unlikely that the cause of the flooding of the lake's flood plains was due to overflow from the lake. Instead, the inundation of these plains is the combined effect of the major rivers overflowing (a common event) and the heavy rainfall on the flood plains and wetlands surrounding the lake. The poor drainage of the area contributed to the problem, causing frequent seasonal flooding. Moreover, population growth has brought large numbers of people and livestock to these flood-prone areas in recent years (Alemu, 2006).

As indicated in figures 10.7 and 10.9, the heavy floods in Dire Dawa and the Omo Valley were associated with intense, short-duration rain storms in areas that already had saturated soil due to seasonally high rainfall in 2006. One difference between the two study areas is the duration of the flooding events. In Dire Dawa, the flood hit with devastating ferocity but lasted for a very short period owing to the small watershed. Even though little flow data exists for 2006 for the lower part of the Omo River, witnesses of the flood reported that the floods lasted for several weeks, consistent with the large catchment of the Omo.

All three case studies reveal that the heavy rains in 2006 were concentrated in specific areas within each basin, although the whole country received precipitation. These and other climatic anomalies (including the 2007 rainfall pattern) indicate that climate change is affecting rainfall patterns in Ethiopia. Thus, the development of adaptation and mitigation plans remains a vital and urgent task.

THE NEED FOR FLOOD PREVENTION AND MANAGEMENT

With the estimated doubling of the Ethiopian population by 2025 and continuing land degradation, the country may experience recurring, even more devastating floods in the foreseeable future. According to the Intergovernment Panel on Climate Change, extreme rainfall events will probably become more prevalent even in the dryer parts of the continent, including much of East Africa (2007). These predictions, as well as the lessons learned from the recent flood events, point out the urgency for institutionalizing flood management and control programs at the national and regional state levels. If the management flood crisis continues on an ad hoc basis of emergency and relief support, seasonal erosion of public and international attention is likely. This constitutes dangers in at least two aspects: 1) public and international support will slowly decrease, particularly in the event of other emerging national and international catastrophes, and 2) there is a danger of society developing a dependency syndrome if floods continue to affect flood-vulnerable districts every few years. It appears that the country has adequate experience, given the long-standing drought and the recent floods, to deal with threats of flooding.

Moges (2007) proposed a conceptual approach of flood management in Ethiopia based on two premises, 1) minimizing the damage of flood water through maximizing the benefits of flood for food security and poverty reduction, and 2) minimizing flood damage through efficient, cost-effective, and sustainable flood management systems that are institutionally manageable and technologically appropriate and flexible. The earlier method focuses on building within the year storage dams and weirs for dual purposes of aversion of flood surges as well as detention of the water for irrigation and other uses during the dry season.

The second approach focuses on nonstructural flood control measures, including development of a Flood Forecasting and Early Warning System (FFEWS) and watershed management. It has always been difficult to permanently displace people from flood prone areas due to economic reasons and people's attachment to their birth place. The economic potential of flood plains and wetlands, such as swamps, is great in Ethiopia, indicating that they can support sizeable populations if they are managed in a sustainable manner (Abebe & Geheb, 2003; Mulugeta, 2004). Therefore, FFEWS is an alternative approach to reduce the damages of flooding by using real-time meteorological information linked with flood forecasting tools and a real-time dissemination mechanism. The minimum requirement of FFEWS is an efficient institutional setup, advanced flood forecasting models, real-time information dissemination mechanisms to reach the end users, and access to real-time meteorological observation and forecasted information. Bangladesh was able to reduce the deaths occurring from floods from 138,882 in 1991 to only 134 in 1997 (United Nations Environmental Program-Division of Early Warning & Assessment System, 2002, cited by Moges, 2007) after the establishment of an efficiently working flood management center.

An equally important, nonstructural measure which can function well with FFEWS is watershed management: Proper management of watersheds, from micro to catchment levels, can drastically reduce the threat of flooding. As described in chapters 7 and 8, reforestation, soil, and water conservation structures such as terraces, sustainable land use practices, and protection of wetlands and other ecosystems with flood regulating properties can effectively reduce or delay runoff. In addition, watershed management can result in increased availability of surface and groundwater resources during the dry seasons as base flow (delayed flow from the soil moisture zone or ground water).

CONCLUSION

Analysis of longitudinal meteorological, hydrological, and satellite derived data (RFE) from three areas in Ethiopia shows that the seasonal and annual rainfall and its areal coverage generally increased sharply in 2006 in the traditionally flood-prone areas. The heavier than normal rainfall during the period June–August rendered the Lake Tana Sub-Basin and the Dire Dawa and lower Omo areas particularly vulnerable when the heavy storms arrived in August. Analysis of monthly and daily precipitation data revealed these relationships. This cumulative effect of prolonged rainfall is revealed especially by the data from the small catchment around Dire Dawa Town. While the data for the large catchment area of the Omo River did not, as expected, reveal a strong temporal association between rainfall patterns and flooding, they provide a measure of flood delays which needs to be considered in flood forecasting models. Although unavailability of daily rainfall data for the Lake Tana Sub-Basin does not permit such an analysis, the relationship between mean annual rainfall, river runoff, and Lake Tana levels is instructive of the interconnectedness of this aquatic ecosystem, which must be considered in the ongoing water development projects in the Lake Tana Sub-Basin and other areas in the Blue Nile Basin selected for dam construction (see chapter 2). The damage caused by the 2006 and other recent floods can be attributed to both increased flood magnitude and increasing encroachment and settlement on flood plains. In view of the continuation of downhill population movements and settlement in the lowlands and the trend toward greater rainfall extremes, it is likely that flood impacts will increase in the future if measures for interventions are not implemented. Institutionalized flood management systems should be considered in developing interventions aimed at reducing flood damages in the country. As it may be difficult to resettle inhabitants of fertile flood plains, the existing Flood Forecasting and Early Warning System (FFEWS), together with enhanced integrated watershed management can reduce flood impacts when effectively integrating real time meteorological data, flood forecasting hydrological models, and real-time dissemination of the forecasted floods to end users. Structural flood measures, such as construction of dams and weirs to store flood surcharge, can effectively reduce flood risks and increase food energy security if they are planned after comprehensive environmental and socioeconomic impact assessments that inform on their optimum size, location, and operation with the objective of ensuring maximum benefits to the whole society and the environment (see chapter 9).

REFERENCES

Abebe, T., Taddese, Y., & Genet, W. (2006, November). *Flooding in Awash River Basin: The case of the 2006 Flood, Addis Ababa, Ethiopia* (Unpublished manuscript). Arba Minch University, Arba Minch, Ethiopia.

Abebe, Y. D., & Geheb, K. (Eds.). (2003). *Wetlands in Ethiopia. Proceedings of a seminar on the resources and status of Ethiopia's wetlands.* Nairobi: International Union for Conservation of Nature and Natural Resources.

Achamyeleh, K. (2003). *Ethiopia: Integrated flood management.* Report prepared for the World Meteorological Organization, Associated Program on Flood Management, Geneva.

Adeba, D. (2007). *Investigation of the cause and impact of flooding of Dire Dawa area* (Unpublished master's thesis). Arba Minch University, Arba Minch, Ethiopia.

Alemu, Y. (2006). *Assessment of causes and impacts of 2006 flooding on agricultural production of Lake Tana's surrounding plains* (Unpublished master's thesis). Arba Minch University, Arba Minch, Ethiopia.

Disaster Prevention and Preparedness Agency. (2006). *Joint government and humanitarian partners: Flash appeal for the 2006 flood disaster in Ethiopia.* Addis Ababa: Author.

Helmer, M., & Hilhorst, D. (2006). Natural disasters and climate change. *Disasters, 30,* 1–4.

Intergovernmental Panel on Climate Change. (2007). *Working Group I, the physical science basis of climate change* (chapter 11). Retrieved from www.ipcc.ch

Interregional Information Network. (2006a, November 28). Kenya: Red Cross estimates 723,000 people affected by floods. *IRIN Newsletter.* Office of the Coordination of Humanitarian Affairs, United Nations.

Interregional Information Network. (2006b, November 30). Somalia: Battered by floods. *IRIN Newsletter*. Office of the Coordination of Humanitarian Affairs, United Nations.

Interregional Information Network. (2007a, August 27). Ethiopia: Flooding causes extensive damage. *IRIN Newsletter*. Office of Coordination of Humanitarian Affairs, United Nations.

Interregional Information Network. (2007b, November 15). Sudan: Floods put 3.5 million people at risk of disease. *IRIN Newsletter*. Office for the Coordination of Humanitarian Affairs, United Nations.

Interregional Information Network. (2007c, August 29). Somalia: Crops destroyed as Shebelle burst banks. *IRIN Newsletter.* Office for the Coordination of Humanitarian Affairs, United Nations.

Ministry of Finance and Economic Development. (2002). *Ethiopia: Sustainable development and poverty reduction. Strategy paper for promoting development and poverty reduction*. Addis Ababa: Federal Democratic Republic of Ethiopia.

Moges, S. A. (2007). *Flood forecasting and early warning system (FFEWS), an alternative technology for flood management system and damage reduction in Ethiopia: A concept note*. Paper presented at the 2007 Lake Abaya Research Symposium, Arba Minch University, Arba Minch, Ethiopia.

Mulugeta, S. (2004). Socio-economic determinants of wetlands and cultivation in Kemise, Illubabor Zone, southwestern Ethiopia. *Eastern Africa Social Science Research Review (EASSRR)*, *20*, 94–114.

Van Aalst, M. K. (2006). The impacts of climate change on the risk of natural disasters. *Disasters*, *30*, 5–18.

Woube, M. (1999). Flooding and sustainable land-water management in the lower Baro-Akobo river basin. *Applied Geography*, *19*, 235–251.

Chapter 11

A Case Study of the 2006 Flood

Dire Dawa

Yohannes Gebre Michael

This chapter examines causes and impacts of the 2006 flood in Dire Dawa Town and prospects of flood prevention. By considering both environmental and anthropogenic factors, particularly poverty, land degradation on steep slopes, uncontrolled slum development, and administrative inefficiencies, and by evaluating measures to prevent floods, this case study complements the hydrological analysis of the 2006 floods in Ethiopia in chapter 10. Specifically, it illustrates the multifactorial nature of floods and the difficulties of preventing them and their impacts in a small catchment. Multiple methodologies were used to collect data on the Dire Dawa flood and the study area. Published and unpublished reports and interviews by the author with urban and rural residents living in the affected areas and officials of the Dire Dawa administration and direct observations by the author of flood damage, land use, settlement patterns, and the physical environment across an altitudinal transect of Dire Dawa constitute the information basis for this study. The author repeatedly visited the area in early 2007 to interview elders, farmers, homeless people, squatters, shop owners, women, and two youth groups in groups or as individuals about the flood and its impacts. The informal interviews, guided by checklists, were carried out in the flooded areas in Dire Dawa Town and in pastoral areas downstream.

PROFILE OF THE DIRE DAWA AREA

Dire Dawa Town and most of the surrounding administrative region is located at the foot of the high Rift Valley escarpment below the Hararghe Highlands in eastern Ethiopia. In Dire Dawa Region and in the vicinity of Harar Town in the nearby highlands, the growing population has long exceeded the carrying capacity of the land and the available water resources, as evidenced by deforestation, high erosion rates, shallow soils, and the recent drying up of lakes Haramaya and Adele (Alemayehu, Furi, & Legesse, 2007; Daba, 2003). Crop failures occur frequently in this arid to semiarid region, where rainfall is unreliable. By diverting flood water to the cultivated river bottom lands and communal domestic and livestock ponds, and by transporting manure from upstream villages to the more arid downstream cultivated areas, a degree of food security is achieved. Similarly, the riverbanks and river bottoms in and around the towns also have multiple functions during the dry and rainy seasons. Both flash floods and overbank river floods are common during the rainy season at intermediate and lower elevations in Dire Dawa's river catchments, as elsewhere in Ethiopia (Ahrens, 1996; Woube, 1999).

The Bio-Physical Environment

Dire Dawa Administrative Region has a total area of only about 2,000 km^2. The foothills of this semiarid escarpment area rise in the south to the more humid Eastern Hararghe Highlands, where all the larger rivers and streams originate. Nearly half of the total area of Dire Dawa Region is classified as intermediate to steep slopes. The steep escarpment and the highlands are intensively cultivated and grazed by livestock and are highly degraded. The mostly shallow soils with low content of organic matter and poor water holding capacity support sparse natural vegetation and few remaining indigenous trees. Rainfall often falls in heavy thunderstorms. Only the small river bottom lands have deeper and more fertile soils. Cultivated land accounts for less than 20% of Dire Dawa Region and bare rock for 36%, with the rest mostly degraded shrub land (Dire Dawa Regional Administration, 2004). Moreover, the conflict between the different land uses (cultivated, grass land, shrub land, and woodland) becomes inevitable with the scarcity of farmland, fodder production, and fuel wood demand.

The area's remaining natural vegetation consists of dispersed patches of degraded drought-resistant grasses, bushes, and shrubs. Expansion of agriculture and settlements and the high demand for fuel wood both in the rural and urban centers account for the severely degraded natural vegetation cover. According to estimates by the Dire Dawa Water, Mines, and Energy Office, 71,257 tons of fuel wood and charcoal were consumed in 2002, which covered only about half of the demand for domestic fuel (Dire Dawa Administrative Council, 2006). To fill this gap, the local population obtains large amounts of fuel wood and charcoal from periodic markets in the highland areas of the neighboring Somali and Oromia regions.

The climate of Dire Dawa Region is arid and semiarid, characterized by high spatial and temporal variability of rainfall, with higher precipitation on the escarpment. The rainfall pattern is bimodal, with the small rains peaking in April and the main rains in July–August. Mean annual rainfall is about 660 mm and the mean annual temperature varies from 20–30°C in the arid and semiarid lowlands along the Rift Valley escarpment (Tadesse & Danachew, 2006; Tefera & Abebe, 2007).

Dire Dawa Region has no perennial rivers. Two major seasonal rivers, the Dechatu and Butugi, passing through the center of the city, and the smaller Lege-Hare and Goro rivers, demarcating the eastern and western boundaries of the urban area, respectively, drain into the Awash catchment. The four rivers originate in the Hararge Highlands around Kulubi, Dengego, and Haromaya towns. The Dechatu River has an estimated watershed area of 215 km^2 (the largest of the four catchments) and is the most susceptible to flooding.

Land Use and Degradation

In the rural communities surrounding Dire Dawa Town, the main source of income is mixed farming (livestock and crop production). For the landless, petty trades, daily labor, and sale of fuel wood and stone, mainly from the common lands, are some of the source of incomes. The main crops produced are sorghum and maize supported with some horticultural production. Due to the unreliability of rainfall, flood water is diverted to the farming plots, and it is also used to harvest water for communal and individual ponds. Some of the local traditional agronomic practices include crop rotation, strip cropping, and intercropping of annuals and perennials such as maize, haricot beans, and *chat (Catha edulis).* Traditional ditches and check dams are maintained on many fields to manage water runoff and soil moisture.

It is estimated that about 70% of Dire Dawa Region is severely affected by land degradation (Jerusalem Children and Community Development Organization, 2005). Field observations around Dire Dawa City revealed that steep slopes, settlement areas, communally owned land, flat bottom land, and river banks are most susceptible to flooding. As indicated above, usually the communally owned land in the rural areas consists of grassland, shrub land, and rock outcrops. The poor, including landless farmers, pastoralists, and urban poor, use these lands for livestock grazing and fuel wood. However, there are no well-defined rules and regulations for the rational use of these common resources due to the conflict between customary rights and recent government rules and regulations. This implies that specific locations need to be prioritized for protection and owners and users of such critical areas need to be actively involved in the planning and implementation processes.

Population and Settlement Patterns in the Flooded Area

The projected population of Dire Dawa Region for 2006 was 398,000, 74.4% of which resided in Dire Dawa City, Ethiopia's second largest city (Central Statistical Authority, 1998). The Oromo and Somali are the two major ethnic groups. The riverbanks and river beds in Dire Dawa Town have multiple functions during both the dry season, when the rivers dry up and become wadis, and during the wet season, when river discharge is normally relatively small, leaving most of the riverbeds dry. Many poor and homeless people established their official residences and business in these flood-prone areas as squatters. This includes street children and dropout farmers and pastoralists affected by drought and famine, probably more than 5,000 people. The river banks and beds also accommodate the major markets for livestock, agricultural and manufactured products for the informal sector, and many vendors of farm products, small shops, and restaurants. In addition, these areas also serve as a playground for the children of the poor, a dumping ground for solid and liquid wastes, an open latrine, and a major source of manually excavated sand and river stones for construction. Due to the lack of proper housing and sanitation facilities and nonenforcement of city sanitary regulations, this densely populated riverine area is highly polluted with human, animal, and chemical waste products, which affect the health of both the local population and the Somali nomadic pastoralists downstream. Only poor squatters live in this undesirable area.

History of Local Flooding and the 2006 Flood

A study by Tadesse and Danachew (2006) based on discussions with elderly informants from rural and urban areas in Dire Dawa indicated that the first well-recognized flash flood in the area occurred in 1945, which resulted in few casualties and little loss of property. The next major flood, in 1977, again caused little damage. They were said to have been followed by more serious floods in 1981, 1997, 2001, 2004, 2005, and 2006, indicating an increase in the frequency and severity of flooding during the last six decades (Tadesse & Danachew, 2006). Informants interviewed by Tefera and Abebe (2007) recalled that the most severe flooding events in Dire Dawa occurred in 1944, 1981, during the small rainy season in 1985, in May 2005, and in August 2006. This implies high variability in the monthly occurrence of intense rains and flooding. The rainfall record for the period 1952–2006 corroborates particularly the reported floods in 1977 and 1981 (Ketema, 2007).

As a result of the 2006 flash flood in Dire Dawa, more than 256 people died (120 died in the 2005 flood) and 244 people became missing. In addition, the flood directly affected more than 10,000 people in the urban area and about 11,000 people in rural areas. Over 230 ha of cultivated land were damaged, more than 1,000 livestock perished, flood control and water conservation structures were destroyed, and property and crop losses were estimated at more than US$8 million each. Flood-related health problems included hunger and malnutrition, and the flood resulted in increased school dropout, increased homelessness, and a decline in household income. The 2006 flash flood also caused damage to infrastructure, estimated at nearly 10 million Birr (mostly roads, bridges, and retaining walls), and damaged or destroyed approximately 1,600 residences (Disaster Prevention and Preparedness Agency, 2006).

Causes of the 2006 Flood

The interaction of environmental and human factors in causing the 2006 Dire Dawa flood has been described by a number of investigators (Kassahun, Tadesse, & Dagnachew, 2007; Ketema, 2007; Tadesse & Danachew, 2006; Tefera & Abebe, 2007; Teruneh, 2007). They generally concluded that the steep, degraded slopes with their shallow soils and sparse vegetation could not absorb the water from the heavy rains, causing the rapid, destructive descent of the water masses into the small catchment into Dire Dawa Town. Overcultivation on steep slopes exacerbated by conflicts between farmers and pastoralists over different land uses and vegetation cover as well as open access to the limited land resources were major underlying problems in the rural areas. In Dire Dawa Town, expansion of squatter settlements in the flood-prone river beds and wadies, weak flood protection structures, lack of adequate environmental knowledge, marginalization of the community in decision making, and absence of a flood prevention and mitigation policy framework increased the impacts of the flood. The disproportionately large flood impact on poor squatters can be viewed as a form of social (and political) injustice, also reported by various investigators in other parts of Africa and South and East Asia (Parker, 2000). Drought and famine conditions and poor coordination among government institutions dealing with flood prevention and land use issues affected both the rural and urban areas.

Inadequate water conservation structures

Field observations in the flooded areas revealed that some of the introduced standard soil and water conservation structures, such as stone bunds, are too long and easily break down during heavy rains, resulting in big rills or gullies. Similarly, in spite of the fact that biological measures are less effective in arid and semiarid areas due to the short growing season, the government and NGOs used food-for-work or cash incentives to construct stone bunds under different land uses. The undesired outcomes of the construction of stone bunds included water splashing, runoff, and evaporation from the stones collected on the plots. Farmers, on the other hand, traditionally left some stones on their plots rationally for multipurpose functions such as minimizing the splash effect, runoff and evaporation, and to obtain better yields. Some local proverbs emphasize that stony plots produce larger crop yields (Gebre Michael & Herweg, 2000).

Land degradation and population growth

Nearly all the research reports on various floods in Dire Dawa and other parts in Ethiopia identified population growth and land degradation as the major causes of flooding (Jerusalem Children and Community Development Organization, 2005; Kassahun et al., 2007; Ketema, 2007; Tefera & Abebe, 2007; Teruneh, 2007), a relationship also found in several other developing countries (Parker, 2000). On the other hand, there is some global and national evidence that population growth and agricultural intensification may permit the improvement rather than deterioration of soil and water resources (Tiften, Mortimore, & Gichuki, 1994).

An inventory of indigenous soil and water conservation technologies from different African countries, including Ethiopia, indicates that the most elaborate soil and water conservation measures have traditionally been applied in the most densely populated areas. Several studies concluded that high population density is one of the preconditions for investment in soil and water conservation measures or land husbandry and that land tenure security is a prerequisite for tree planting (Gebre Michael, 1999; Reij, Scoons, & Toulmin, 1996), relationships which were not observed in Dire Dawa.

Heavy rainfall

The analysis of rainfall patterns in the Dire Dawa catchment in chapter 10 shows that daily rainfall on August 5, 2006, was more than twice as heavy at the upstream stations than during the other days during that month, when rains were fairly normal. According to daily rainfall for 1952–2006 in Dire Dawa Town, the most intensive and heavy rainfalls were recorded in 1956, 1977, 1981, 1983, 1988, and 1998 (Ketema, 2007), indicating that the 2006 flood was not an anomaly. Thus, it was the concentration of heavy rainfall on one day that was instrumental in the 2006 flood, a pattern sharply different from the one observed in the large Omo-Gibe Basin (chapter 10).

FLOOD MITIGATION STRATEGIES AND CHALLENGES

Existing Strategies

Various strategies to control flooding have been proposed in Ethiopia. Most of them revolve around awareness creation, implementation of watershed protection programs, establishment of soil and water conservation and flood protection structures, flood mapping, policy reorientations, and strengthening of partnerships among the sectoral offices and regional states (Jerusalem Children and Community Development Organization, 2005; Kassahun et al., 2007; Ketema 2007; Tefera & Abebe, 2007). These strategies have reduced flooding in some localities but were largely ineffective in preventing large-scale inundations.

The Proposed Watershed Approach

Various studies of the 2006 floods in Ethiopia have underlined the need to use the watershed approach as a priority to alleviate flooding problems (Jerusalem Children and Community Development Organization, 2005; Kassahun et al., 2007; Teruneh, 2007). However, the watershed approach faces a number of challenges in Ethiopia. First, the watershed and natural hydrological boundaries do not coincide and are thus incompatible with administrative boundaries (*kebele*, district, and region). Hence, mobilization of labor and limited resources, including safety net activities, create leadership and other administrative conflicts of interest. Second, communities usually equate the watershed approach with the protection of natural resources, including closures and afforestation and limited efforts toward utilization and management, making them reluctant to participate. Third, standard soil and water conservation technologies are introduced under this approach, with incentives limiting the range of choices for the differentiated community. Fourth, the incentives of food-for-work and cash-for-work, which are commonly used in watershed conservation and rehabilitation programs, may have some undesired outcomes. Most importantly, the grain or cash tend to stimulate powerful individuals to impose inappropriate technologies, and the incentive might be considered as payment and contribute to loss of land ownership, exacerbating the dependency syndrome (Gebre Michael & Waters-Bayer, 2007).

Can Flood Risk Be Reduced?

High risk of flooding in Dire Dawa and elsewhere in Ethiopia will be difficult to reduce in the near term due to several factors. First, Dire Dawa area and surrounding communities experience recurring droughts and famines which generate high rates of rural-urban migration and squatter settlements, with many squatters settling in the undesirable beds of the flood-prone intermittent river and wadis within the urban area. Secondly, the coordination of different stakeholders (different sectors and bureaus of government at the regional, *woreda,* and *kebele* levels and NGOs) in sustainable resource management is process oriented and time consuming and, thus, an expensive, slow-moving venture. Third, flood prediction and mitigation is difficult in the absence of extensive multi-sectoral information and the scarcity of basic meteorological data.

CONCLUSION

This case study illustrates that although the flood plains of Dire Dawa have been used by different stakeholders, it was the poor and disenfranchised who suffered most during the 2006 flood. By living in the flood-prone river bottoms, shunned by people who can afford to live away from wadis, the squatters were the most vulnerable population group. Although flooding occurs periodically in the Dire Dawa area, there are indications that its magnitude and impacts on human lives and properties have increased in recent years. Because multifactorial land degradation appears to be a major cause of flooding, local flood prediction and mitigation will be difficult as it needs extensive and diverse information, including meteorological data and a holistic, integrated management approach to address the problem. The solution to the flood problem needs to be flexible to address the needs and priorities of the affected population living in an environment characterized by scarce resources and extreme weather conditions. This case study also suggests that flood prevention and mitigation programs may benefit from the development of a strategy that harmonizes indigenous and modern soil and water conservation technologies and resource management. This will require coordinated efforts across hydrological and administrative boundaries in the watershed and customary and government laws and regulations, including city zoning ordinances and flood prevention measures, a challenge faced by many integrated water resources management efforts (Lenton & Muller, 2009).

References

Ahrens, J. D. (1996, August 26–27). *Koka Dam-River Awash floods.* Report prepared for the UN Development Programme, Emergency Unit for Ethiopia, Addis Ababa.

Alemayehu, T., Furi, W., & Legesse, D. (2007). Impact of water over-exploitation on highland lakes in eastern Ethiopia. *Environmental Geology, 52*, 147–154.

Central Statistical Authority. (1998). *The 1994 population and housing census of Ethiopia: Results at country level.* (Vol. 1, Statistical report). Addis Ababa: Author.

Daba, S. (2003). An investigation of the physical and socioeconomic determinants of soil erosion in the Hararghe Highlands, Eastern Africa. *Land Degradation and Development, 14*, 69–81.

Dire Dawa Administrative Council, Rural Development Bureau. (2006). Unpublished data. Dire Dawa.

Dire Dawa Regional Administration. (2004). *Dire Dawa Statistical Abstract (Dire Dawa).* Dire Dawa: Author.

Disaster Prevention and Preparedness Agency. (2006). *Flash appeal for the 2006 flood disaster in Ethiopia.* Addis Ababa: Author.

Gebre Michael, Y. (1999). *The use, maintenance and development of soil and water conservation measures by small-scale farming households in different agro-climatic zones of northern Shewa and southern Wello, Ethiopia* (SCRP, Research Report 44). Berne: University of Berne, Center for Development and Environment.

Gebre Michael, Y., & Herweg, K. (2000). *From indigenous knowledge to participatory technology development.* Berne: University of Berne, Center for Development and Environment.

Gebre Michael, Y., & Waters-Bayer, J. (2007). *Trees are our backbone: Integrating environment and local development in Tigray Region of Ethiopia* (IIED Dryland Issue Paper No.145, p. 26). London: IIED.

Jerusalem Children and Community Development Organization. (2005). *Changing environmental threats into sustainable livelihood opportunities: Lessons and experiences on community-based environmental rehabilitation initiatives of Jerusalem Children and Community Development Organization.* Paper prepared for the Second International Work Conference on Climate Change and Disaster Risk Reduction, Dire Dawa.

Kassahun, D., Tadesse, Y., & Dagnachew, S. (2007). Flood disaster impacts in Dire Dawa: Some recommendations for future risk reduction in Dire Dawa. In D. Kassahun & W. Bewket (Eds.), *Proceedings of the Workshop Options for Flood Risk Mitigations in Dire Dawa* (pp. 27–46). Dire Dawa: Catholic Relief Services.

Ketema, T. (2007). Hydro meteorological analysis and its implication for flood risk reduction in Dire Dawa. In D. Kassahun & Bewket, W. (Eds.), *Proceedings of the Workshop Options for Flood Risk Mitigations in Dire Dawa* (pp. 5–17). Dire Dawa: Catholic Relief Services.

Lenton, R., & Muller, M. (2009). *Integrated water resources management in practice: Better water management for development.* London: Earthscan.

Parker, D. J. (Ed.). (2000). *Floods.* London: Routledge.

Reij, C., Scoons, I., & Toulmin, C. (1996). *Sustaining the soil: Indigenous soil and water conservation in Africa.* London: Earthscan.

Tadesse, Y., & Danachew, S. (2006). *Flooding at Dire Dawa: A case study of the 2006 flash floods* (Unpublished manuscript). Addis Ababa.

Tefera, B.,& Abebe, K. (2007). A study of flood risk mitigation in East Hararghe Zone, D. Dawa Administrative Council and Shinile Zone. In D. Kassahun & W. Bewket (Eds.), *Proceedings of the Workshop Options for Flood Risk Mitigations in Dire Dawa* (pp. 47–68). Dire Dawa: Catholic Relief Services.

Teruneh, Y. (2007). Storm water management issues in the current Dire Dawa integrated development planning in Dire Dawa. In D. Kassahun & W. Bewket (Eds.), *Proceedings of the Workshop Options for Flood Risk Mitigations in Dire Dawa* (pp. 69–87). Dire Dawa: Catholic Relief Services.

Tiften, M., Mortimore, M., & Gichuki, F. (1994). *More people, less erosion: Environmental recovery in Kenya*. London: John Wiley.

Woube, M. (1999). Flooding and sustainable land-water management in the lower Baro-Akobo river basin, Ethiopia. *Applied Geography, 19*, 235–251.

Chapter 12

Water Pollution From Industrial and Agricultural Sources

Worku Legesse and Helmut Kloos

Introduction

Increasing populations, industrial establishments, and irrigated agriculture, among others, are competing for dwindling freshwater sources in practically all countries. Both manufacturing industries and agricultural activities contribute significantly to the pollution of surface and groundwater resources worldwide, water insecurity, diseases, and ecological upsets. It is now generally agreed that adequate freshwater supplies and protection for aquatic habitats can be provided in the 21st century only with fundamental changes in the way society uses, manages, and values freshwater (Postel, 2002). The urgency of water conservation, water reuse, and detoxification is further increased by global climatic change

(Payet & Obura, 2004). The production of industrial and agricultural chemicals has increasingly shifted from the developed to the developing world during the 1990s, when the industrial growth rate was nearly six times higher in the latter (World Resources Institute et al., 1998).

Disturbing relationships between quality and quantity of water and problems due to decreasing and inequitable distribution of freshwater are being reported from countries and communities experiencing water pollution. In addition to rapidly increasing consumption of freshwater by people, their livestock, agriculture, and industries, the similarly growing amounts of wastewater tend to decrease available safe water supplies for all consumers, often contributing to acute infectious and chronic noninfectious diseases, including cancer, neurological dysfunction, mutagenic effects, and birth defects (Wargo, 1996). Pollution of water resources also is impacting freshwater ecosystems, causing soil degradation and damage to natural vegetation and crop lands, loss of animal species, and in oceanic ecosystems devastating losses of marine life. There is growing consensus that human health and wellbeing is inextricably related to the health of the natural environment as reflected by the maintenance of biodiversity, and a basic tenet of the new science of environmental economics is that provision of adequate safe water for environmental protection and a healthy ecosystem is cost effective in the long term (United Nations, 2004).

Although Ethiopia's economy is predominantly based on subsistence agriculture, water pollution by chemicals and heavy metals from industries, including manufacturing and agro-industrial firms, is a potential threat to water security. In several widely distributed documents, Ethiopia, unlike Egypt, Kenya, Ghana, Nigeria, Uganda, and South Africa, is still depicted as a country subject primarily to soil degradation and water shortages, without significant industrial and agricultural pollution (Marcoux, 1998; *The World Factbook*, 2004). However, although only four African countries applied smaller amounts of pesticides per hectare of cropland than Ethiopia in 1999 (Center for Environmental Systems Research, 2001; World Bank, 2001), recent extensive water surveys in the Awash River Basin and in the most urbanized and industrialized region of the country around Addis Ababa revealed high levels of pollutants in streams and lakes that originated from industrial and agricultural sources (Agizew, 1999; Alemayehu, Walenigus, & Tadesse, 2003; Kumie & Kloos, 2006; Teklehaimanot, 2003). Population growth and economic development since the 1960s, low awareness of the health and ecological impacts of polluted water, and weak enforcement of anti-pollution regulations have exacerbated water pollution by agricultural and industrial enterprises in Ethiopia. The objective of this chapter is to review the occurrence of water pollution by chemicals and heavy metals from industrial and agricultural sources in Ethiopia within a development policy context and to examine some treatment and mitigation options.

Industrial Development in Ethiopia

The beginnings of the manufacturing industry in Ethiopia go back to the 1890s but grew slowly until the mid 1980s, after which half of the country's manufacturing plants were established. The 1991 war with Eritrea caused a temporary setback in industrial expansion, which resumed in 1993, and there were 909 medium and large manufacturing firms in 2001–2002 (figure 12.1). Private firms have become more numerous than public firms since the liberalization of the economy in 1992–1993, and the growth of the cottage and handicraft industry, numbering almost 900,000 undertakings in 1996, also accelerated (Central Statistical Authority, 2003a). The manufacturing industry is concentrated in the Addis Ababa metropolitan area, but factories are also becoming more numerous in provincial towns and on large commercial farms. In 2000–2001, 485 (61%) of the country's medium and large manufacturing firms were in Addis Ababa, most of them in the Akaki-Kaliti industrial center. In addition to leather, food, cotton milling, nonmetallic mineral, and textile industries, which also exist in smaller towns, Addis Ababa has most of the chemical, printing, metal, machinery, and equipment manufacturing industries in Ethiopia (Central Statistical Authority, 2002a, 2002b, 2003a, 2003b; Kumie & Kloos, 2006).

FIGURE 12.1. Number of manufacturing firms in Ethiopia between 1975/1976 and 2001/2002.

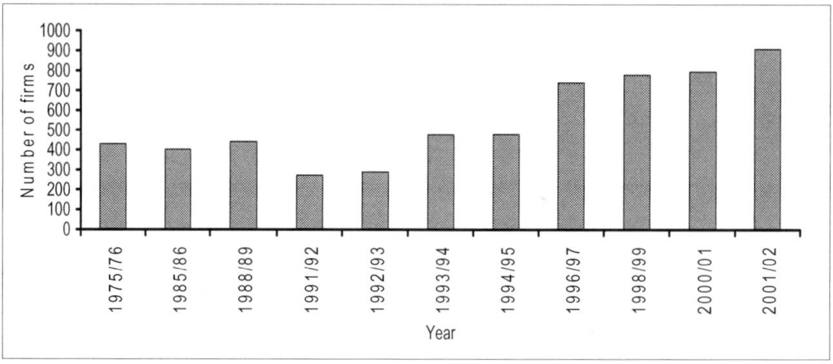

Source. Based on Environmental Protection Agency (2001b) and various Statistical Bulletins and Statistical Abstracts of the Central Statistical Authority, Addis Ababa.

In 2001–2002, 31.0% of the 909 manufacturing firms produced food and beverages, 11.0% produced nonmetallic mineral products, 9.6% produced iron and steel, metal products, machinery, and motor vehicles, 10.3% produced wood and paper products, 8.1% produced chemical products and plastics, 6.6% produced textiles, and 5.3% were involved in tanning and the manufacture of leather products. Most plants (41.3%) were small in size, employing 10–19 workers each, 25.3% were medium size (20–49 workers), and 33.4% were large size (50 or more workers) (Federal Democratic Republic of Ethiopia, 2004).

INDUSTRIAL WATER POLLUTION

Available data indicate that water pollution from industrial sources in Ethiopia is most intense in Addis Ababa and less so in smaller cities and in mining areas. The widespread practice of industrial and commercial firms to discard untreated liquid waste into streams, together with inadequate sewage systems for domestic waste, is causing an increasing number of streams, rivers, and lakes to become polluted. Little is known about the polluting potential and impact of the Ethiopian mining industry, but there is some evidence that it may be locally significant, particularly in gold mines (Japan International Cooperation Agency, 1999).

Addis Ababa

Most studies on water pollution in Ethiopia have been carried out in Addis Ababa and along the rivers flowing from this city into the Awash River. Illegal dumping of waste increased during the 1980s, and by the late 1990s, 90% of all factories engaged in this practice. Even those firms that reportedly treated water provided only preliminary treatment (Kumie & Kloos, 2006; United Nations Industrial Development Organization, 2001). The concentration of industries in Addis Ababa, together with the domestic waste and sewage generated by its population of 3 million, has resulted in a major surface and groundwater pollution problem in the metropolitan area and in the upper Awash River catchment. The limited capacity of the Addis Ababa sewage plant (built for only 50,000 people) can treat only a small fraction of the waste and sewage of Addis Ababa's population. This situation prevails in most African countries, where the construction of sewage plants has not kept pace with urbanization and industrialization (Qadir, Sharma, Bruggeman, Choukr-Allah, & Karejeh, 2007).

Tanneries, chemical firms, and beverage and food producers were the major polluters in terms of dissolved substances, nitrate, phosphate, ammonia, BOD_5 (biological oxygen demand, a measure of organic chemical pollution and a cause of oxygen depletion), and suspended solids. Tanneries also emitted high levels of ammonia, chromium, and lead that surpassed EPA's maximum concentration levels (MCL), and textile, metal, and nonmetal industries surpassed intermediate levels (table 12.1; United Nations Industrial Development Organization; 2001). High concentrations of nickel and chromium in subsurface soils and deep wells in the Akaki/Kaliti industrial area and the susceptibility of the local geology and hydrology to rapid circulation of pollutants indicate

TABLE 12.1. Mean concentrations of selected pollutants in effluents in Ethiopia in 2001, by industrial subsector, and provisional EPA industrial effluent standards.

Subsector	BOD_5	Ammonia	Nitrate	Phosphate	pH	SS*	Cadmium	Chromium	Lead
Tannery									
Beverage	913	46.1	4.8	18.5	8.2	1212.1	0.1	0.1	0.3
Food	565.8	44.4	17.1	31.0	8.5	2031.3	0.1	0.1	<0.1
Chemical	249.2	11.5	38.0	9.7	9.3	5905.8	0.2	0.2	0.7
Textile	45.8	0.7	4.1	2.0	9.1	70.3	<.1	<0.1	0.2
Metals and nonmetals	14.6	0.5	3.8	1.0	8.0	222.4	<0.1	<0.1	0.1
EPA effluent standards	30**	4.5	50	0.7	6.9	50	<0.1	<0.1	<0.1

Note. All values exce pt pH are in mg/L, *Suspended solids, **WHO Standard (World Health Organization, 1984).
Sources. Environmental Protection Agency (2001a, 2001c).

that metal and chemical pollutants are leached into the groundwater (Alemayehu et al., 2003; UNEP/UNESCO/UN-Habitat/ECA, 2003).

Spatial distribution of pollutants

Several surveys along the Little Akaki River in Addis Ababa revealed a strong relationship between the location of industries and levels of chemical and heavy metal pollution. Pollution levels were lowest in the Little Akaki River upstream of the city boundary, where no factories are located, highest in the Akaki/Kaliti industrial area in the southern part of Addis Ababa, and declined with distance further downstream, outside the urban area. Although some of the pollutants were below the maximum acceptable concentrations (MAC) suggested by WHO and the European Union, most chemicals, including BOD_5, COD (chemical oxygen demand, a indirect measure of the amount of organic compounds in water), and the nutrient-rich pollutants (ammonia, nitrite, and nitrate) exceeded the limits by wide margins at all sites in the inner city and in the industrial area downstream. Lead and chromium levels exceeded the standards two to three fold (table 12.2). Several other chemicals, including dioxin-like compounds, estrogenic compounds (Redda Teklehaimanot, 2003), silicon dioxide, and total dissolved solids (Beyene Addis, et al., 2009), were elevated but below MACs at several sites in Addis Ababa. Ali (2002) reported a similar distribution of pollutants in the Little Akaki River and an increase in their concentration during the dry season, when river levels are lowest. In the Awash River, which receives all runoff from the Addis Ababa area, the concentration of all the chemicals and heavy metals associated with manufacturing industries in Addis Ababa was below the maximum concentration limits (MCLs) and declined further downstream (table 12.2).

Pollution impacts

Water pollution in Addis Ababa has eliminated fish and nearly all other macroaquatic fauna in local rivers and streams. The spatial and temporal distribution of macroinvertebrates and diatoms in the Akaki, Little Akaki, and Kebena rivers in Addis Ababa were strongly correlated with

TABLE 12.2. Concentration of chemical pollutants and heavy metals in the three major rivers (Little Akaki, Akaki, and Kebena) in Addis Ababa and in the Awash River in the Rift Valley.

Pollutants	Concentration in Addis Ababa* rivers	Concentration in the Awash River**	Source	MCL***
COD	52–400 mg/L	28–944 mg/L	Redda Teklehaimanot (2003)	25 mg/L
BOD_5	98–1,250 mg/L	ND#	Beyene, Addis, et al. (2009)	30 mg/L
Nitrite	7.7–17.2 mg/L	0.6–4.3 mg/L	Redda Teklehaimanot (2003)	1 mg/L
Nitrate	7.7–81.4 mg/L	ND	Beyene, Addis, et al. (2009)	10 mg/L
Ammonia	5.6–19.0 mg/L	0.2–0.9 mg/L	Redda Teklehaimanot (2003)	1.5 mg/L
Ammonia	8.8–80.6 mg/L	ND	United Nations Industrial Development Organization (2001)	1.5 mg/L
Chromium	99 μg/L	6–40 μg/L	Redda Teklehaimanot (2003)	50 μ/L
Lead	30 μg/L	30 μg/L	Redda Teklehaimanot (2003)	10 μg/L

*Data are for the Akaki, Little Akaki, and Kebena rivers. Data from sampling sites at the headwaters of the rivers at the periphery of the city were excluded.
**Only river sampling sites downstream of Koka I Dam were included.
***Maximum concentrations limits are the EURO Directives (Reimann et al., 2003), WHO standards (World Health Organization, 1984), and the limits set for water bodies for agricultural use in China (Ma & Ortolano, 2000); #No data available.

the concentration of various pollutants. These aquatic organisms were absent from the most polluted, deoxygenated sections of these rivers, where only a few pollution-tolerant species can survive (Beyene, Addis, et al., 2009). The use of stream and lake water by urban and rural human and livestock populations and its effects on stream ecology downstream of towns extends the pollution problem far beyond the urban area (Redda Teklehaimanot, 2003). Aba Samuel Reservoir on the Akaki River about 25 km downstream of Addis Ababa has similar pollution levels that resulted in rapid eutrophication in recent years as indicated by the spread of water hyacinth and the extermination of fish (Gizaw, Legesse, Haddis, Deboch, & Birke, 2004). Similarly, a pond outside Jimma town, until recently a prime habitat of hippopotami, has experienced increasing eutrophication caused by a local soap factory and the municipal abattoir (Wubshet, 1998).

The public health impact of chemical river water pollution has not been assessed but may be considerable locally. Although the highly polluted rivers and streams within the Addis Ababa metropolitan area are not being used for drinking and cooking, contaminated surface and groundwater is widely used by poorer segments of the population for domestic, hygiene, and medicinal purposes and by small commercial growers for food production. This includes the use of numerous nitrate and coliform-contaminated springs, including holy springs (*tsebel*), for medicinal purposes by the population (Alemayehu et al., 2003; Kloos et al., 1987) and irrigation of vegetables using highly polluted river water (Itanna, 2002). An estimated 30% of the vegetables marketed in Addis Ababa come from small irrigation plots along the two Akaki rivers in Addis Ababa (Bekele, Tadesse, & Peden, 2003). The extent and potential health impacts of groundwater contamination are not known. Health risks associated with increasing reliance of the population of Addis Ababa and other towns on groundwater sources for human consumption found chemically and biologically contaminated (Endale, 1999; Tolke, 2007; UNEP/UNESCO/UN-Habitat/ECA, 2003) have also not been assessed.

Other Urban Areas

In none of the smaller towns do communal sewage plants exist. Recent studies reveal that water pollution in medium-sized towns impacts stream ecology and water quality in regional towns (Adem, Haddis, & Kassie, 2004; Beyene, Legesse, Triest, & Kloos, 2009; Endale, 1999;

Tilahun, 2004). Where wastewater treatment facilities exist, they are often inefficient, as at a brewery in Ilubabor Region (Tefera, 2001; Tilahun, 2004). Similarly, Gebre-Mariam and Desta (2002) found high level of heavy metal concentration in the effluent from a textile factory in Awasa Town and in the aquatic biota, which Adem (2004) attributed to the malfunctioning of the chemical rehabilitation ponds.

Significant industrial and domestic pollution impacts on aquatic macroinvertebrates similar to those in Addis Ababa were reported for rivers passing through Dessie, Debre Berhan, and Kombolcha towns in Amhara Region (Beyene Legesse, et al., 2009; Seyoum, 2007) and Jimma town in Oromia Region (Van der Bruggen, Smets, & Haddis, 2009). High BOD_5, COD, nitrate, nitrite, and ammonia pollution levels have been major factors in the eutrophication of the now defunct Aba Samuel Reservoir on the Akaki River downstream of Addis Ababa. In Lake Awasa, effluents from textile factories have been associated with high trace metal levels, particularly mercury, in predator species of commercial fish (Ataro, Wondimu, & Chandravansh, 2003) and fish mortality (Adem et al., 2004). The effects of the pollutants mentioned in this section on human health, including cancer, metemoglobinemia in infants, and teratogenic and mutagenic effects, have been widely described in the literature (Koren & Biscesi, 2003).

AGRICULTURAL WATER POLLUTION

The economic development policy of the Federal Democratic Republic of Ethiopia includes, among others, provisions for increased food and fiber production and to conserve, develop, and utilize natural resources in a sustainable manner (Ministry of Finance and Economic Development, 2002). Although the potential benefits of this program in terms of food and fiber production from expanded and intensified agricultural development are well articulated, its impact on environmental pollution remains to be assessed. One study indicated that environmental and public health impacts of pollution from agro-chemicals in the Awash Valley, the most intensively farmed area of Ethiopia, may be considerable (Redda Teklehaimanot, 2003).

Pesticide and Herbicide Use and Storage

Extensive irrigated agriculture confers considerable risk of water pollution by pesticides and herbicides in Ethiopia, similar to the situation in other African countries. State farms in Ethiopia, mostly large irrigation schemes, are the major user of pesticides, using 70% of the 1,100 tons applied annually. Nearly 70% of the pesticides used on the medium and large state farms were insecticides, 22.5% herbicides, 7% fungicides, and 0.7% rodenticides (Michael, 2002). Pesticides were also applied on 3.4 million ha of subsistence farms, and fertilizers were applied on 731,000 ha in 2001–2002 (Central Statistical Authority, 2004), but amounts applied were relatively small due to their high costs and farmers' lack of access to land, credit, and irrigation (Suleiman, 2001). Pesticides used include mostly chlorinated hydrocarbons and herbicides, which are being used by both rural and urban farmers in Ethiopia (Bizualem, 1999).

Agro-chemical water pollution in Ethiopia is most serious in the Awash Valley. The three commercial sugar plantations (Wonji, Wonji-Shoa, and Metahara) were found to emit untreated waste from their milling factories into the upper Awash River. Chemical pollution linked to farm operations (either sugar cane processing or pesticide application) was associated with episodes of fish mortality in the Awash River in the vicinity of one irrigation scheme in the upper Awash Basin (Redda Teklehaimanot, 2003). Jile, Kereyu, Arsi Oromo, and Afar pastoralists and subsistence farmers used Awash River water for all domestic purposes, including bathing, washing clothes, watering livestock, cooking, and drinking, as well as small-scale irrigation. Reports of acute illness and occasional deaths by local farm laborers on several irrigation schemes throughout the Awash Valley in the 1970s and early 1980s were associated with pesticide poisoning (H. Kloos, unpublished data).

The handling and storage of obsolete pesticides for agricultural use and malaria and locust control presents significant public health and environmental hazards (Wargo, 1996). Few users in Ethiopia have adequate knowledge about the toxicity, safe handling, and application of these chemicals, and most containers are reused for storage of water and household items, especially in rural areas. Storage of obsolete pesticides and herbicides in containers prone to leak in unsafe environments, such as humid buildings with dirt floors, exacerbates the risk of environmental contamination. According to the Ethiopian Deputy Minister of Agriculture and senior officials of the Food and Agriculture Organization of the United Nations, Ethiopia is among the countries affected most by obsolete pesticides in Africa, with hundreds of thousands of people estimated to be at risk. Clean up of the more than 2,800 tons of pesticides stored at 949 sites nation-wide has been hampered by financial constraints (Brough, 2003). Leakage of pesticides stored on commercial irrigation farms was observed in the mid-1970s in Awara Melka cotton farm (H. Kloos, unpublished data) and more recently in Melka Sadi banana plantation in the Awash Valley (Michael, 2002). Leakage of pesticides that are no longer used because of obsolescence or government restrictions, including DDT and malathion, has also been reported from smaller farms in the highlands, but little is known about their public health hazards and ecological impacts (Wassie, 2004).

Although the Ethiopian government has been shipping some of the obsolete pesticides abroad for detoxification (Michael, 2002), the high risk of pesticide poisoning conferred by generally poor management of these chemicals and use of canal and river water for human consumption in some less developed farms persists. This is further indicated by the low level of awareness of the health risks the handling of pesticides entail among the farm labor populations, including pesticide sprayers (Mekonnen & Agonafer, 2002).

Fertilizers

Various studies in Ethiopia (Gebre-Mariam, 1998) and other African countries (Vordzogbe, 2003) reveal that chemical pollution of rivers and lakes is due not only to pesticides and herbicides but also fertilizers, often from farms located in uplands far from large surface waters. Nitrogen in manure and artificial fertilizer is a major cause of eutrophication and ecological upsets in surface waters. Environmental studies of nitrogen in various countries (van Drecht, Bouman, Knoop, Meinardi, & Beuse, 2001) indicate that its movement through plants and groundwater and its impact on aquatic systems may vary widely in different localities in Ethiopia due to the role of local hydrology, climate, geology, soils, and land use.

Coffee processing waste

Although the cultivation and processing of coffee, the major export commodity of Ethiopia, continue to expand, little is known about the environmental pollution that has been associated with wet coffee processing (Chanakya & De Alwis, 2004). Coffee processing, involving removing the coffee hull and washing the green coffee beans, is a major source of stream and river pollution in the coffee growing areas of southwestern and southern Ethiopia, where hundreds of processing plants are discharging their untreated waste into rivers and streams. Significantly higher turbidity, ammonia, nitrate, phosphate, and BOD_5 (biological oxygen demand) and lower pH values were recorded in streams below than above processing plants in Jimma Zone and Sidamo Zone in the western and southern highlands, respectively. In Jimma Zone, mean turbidity, ammonia, nitrate, phosphate, and BOD_5 levels in the 10 streams tested downstream of coffee processing plants were above the effluent standards set by the EPA. Density and diversity of bottom-dwelling aquatic fauna were correspondingly poorer at the downstream sites (Endris et al., 2008). The use of these waters by local populations for their daily needs has been associated with intestinal, respiratory and skin diseases (Haddis & Devi, 2008). Various methods and techniques used in several other coffee producing countries, such as reducing water use, reusing wastewater, using water-efficient machinery, and preparing coffee waste for feeds, beverages, biogas, caffeine, protein, compost, and several other products, have not been applied in Ethiopia (Endris et al., 2008; Rathinavelu & Graziosi, 2005).

Pollution at the watershed level

Little is known about land use water quality relationships in developing countries, but there is increasing evidence that accelerated deforestation, intensified farming, livestock grazing, soil erosion, and land degradation are contributing to surface water pollution. One strong indication of agricultural water pollution is high nitrogen and phosphate levels found in springs in southern Ethiopia (McKenzie, Siegel, Patterson, & McKenzie, 2001). Moreover, silt and nutrient loads (including chemical fertilizers) and pesticide levels of runoff have been associated with increasing pollution levels and algal blooms in rivers draining such areas and in lakes in Ethiopia and other developing countries, especially in densely populated rural areas (Gebre-Mariam, 1998; Grepperud, 1996). Various s-triazine herbicides were identified in lakes Chamo and Abaya, although at levels below EU directives (Megersa, Solomon, Chandravanshi, & Jönsson, 2000).

Contamination of commercial fish and related public health effects will assume greater importance with increasing demand for this food source and plans to sharply increase fisheries in Ethiopian freshwaters. A recent plan of the Ministry of Agriculture calls for an increase in fish catches from 15,000 tons in 2004 to 100,000 tons in 2010. This development will require regulations of fisheries (now absent in eight of the nine regions) (Integrated Regional Information Network, 2004) and monitoring of synthetic chemicals in marketed fish.

Little information is available on groundwater contamination from agricultural sources in Ethiopia. A similar lack of information characterizes the monitoring and surveillance of aquifers in rural areas in other African countries (Xu & Usher, 2006). The rapid development of towns drawing their groundwater supplies in the vicinity of irrigation schemes in the Ethiopian lowlands warrants the attention of planners, geologists, and engineers. The detoxification of drinking water supplies from pesticides using carbon filtration systems, a preferred method for removing pesticides and other organic compounds from groundwater in industrialized countries, may not be affordable for Ethiopian communities owing to their high cost.

PREVENTION AND TREATMENT OF POLLUTED WATER

Efforts dealing with prevention of water pollution and treatment of industrial and agricultural wastewater are in their infancy in Ethiopia.

Prevention, by far the cheapest and effective approach to deal with water pollution if systematically implemented, requires collaboration and cooperation at all levels of society, from individual workers and farmers to managers of companies and farms, as well as government regulating organizations and planners. Changing prevailing practices toward lower use and discard of pollutants at the source, such as using fewer polluting chemicals and metals and recycling of water, have been recommended, but few measures have been implemented in Ethiopia thus far. This country participated in an eight-country program sponsored by the United Nations Industrial Development Organization (UNIDO) that aimed at industrial waste purification and the long-term problems associated with the generation and treatment of industrial wastewater (Mebratu, 1990), but no information is available on the outcome of this program.

Legislative Framework

The Ethiopian government has issued several policies and legislative frameworks dealing with various aspects of water pollution, management, and sustainable development issues. They include the Constitution of the Federal Democratic Republic of Ethiopia, the Environmental Policy (Environmental Protection Authority, 1997), and four proclamations dealing with water quality control, one of which established the EPA. A draft of the Industrial Pollution Prevention Regulation and Effluent Standards, which contains provisions for the prevention of pollution and restrictions on discharging industrial waste, was being prepared in 2008 by the EPA in consultation with various stakeholders (M. Ali, personal communication). This document contains provisions related to the prevention of pollution and restrictions on discharging industrial releases into the environment. As a signatory of the Stockholm Convention, Ethiopia has committed itself to control the use and release into the environment of highly toxic persistent organic pollutants (Federal Democratic Republic of Ethiopia, 2006).

Pollution in rivers crossing regional boundaries will require interregional cooperation and coordination of preventive and pollution control programs and activities to minimize downstream impacts. The problem of downstream pollution by upstream sources is illustrated by the Awash River, which receives high levels of chemical, heavy metal, and organic pollutants from industries and residences in Addis Ababa and agrochemicals from the commercial farms in the upper part of the Awash Valley before entering Afar Region (Teklehaimanot, 2003). While the existing policies and legislative framework provide comprehensive guidelines and regulations for the implementation of pollution prevention measures, they are rarely enforced due to administrative weakness and lack of political will (Kumie & Kloos, 2006), as evidenced by both the pollution of surface waters with urban areas and inadequate environmental and socioeconomic impact assessment of various dam projects (see chapter 9).

Community Participation

It is increasingly agreed that any efforts in water pollution remediation must focus on both the sources of pollution and water courses and aquifers receiving the contaminants and that both regular monitoring of pollution levels and awareness of and participation by communities are crucial for the success of remediation efforts. There are encouraging environmental policy and community organizational developments that may facilitate remediation efforts at the institutional, community, and regional levels. Ethiopia's environmental conservation policies are now calling for the active participation of local communities in management of water resources with the objective of increasing community inputs in identifying pollution problems and implementing and evaluating prevention and control measures and programs. The Ethiopian government started to embrace the concept of community participation in its environmental policies after the failure of the land use and conservation initiatives in the 1970s that used the top-down approach (Federal Democratic Republic of Ethiopia, 1997). Nevertheless, appropriate institutional frameworks and collaborative antipollution programs involving community, *woreda*, regional leaders, innovative education programs, and the communities themselves, as elsewhere in Africa (Makule, 2000; Vordzogbe, 2003), have not been developed in Ethiopia.

Potential Water Treatment Methods

A number of treatment methods for chemical pollutants that have been tested in Ethiopia or other African countries may be employed if the necessary technical and managerial capacities can be generated.

Recycling of wastewater and toxic components

This method represents an alternative to the typical single-use system of urban areas, characterized by discarding "waste" water into the urban drainage/sewage system or surface waters. Recycling reduces the pollution risk in downstream communities and converts waste into valuable materials, including fertilizer (Gumbo, 2000; Postel, 2002). It has been suggested that water from some polluted streams may be reused after aerobic or anaerobic treatment (Rivela, Mendez, Bornhardt, & Vidal, 2004). Recycling plants already operate in South Africa (Pillay, Friedrich, & Buckley, 2002) and Egypt, where the use of waste sludge and sewage to fertilize and irrigate soils was associated with higher cobalt levels in the treated soil (Abdel-Sabour, 2003), and in Morocco, where a chromium recycling plant designed by tanning firms and local utility companies was able to clean chromium from rivers and sell it to tanneries (Chemonics International Consortium, 2008).

Oxidation ponds and phytoremediation

Results from wastewater treatment field trials in African countries involving oxidation ponds and wetlands through phytoremediation technology show mixed results. Whereas wetlands near Cape Town in South Africa were able to retain and decrease the concentration of the pesticide chlorpyrifos in agricultural runoff (Moore, Schultz, Cooper, Smith, & Rodgers, 2002) and in experimental containers in Kenya (Abira, Nirigacha, & van Bruggen, 2003), wetlands may also become a source of contaminants by accumulating toxins in plants (Weis & Weis, 2004). Similarly, oxidation ponds at a pulp and paper mill in Kenya failed to reduce metal loadings of cadmium, copper, nickel, lead, and zinc below the limits set by EC for drinking water (Achoka, 2002). In a pilot-scale experiment in an advanced integrated wastewater pond system at a tannery near

Addis Ababa, high ammonia and organics removal and somewhat lower phosphorous removal efficiencies were reported, except for high loading conditions, which impeded removal rates (Tadesse et al., 2003).

Soil filtration

A simple, cost-effective method of treating polluted water through soil filtration has been developed and tested in Addis Ababa. Surface application of water from the Akaki River and the nearby wastewater treatment plant to the Akaki Well Field, an area of deep wells increasingly used by Addis Ababa to meet domestic water supplies, resulted in significant improvement in 15 of the 17 chemical and biological parameters tested. These beneficial results of this managed aquifer recharge method were attributed mainly to physico-chemical and biodegradation processes associated with the local vertisol soils (Abiye, Sulieman, & Ayalew, 2008). The widespread distribution of the typically deep vertisols in the Ethiopian highlands (see chapter 8) may make this treatment method particularly attractive for communities. Reductions between 65% and 85% were achieved for nitrate, nitrite, ammonia, and phosphate in a field experiment in a highland wetland near Jimma Town whose ecological function was, however, threatened by the misuse of this ecosystem (Mohamednur, 2007).

Bioremediation

Bioremediation is a developing technology that uses microorganisms to degrade contaminants into less toxic or nontoxic compounds or recover them for reuse, a promising technology in cleaning up waters and soils containing organic and inorganic contaminants and in recycling heavy metals. Due to the high cost and often slow results achieved with the more traditional remediation technologies, bioremediation has become a popular and cost-effective alternative (Kalogerakis, Alvarez, & Psillakis, 2005). The utilization of heavy metal and organic compound-reducing microorganisms is increasingly used in India and some other developing countries (Kamaludeen, Arankumar, Avudainayagam, & Ramasamy, 2003).

Other technologies

Recent solar experiments in Ethiopia resulting in the successful drying of high-performance starches from local plant products and the prospect of using them in making materials for sequestering heavy metals from polluted water represents a promising application of appropriate technology (Asfaw, License, Engida, & Poliakoff, 2007). These and other innovative wastewater treatment systems rather than conventional models that involve end-of-pipe treatment using costly chemicals may prove appropriate for Addis Ababa and other Ethiopian and African cities. The African 10-Year Framework Program on Sustainable Consumption and Production launched in Addis Ababa in 2006 is promising, as it calls for environmental, economic, and social sustainability of treatment systems that would include effective recycling of wastewater and improved production techniques, among others, (United Nations Economic Commission for Africa & United Nations Environmental Program, 2006).

Conclusion

In the competition for clean water supplies between growing industrial and agricultural needs on one hand and those of Ethiopian communities and ecosystems on the other, there is a real danger for the latter to lose in an environment of poverty, weak enforcement of water legislation, and a general lack of efforts at the community level to address this problem. Protection of water resources and other natural resources has traditionally received inadequate attention at all levels in Ethiopia and other African countries, where the first priority is to meet immediate, basic needs of the population, as noted in a recent United Nations (2004) report. Thus, while the recent water policy, proclamations, and initiatives may create the necessary legal framework and infrastructure, an environment conducive for the successful implementation of effective preventive and remedial programs will also require improvements in the standard of living, greater awareness of the pollution problem, and political will. Although manmade chemical water pollution is still fairly rare in most Ethiopian river basins and groundwater compared to the situation in more industrialized and urbanized developing countries, monitoring of all major human water sources is warranted as part of an effort to build capacity to face the greater pollution problems in the future. High pollution levels in several rivers in and around Addis Ababa and in several regional towns already far exceed the Ethiopian provisional standards proposed by EPA (Beyene, Addis, et al., 2009; Teklehaimanot, 2003), and pesticides, herbicides, and coffee processing also threaten water quality. The distribution, types, and intensity of pollution described in this review, together with continuing heavy dependence by the Ethiopian population on untreated surface water indicate that all towns with manufacturing industries and all commercial farms, both of the irrigation and rain-fed types, must be considered as potential sources of significant chemical pollution.

Further studies are needed to determine the fate of individual pollutants in surface water and groundwater in different localities, their bioaccumulation and impact on aquatic fauna and flora, stream, river, and lake ecosystems, and on public health, as well as effective and sustainable pollution prevention measures and wastewater treatment systems. The findings that the transport and deposition of agricultural fertilizers and pesticides in surface water supplies is linked with soil erosion and runoff and that industrial waste adversely affects large river systems such as the Awash over long distances downstream point to the diffusion of these chemicals far beyond their places of origin. A better understanding of their movements, fate, and broadly based ecological and public health impacts may contribute to integrated land use planning, management, and regulation of anthropogenic chemicals at the community and river and lake basin levels. There is also a need to carry out carefully designed cross-sectional and longitudinal epidemiological studies to examine the role of industrial and agricultural chemicals in the prevalence and incidence of acute and chronic noninfectious diseases. Because environmental toxins, singly and in combination, tend to enhance carcinogenic, teratogenic, and mutagenic processes and also interact with infection, biomedical research can effectively contribute to risk assessment (Kirsch-Volders, 1984). Parallel studies of the knowledge and perception of the presence and health impacts of pollutants in places of residence and work may provide a solid basis for the development of health education programs aimed at reducing exposure risk through adherence to accepted use, handling, and discarding of chemicals, as well as the use of protective clothing. The treatment of wastewater may contribute to reducing water scarcity and increasing water quality. Greater public awareness and enforcement of existing antipollution laws and further toxicological and ecological studies will be essential for the effective prevention and treatment of chemical pollution by industries and commercial farms.

REFERENCES

Abdel-Sabour, M. F. (2003). Impact of wastewater reuse on cobalt status in the Egyptian environment. *Journal of Environmental Science, 15*, 388–395.

Abira, M. A., Nirigacha, H. W., & van Bruggen, J. J. (2003). Preliminary investigation of the potential of four tropical emergent macrophytes for treatment of pre-treated pulp and paper mill wastewater in Kenya. *Water Science and Technology, 48*, 223–231.

Abiye, T. A., Sulieman, H., & Ayalew, M. (2008). Use of treated wastewater for managed aquifer recharge in highly populated urban centers: A case study in Addis Ababa, Ethiopia. *Environmental Geology, 58*, 55–59. doi: 10.1007/s00254-008-1490-y

Achoka, J. D. (2002). The efficiency of oxidation ponds at the Kraft pulp and paper mill at Webuye in Kenya. *Water Research, 3*, 1203–1212.

Adem, J. (2004). Assessment of the quality and environmental effects of wastewater from Awasa Textile Factory (Unpublished master's thesis). Jimma University, Jimma, Ethiopia.

Agizew, N. (1999). Evaluation of water quality and suggested improvement measures: The case of the Little Akaki River, Addis Ababa, Ethiopia (Unpublished master's thesis). Royal Institute of Technology, Stockholm, Sweden.

Alemayehu, T., Walenigus, S., & Tadesse, Y. (2003). *Surface and ground water pollution status in Addis Ababa, Ethiopia.* Report prepared for the Department of Geology and Geophysics, Addis Ababa University, and Addis Ababa Water Supply and Sewerage Authority.

Ali, M. (2002). *Industrial pollution and its impact on the Little Akaki River.* A research report submitted in partial fulfillment of the requirements for the master's thesis. Loughborough University, Loughborough, U.K.

Asfaw, N., License, P., Engida, T., & Poliakoff, M. (2007). Empowering green chemists in Ethiopia. *Science, 316*, 1849–1850.

Ataro, A., Wondimu, T., & Chandravansh, B. S. (2003). Trace metals in selected fish species from lakes Awasa and Ziway, Ethiopia. *SINET: The Ethiopian Journal of Science, 26*, 103–114.

Bekele, T., Tadesse, G., & Peden, D. (2003, May). *Community-based traditional irrigation schemes performance: A case study of Upper Awash River Basin of Addis Ababa sub-catchments.* Paper presented at the Workshop on Community-Based Irrigation Management at ILRI, Addis Ababa.

Beyene, A., Addis, T., Kifle, D., Legesse, W., Kloos, H., & Triest, L. (2009). Comparative study of diatoms and macroinvertebrates as indicators of severe water pollution: Case study of the Kebena and Akaki rivers in Addis Ababa, Ethiopia. *Ecological Indicators, 9*, 381–392.

Beyene, A., Legesse, W., Triest, L., & Kloos, H. (2009). Urban impact on ecological integrity of nearby rivers in developing countries: The Borkena River in highland Ethiopia. *Environmental Monitoring and Assessment, 153*, 461–476.

Bizualem, B. (1999). Assessment of knowledge, attitude and practice of residents toward the public health impact of pesticides in Jimma town and Geruke-Jimmate rural community (Unpublished undergraduate thesis). Jimma University, Jimma, Ethiopia.

Brough, D. (2003). Ethiopia says pesticide dumps are "time bomb." Report to Planet Ark, Reuters News Service. Retrieved from http://www.planteark.com/avantgo/dailynewsstory.cfm?newsid=10580

Center for Environmental Systems Research. (2001). *Water gap 2.1.* Kassel, Germany: University of Kassel.

Central Statistical Authority. (2002a). *Report on large and medium scale manufacturing and electricity survey in industries in March 2002* (Statistical Bulletin No. 205). Addis Ababa: Federal Democratic Republic of Ethiopia.

Central Statistical Authority. (2002b). *Report on large and medium scale manufacturing and electricity industries* (Statistical Bulletin No. 256). Addis Ababa: Author.

Central Statistical Authority. (2003a). *Report on cottage/handicraft manufacturing industries survey November 2002.* Addis Ababa: Federal Democratic Republic of Ethiopia.

Central Statistical Authority. (2003b). *Report on small-scale manufacturing industries survey December 2002.* Addis Ababa: Federal Democratic Republic of Ethiopia.

Central Statistical Authority (2004). *Statistical Abstract 2003.* Addis Ababa: Author.

Chanakya, H. N., & De Alwis, A. A. P. (2004). Environmental issues and management in primary coffee processing. *Process Safety and Environmental Protection, 8*(B4), 1–10.

Chemonics International Consortium. (2008, March 3). Moroccans secure precious natural resource. In *Water IQC II.* Washington DC: International Water Management Institute. Retrieved from http://ridgetoreef.com/en/Article.aspx?id=9

Endale, H. (1999). *Assessment of the polluting effects of wastewaters from Jimma Council Abattoir and Soap Factory on the nearby hand-dug well water sources* (Unpublished undergraduate thesis). Department of Environmental Sciences, Jimma University, Jimma Ethiopia.

Endris, S., Legesse, W., Kebede, T., W/Senbet, W., & Sualeh, A. (2008). Environmental contamination from by-products of wet coffee processing in southwestern Ethiopia: Current status and suggested management options. *SINET: The Ethiopian Journal of Science, 31*, 69–74.

Environmental Protection Authority. (1997). *Environmental Policy.* Addis Ababa: Federal Democratic Republic of Ethiopia.

Environmental Protection Agency. (2001a). *Industrial pollution prevention regulation* (Vol. 3). Addis Ababa: Author.

Environmental Protection Agency. (2001b). *Situation analysis: The industrial sector* (Vol. 1). Addis Ababa: Author.

Environmental Protection Agency. (2001c). Unpublished industrial effluent data. Addis Ababa: Author.

Federal Democratic Republic of Ethiopia. (2006). *National implementation plan for the Stockholm convention*. Addis Ababa: Author.

Gebre-Mariam, Z. (1998, February). *Emerging water management issues: Human interactions and water quality in the Horn of Africa*. Paper presented at the symposium of the American Association for the Advancement of Science, Science in Africa: Emerging Water Management Issues, the Advancement of Science in Africa: Emerging Water Management Issues, Philadelphia, PA.

Gebre-Mariam, Z., & Desta, D. (2002). The chemical composition of the effluent from Awassa Textile Factory and its effects on aquatic biota. *SINET: The Ethiopian Journal of Science, 25*, 263–274.

Grepperud, S. (1996). Population pressure and land degradation: The case of Ethiopia. *Journal of Environmental Economics and Management, 30*, 18–33.

Gumbo, B. (2000). Re-engineering the urban drainage system for resource recovery protection of drinking water. *Schriftenreihe über Verhandlungen der Wasser, Boden und Lufthygiene, 105*, 15–21.

Haddis, A., & Devi, R. (2008). Effect of effluent generated from coffee processing plant on the water bodies and human health in its vicinity. *Journal of Hazardous Materials, 152*, 259–262.

Integrated Regional Information Network. (2004, December). Ethiopia: ADF provides US $1.4 million to improve fish exploitation. *IRIN News*. Office for the Coordination of Humanitarian Affairs, United Nations.

Itanna, F. (2002). Metals in leafy vegetables grown in Addis Ababa and toxicological implications. *Ethiopian Journal of Health Development, 16*, 295–302.

Japan International Cooperation Agency. (1999). *Country profile on environment: Ethiopia*. Tokyo: Author.

Kalogerakis, N., Alvarez, P., & Psillakis, E. (2005). Special issue: Recent advances in bioremediation. *Environment International, 31*, 147.

Kamaludeen, S. P., Arankumar, K. R., Avudainayagam, S., & Ramasamy, K. (2003). Bioremediation of chromium contaminated environments. *Indian Journal of Experimental Medicine, 41*, 927–985.

Kirsch-Volters, M. (1984). *Mutagenicity, carcinogenicity, and teratogenicity of industrial pollutants*. New York: Plenum Press.

Kloos, H., Etea, A., Degefa, A., Aga, H., Solomon, B., ... & Belemo, G. (1987). Illness and health behavior in Addis Ababa and rural central Ethiopia. *Social Science and Medicine, 25*, 1003–1019.

Koren, H., & Biscesi, M. (2003). *Handbook of environmental health: Biological, chemical, and physical agents of environmentally related disease* (4th ed., Vol. 2). Boca Raton: CRC Press.

Kumie, A., & Kloos, H. (2006). Occupational health and industrial pollution. In Y. Berhane, D. Haile Mariam, & H. Kloos (Eds.), *The epidemiology and ecology of health and disease in Ethiopia* (pp. 171–195). Addis Ababa: Shama Books.

Ma, X., & Ortolano, L. (2000). *Environmental relations in China institutions: Enforcement and compliance*. Manham, MD: Rowman and Littlefield.

Makule, D. E. (2000). Pollution of water sources due to poor waste management—The city of Dar-es-Salaam. *Schriftenreihe über Verhandlungen der Wasser, Boden und Lufthygiene, 105*, 117–121.

Marcoux, A. (1998). *Population change, natural resources, environmental linkages in East and Central Africa*. Paper prepared for the Sustainable Development Department, Food and Agriculture Organization, Rome.

McKenzie, J. M., Siegel, D. I., Patterson, W., & McKenzie, D. J. (2001). A geochemical survey of spring water from the main Ethiopian rift valley, southern Ethiopia: Implications for well-head protection. *Hydrology Journal*, *9*, 265–272.

Mebratu, D. (1990, October). *Program on purification of industrial waste water. Country paper: Ethiopia*. Paper presented at the regional meeting on Management of Industrial Waste Water, United Nations Industrial Development Organization, Paris.

Megersa, N., Solomon, T., Chandravanshi, B. S., & Jönsson, Å. (2000). Sample clean-up enrichment and determination of s-triazine from southern Ethiopia using supported liquid membrane extraction. *Bulletin of the Chemical Society of Ethiopia*, *14*, 9–24.

Mekonnen, Y., & Agonafer, T. (2002). Pesticide sprayers' knowledge, attitude and practices of pesticide use on agricultural farms in Ethiopia. *Occupational Health, 52*, 311–315.

Michael, B. (2002). *Health status of farm workers serving in selected state farms of agricultural development organizations in the main rift of Ethiopia* (Unpublished manuscript). Department of Biology, Addis Ababa University, Addis Ababa.

Ministry of Finance and Economic Development. (2002). *Sustainable development and poverty reduction program*. Addis Ababa: Federal Democratic Republic of Ethiopia.

Mohamednur, B. (2007). *Assessment of pollution filtering capacity of natural wetland from Aweytu and Kitto streams* (Unpublished master's thesis). Addis Ababa University, Addis Ababa.

Moore, M. T., Schultz, R., Cooper, C. M., Smith, S., Jr., & Rodgers, J. H., Jr. (2002). Mitigation of cholopyrifos runoff using constructed wetlands. *Chemosphere*, *46*, 827–835.

Payet, R., & Obura, D. (2004). The negative impacts of human activities in the eastern African region: An international water perspective. *Ambio*, *33*, 24–33.

Pillay, S. D., Friedrich, E., & Buckley, C. A. (2002). Life cycle assessment of an industrial water recycling plant. *Water Science and Technology*, *46*(9).

Postel, S. (2002). Rivers of life: The challenge of restoring health to freshwater ecosystems. *Water Science and Technology*, *45*, 3–8.

Qadir, M., Sharma, B. R., Bruggeman, A., Choukr-Allah, R., & Karejeh, F. (2007). Non-conventional water resources and opportunities for water augmentation to achieve food security in water-scarce countries. *Agricultural Water Management*, *87*, 2–22.

Rathinavelu, A., & Graziosi, A. (2005). *Potential alternative use of coffee wastes and by-products.* Trieste: International Coffee Organization.

Reimann, C., Bjorvatn, K., Frengstad, B., Melaku, Z., Tekle-Haimanot, R., & Siewers, U. (2003). Drinking water quality in the Ethiopian section of the East African Rift Valley. I. Data and health aspects. *Science of the Total Environment, 311*, 65–80.

Rivela, B., Mendez, R., Bornhardt, C., & Vidal, G. (2004). Towards a cleaner production in developing countries: A case study in a Chilean tannery. *Waste Management Research*, *22*, 121–141.

Seyoum, E. (2007). Assessment of the physicochemical parameters of River Beressa in Debre Berhan Town (North Shoa) for suitability of drinking water. Unpublished master's thesis, Addis Ababa University, Addis Ababa.

Suleiman, A. (2001). Supply response of Ethiopian farmers to price and non-price factors: A micro-economic analysis. *Proceedings of the EAF International Conference on Contemporary Development Issues in Ethiopia* (Vol. 1, pp. 186–201). Kalamazoo, MI: Center for African Development Policy Research, Western Michigan State University at Kalamazoo.

Tefera, W. (2001). *Treatment process and efficiency of liquid waste treatment system of Bedele Brewery, southwestern Ethiopia* (Unpublished undergraduate thesis). Jimma University, Jimma, Ethiopia.

Teklehaimanot, R. R. (2003). *A literature based study into the expected pollution of the Awash River, Ethiopia (TOX-80316) and Preliminary* HKWL: change to *"(TOX-80316): Preliminary" toxic risk evaluation of the Awash River, Ethiopia (TOX-80328)* (Unpublished master's thesis). Wageningen University, Wageningen, The Netherlands.

Tilahun, T. (2004). *Physicochemical and biological characterization of Sorba Stream and Elike River: A tool for impact assessment of Bedele Brewery effluent discharge, southwestern Ethiopia* (Unpublished undergraduate thesis). Jimma University, Jimma, Ethiopia.

Tolke, A. A. (2007). *Influence of uncontrolled land use and urbanization on the hydrochemistry and contamination of groundwater from Wolaita Soddo town and its environ, south Ethiopia.* Paper presented at the 2007 Lake Abaya Research Symposium, Arba Minch University, Arba Minch, Ethiopia.

UNEP/UNESCO/UN-Habitat/ECA. (2003). *Groundwater vulnerability mapping of the Addis Ababa water supply aquifers, Ethiopia.* Report prepared by Addis Ababa University, Department of Geology and Geophysics, and Addis Ababa Water and Sewerage Authority, Addis Ababa.

United Nations. (2004). *Water and the environment. United Nations Series Hydrology for the Environment, Life and Policy.* New York: Author.

United Nations Economic Commission for Africa and United Nations Environmental Programme. (2006). *Report of the High-Level Launch for the African 10-Year Framework Programme on Sustainable Consumption and Production.* Report of the Fourth African Roundtable on Sustainable Consumption and Production. Addis Ababa: Author.

United Nations Industrial Development Organization, Integrated Programme for Private Sector Development-Enhanced Competitiveness and Environmentally Friendly Production. (2001). Ecologically Sustainable Industrial Development Project-US/ETHG/99/068 Ethiopia (Vol. 1). *Situational analysis of the industrial sector*. Addis Ababa: Author.

Van der Bruggen, B., Smets, I. Y. & Haddis, A. (2009). Impact of wastewater discharge in Jimma, Ethiopia, and remediation possibilities. *Desalination, 248*, 603–609.

van Drecht, G., Bouman, A. F., Knoop, J. M., Meinardi, C., & Beuse, A. (2001). Global pollution of surface waters from point and nonpoint sources of nitrogen. *Scientific World Journal*, *6*, 632–641.

Vordzogbe, S. D. (2003, December). *Managing water risks in Africa.* Paper presented at the Pan-African Implementation and Partnership Conference on Water, United Nations, Economic Commission for Africa, Addis Ababa.

Wargo, J. (1996). *Our children's toxic legacy: How science and law fail to protect us from pesticides.* New Haven and London: Yale University Press.

Wassie, F. (2004). *Soil pollution by obsolete pesticides in Ethiopia* (Unpublished master's thesis). Gent University, Gent, Belgium.

Weis, J. S., & Weis, P. (2004). Metal uptake, transport and release by wetland plants: Implications for phytoremediation and restoration. *Environment International*, *30*, 685–700.

World Bank. (2001).World Development Indicators. Washington, DC: Author.

The World Fact Book. (2004). Field listing-environment-current issues. Retrieved from http://www.brainyatlas.com/fields/2032/fml

World Health Organization. (1984). *Guidelines for drinking water quality* (2nd ed.). Geneva: Author.

World Resources Institute, United Nations Environmental Program, United Nations Development Program, & World Bank. (1998). *World resources 1998–99: Environmental change and human health*. Washington, DC: Authors.

Wubshet, M. (1998). *Assessment of knowledge, attitude and practice of residents towards the public health impact of pesticides in Jimma town and Geruke-Jimmate rural community* (Unpublished undergraduate thesis). Jimma University, Jimma, Ethiopia.

Xu, Y., & Usher, B. (Eds.). (2006). *Groundwater pollution in Africa*. London and Leiden: Taylor & Francis.

CHAPTER 13

WATER-RELATED, VECTOR-BORNE, AND GEOCHEMICAL DISEASES

Helmut Kloos and Worku Legesse

Water resources development may have both positive and negative impacts on human health. On the positive side, irrigation can play a major role in increasing agricultural productivity, reducing poverty, and sustaining rural livelihoods. Similarly, adequate supply of potable domestic water can help maintain sanitation and hygiene standards that prevent a wide range of water-borne and water-washed diarrheal, skin, and eye diseases, poliomyelitis, typhoid fever, and several types of hepatitis and intestinal infections. On the other hand, the development of water resources for irrigation and hydropower generation often creates habitats for insect vectors transmitting the water-related diseases such as malaria, dengue fever, onchocerciasis, lymphatic filariasis, and yellow fever, as well as aquatic intermediate hosts transmitting schistosomiasis, Guinea worm infection, and fascioliasis, the latter of which infects both humans and livestock (Erlanger et al., 2005; Hunter, Rey, Chu, Adekolu-John, & Mott, 1993; Mara & Feachem, 1999). The risk of the transmission of soil-transmitted helminthiases, including hookworm infection and trichuriasis, has also increased in many irrigation schemes and dam sites because of the congregation of large numbers of people, which increases the risk of soil contamination with human excreta, and the increase in soil moisture, which enhances the chances of survival of the larval stages (Alemayehu et al., 1998; Kloos, DeSole, & Lemma, 1981).

Climate change appears to be increasing the health risks associated with water resources and their development. Increases in climate-sensitive infectious diseases, including malaria, dengue fever, cholera, and diarrhea, have been reported from various developing countries. Drought-induced crop failures resulting in malnutrition during famines tend to reduce host resistance to infections, cause the spread of disease agents through human migrations, and deteriorate health services in affected areas (Khasnis & Nettleman, 2005). A recent WHO study estimated that global warming since the 1970s was causing over 150,000 excess deaths annually by 2000 (Integrated Regional Information Network, 2008b). Closer collaboration between climate modelers and health planners may provide the concrete evidence needed to reliably determine the magnitude of the climate/disease link and to best inform policy makers (Huntingford, Hemming, Gash, Gedney, & Nuttal, 2007).

The transmission potential of vector-borne diseases in water projects and their control is best illustrated by the interaction of the three components of their generalized life cycle: the disease agent, the vector (or intermediate host), and the human host. The survival, density, and seasonal occurrence of mosquito vectors and snail intermediate hosts and, thus, transmission patterns, are largely determined by the presence of suitable surface waters for breeding. Mosquitoes and snails occur widely in the African environment and rapidly invade and breed in reservoirs and irrigation canal systems, which constitute particularly suitable habitats. These water bodies, thus, provide many opportunities for the transmission of malaria by mosquitoes and of schistosomiasis by aquatic snails. Among the numerous mosquito species and aquatic snails endemic in Africa, species able to transmit human parasites have become well adapted to many reservoirs and irrigation systems (Hunter et al., 1993).

The magnitude of the health risk of dams and irrigation schemes is indicated firstly by the fact that an estimated 40,000 large dams and 800,000 small dams had been built and 272 million ha of land were under irrigation worldwide by 2004. Secondly, two of the most important vector-borne diseases—malaria and schistosomiasis—tend to become highly prevalent around water development projects. Nearly 10 million people live near large dams or irrigation schemes in the hyperendemic region, with 87.9% of the global malaria burden (Keiser et al., 2005). Most information on health impacts of irrigation schemes comes from large commercial farms, but, contrary to widespread belief, community-based and small-scale irrigation schemes are also affected by environmental and health problems (McCartney, Boelee, Cofie, & Mutero, 2007).

Comprehensive planning, implementation, and efficient, ecologically sensitive operation of water resources projects can prevent or at least help to mitigate the occurrence and impact of water-related diseases, both around manmade lakes and in irrigation schemes (Hunter et al., 1993; Jobin, 1999), and inadequate consideration of environmental and health issues is now widely associated with project failure (McCartney et. al., 2007). In spite of the magnitude of dam-related health problems, the 2000 World Commission on Dams report allocated only 2 of its 448 pages to health. The failure to include resources for the provision of health services in many plans for dam construction also indicates that dams are still not considered as investments in health, contrary to the concept of "human capital," which "includes health as an input to all economic development" (Sleigh & Jackson, 2001, p. 570).

The objectives of this chapter are to summarize major negative health impacts of water resources development in Africa, to examine the role of different types of water resources development programs and activities in Ethiopia in the transmission of endemic vector-borne and geochemical diseases, and to evaluate some preventive strategies. The focus will be on dam and irrigation development, with preliminary discussions of health issues of resettlement projects and the less well understood wetlands. Chemical and organic water pollution from industrial, agricultural, and domestic sources are discussed in chapter 12, and bacterial and viral intestinal diseases are discussed in the book by Berhane, Haile Mariam, and Kloos (2006).

HEALTH ASPECTS OF WATER RESOURCES DEVELOPMENT IN AFRICA

Increases in the prevalence of schistosomiasis, malaria, lymphatic filariasis, dracunculiasis, and chronic respiratory and diarrheal diseases in irrigation schemes are reliable indicators of the human-induced environmental and ecological changes taking place in these systems because all these diseases are carried solely by humans, without animal reservoir hosts being involved. A combination of population dynamics (heavy influx and seasonal migration of poor and typically infected farm labor populations), working conditions (intense contact with soil and water), living conditions (inadequate water and sanitation facilities and overcrowding in labor camps), and inadequate health services that often prevail in irrigation schemes facilitate the transmission of schistosomiasis, malaria, lymphatic filariasis, soil-transmitted intestinal parasites, diarrheal disease, tuberculosis, and other infectious diseases, which in turn exacerbate the incidence of nutritional disorders and result in a consequent increase in susceptibility to infection. These conditions have been reported from most African countries developing land for irrigation or hydropower projects. The effectiveness of control programs is diminished because national disease control programs in Africa most often target only malaria, while few of them target schistosomiasis, filariasis, or the intransigent diarrheal diseases, and intersectoral collaboration is usually inadequate (Hunter et al., 1993).

Dams, Irrigation Projects, and Wetlands

The ecological changes normally occurring in tropical countries after the completion of manmade lakes and irrigation systems include a wide range of aquatic, edaphic, biological, and human population changes in and around these projects which usually result in the transmission and spread of vector-borne diseases in the absence of adequate prevention and mitigation efforts. These relationships influence the spatial and temporal interaction between vector, parasite, host, and physical and human environment, making disease risk place-specific but difficult to predict within endemic areas. Environmental degradation and health risks may appear even before ecological changes have been recognized and preparations have been made to overcome them. The rate of ecological change and the severity of health impacts in different areas vary considerably and are not well understood but are often related to degradation of the environment, hydrological changes, the technical characteristics and size of projects, and the socioeconomic status and economy of the local population intersecting with water use. Poor, rural agricultural people heavily depending on the newly created water bodies and the influx of migrant populations tend to result in rapid spread of vector-borne diseases in the absence of appropriate engineering and adequate preventive and curative health services. Large dams not only provide the services stated in project protocols, including hydropower production, flood control, and dry-season water supply for commercial irrigation schemes but also often encourage subsistence farmers to irrigate their plots and fishermen to settle around reservoirs. Many dams and irrigation systems attract large populations from other areas who may introduce and spread water-related diseases in the absence of adequate health services. As stated above, the various ecological changes and health impacts associated particularly with dam construction are not confined to the immediate areas around new lakes and dam sites, but impact surrounding areas, especially downstream communities (Hunter et al., 1993; McCartney et al., 2007).

Natural wetlands, particularly swamps and marshes, have traditionally been viewed by governments in industrialized countries as useless areas ridden with malaria and other infectious disease and were consequently drained and developed to agriculture and settlement (Kitron & Spielman, 1989). It is now accepted, however, that wetlands are highly productive, economically and ecologically important ecosystems and support large human and animal populations in developing countries (see chapter 2). There is increasing evidence that control of mosquito vectors can be achieved by preserving and wisely managing wetlands without damaging environmental modifications (Zimmerman, 2001).

In order to prevent water-related diseases from spreading in water development projects, ecological and health impact studies should be carried out and, more importantly, acted upon prior to the implementation of such projects. Studies of the health effects of water development schemes in the tropics were first carried out in the early 1970s (Hubendick, 1973), but few comprehensive studies have been carried out in Africa, due to economic, political, and administrative factors, including lack of transparency of planning procedures and transactions (Hunter et al., 1993; McCartney et al., 2007).

The Need for Health Risk Analyses

Few detailed health risk analyses of water development projects and no comprehensive, longitudinal health impact analyses within the context of social and environmental assessments have been carried out in Africa. One of the few studies quantifying the risk of water resources development on vector-borne disease transmission found a significantly higher risk for schistosomiasis among people living in irrigation schemes in 24 African countries (Steinmann, Keiser, Bos, Tanner, & Utzinger, 2006). To capture the full burden of water-related diseases, longitudinal assessments should also consider the secondary health impacts of many water-related infectious diseases, such as malaria causing haemolytic anemia and Burkitt's lymphoma, urinary schistosomiasis increasing the risk of bladder cancer, and diarrhea exacerbating malnutrition (Mara & Clapham 1997; Mara & Feachem, 1999).

As an outcome of the long-standing debate over the trade-offs in development, ecology, and health, governments and nongovernmental organizations in Ethiopia and elsewhere are increasingly interested and also required by donors to include environmental impact studies with the objective of preventing adverse effects or mitigating them if and when they arise. Any preventive and curative interventions need to be included in development projects from the beginning. In spite of the fact that these guidelines have been set out for the last three decades and in spite of the rhetoric, little has come out of the prescribed planning models and intersectoral collaborations (Brewster, 1999; Hunter et al., 1993), which tend to be blessed by everyone and funded by no one (Millenium Ecosystem Assessment, 2005). Existing and feasible integrated disease prevention and control approaches have been described for numerous dam and irrigation projects in Africa by Jobin (1999), Hunter et al. (1993), and McCartney et al. (2007).

A rigorous, comprehensive evaluation of health risks requires the inclusion of both positive and negative factors in health impact studies undertaken among well-defined indigenous and migrant population groups at different times starting from the planning stage. This would have to include examination of causal relationships of environmental and human impacts at the project, river basin, or regional levels in comparison to alternative water development scenarios (McCartney et al., 2007). Such evaluations and health impact assessments in general lack the capacity and capability to manage the need for health impact assessment building, prompting WHO to start developing comprehensive capacity methodology for development projects (Danish Bilharzia Laboratory, 2007a, 2007b).

The Neglect of Downstream Impact Assessments

An important but often neglected aspect of the construction of large dams is their health impact on downstream and relocated populations due to changes in river and wetland ecology. Downstream impacts have been unrecognized, misunderstood, and underestimated by planners, largely because they occur in remote areas, far from dam sites, causing them to be easily ignored. Even when recognized, downstream impacts are daunting in their complexity (Adams, 2000). Downstream impacts of dams are complex because of the dynamic changes caused by changing river flow regimes on both the biophysical and human environments. Another major factor in their neglect is that impacted sites are usually outside the dam project and planning areas and typically difficult to reach, rendering the downstream places of marginal political importance. Economic, social, and health impacts tend to be closely linked and self-reinforcing, often reducing agricultural, pastoral, and fish production, exacerbating poverty, food insecurity, poor health, and culture loss and resulting in out-migration (Adams, 2000).

Examples of Health Impacts in Africa

Several studies of health impacts carried out around and downstream of large manmade lakes have reported infectious and nutritional diseases and mental illnesses. As a result of water impoundment, abstractions, and drainage of wetlands, new vector habitats and disease transmission sites are often created after rivers change flow regimes.

Aswan Dam

The ecology of the two forms of schistosomiasis endemic in Egypt significantly changed after the construction of Aswan Dam. The elimination of the seasonal flood and high silt load downstream of the dam promoted the spread of the aquatic snail host of the *Schistosoma mansoni* parasite in irrigated areas and resulted in its establishment for the first time in the Nile downstream of the dam. Snail host populations of *S. haematobium*, on the other hand, decreased in the Nile (El Khoby et al., 2000). These changes underscore the need to study the ecology of vector-borne diseases in water projects.

Kariba Dam

The longitudinal (and ongoing) anthropological and human ecological studies by Scudder and Colson (1982) and Colson (1971) on the impact of Kariba Dam, constructed in the late 1950s on the Zambezi River, on the lives of the indigenous Gwembe Tonga people describe in detail the numerous health and social problems they experienced during and after their forced relocation to infertile areas surrounding the lake. The dam was built for hydroelectric production only, without consideration of its social and health impacts on the 57,000 displaced Gwembe Tonga and on wildlife, the ecology of the new lake, and river and delta fisheries; only hydropower production, irrigation agriculture, and tourism have been benefits of the dam (Scudder, 2005; Soils Incorporated, 2000). Malaria and schistosomiasis spread rapidly and severely afflicted the population living around the reservoir. Sociocultural stress and malnutrition among the resettlers also were surprisingly common. Moreover, the spread of tsetse flies flushed out of their riverine habitat by the reservoir and consequent cattle mortality were most severe after resettlement and could only be reduced over the years through adaptation and sustained government-sponsored control programs. Ecological impacts in the lake in the form of invading aquatic weeds, deoxygenation of lake water caused by the submergence and decomposition of forests and consequent fish kills in the lake, and significant downstream reduction in fisheries were more persistent, and new dam-related health, social, and ecological problems have arisen in recent years (Colson, 1971; Hira, 1970; Scudder & Colson, 1982). In particular, poorly planned water releases from Kariba Lake and the newer downstream Cahora Bassa Dam during the rainy seasons in 2002 resulted in severe flooding causing the displacement of more than 400,000 people and a number of deaths on the flood plains of the lower Zambezi (Scudder, 2005). Releases from Cahora Bassa during heavy rains in early 2008 caused renewed flooding and the resettlement of 95,000 people, with another 200,000 reluctant to leave their homes at risk (Integrated Regional Information Network, 2008a).

Akosombo Dam

Lake Volta in Ghana, created after the completion of Akosombo Dam in 1964, has been associated with health problems similar to those at Kariba Dam. Although negative health predictions had been made and a resettlement authority established, both the resettled population and the Ewe fisherman settling around the lake were severely afflicted with malaria, schistosomiasis, diarrheal diseases, and sleeping sickness (Jobin, 1999).

VECTOR-BORNE DISEASES AND WATER RESOURCES DEVELOPMENT IN ETHIOPIA

Failure to follow the national guidelines and recommendations and those of the World Commission on Dams, among other reasons, was recently identified during the construction of the Gilgel Gibe Hydroelectric I, II, and III Hydroelectric Projects in Ethiopia. The government's attempt to fast track these and other water development projects in its Master Plan and weakness of the Ethiopian regulatory agencies (Environmental Protection Agency and the Ministry of Water Resources) have been associated with this noncompliance (Campagna per la riforma della Banco Mondiale, 2008).

Malaria

Of all vector-borne diseases endemic in Ethiopia, malaria has traditionally had the greatest impact on the health and socioeconomic fabric of the population. Over 5 million episodes of malaria are estimated to occur in Ethiopia annually. About three quarters of the country's total area is malarious, with about two thirds of the population at risk of infection (Adhanom, Deressa, Witten, Getachew, & Seboxa, 2006). These estimates were based on altitude, rainfall, and the history of malaria occurrence, which is endemic at altitudes below 2,000 m but with occasional epidemic outbreaks at higher altitudes up to 2,400 m. Mean annual temperatures above 2,500 m are 10–15°C, too low for development and survival of the *Plasmodium* parasite (Tulu, 1993). The incidence and altitudinal occurrence of malaria transmission in the Ethiopian and East African highlands are expected to shift to higher elevations due to increasing temperatures linked to global warming (Boko et al., 2007; Lindblade, Walker, Onapa, Katungu, & Wilson, 1999). In the Ethiopian highlands, four consecutive abnormally high transmission seasons were recorded between 2002 and 2005, with explosive epidemic malaria occurring in 2003 up to 2,500 m altitude (Negash et al., 2005). Malaria epidemics among highland populations can result in particularly high morbidity and mortality rates because of their lack of acquired immunity

(Lindblade et al., 1999). Climatic change is also likely to result in the extension of the areas of endemic schistosomiasis, trypanosomiasis, onchocerciasis, Bancroftian filariasis, and yellow fever, all of which were traditionally absent from the Ethiopian highlands between 1,800–2,000 m (Berhane et al., 2006). These evolving transmission patterns put into question the suggestion to locate water development schemes in hitherto "safe" highland areas (Ghebreyesus et al., 2002).

A review of the literature shows that malaria epidemics may also be caused by unusually heavy rainfall episodes, massive environmental changes due to deforestation, flooding, influx of nonimmune populations into endemic areas, irrigation projects, high incidence of other diseases or poor nutritional status, interruption of effective surveillance and control, and parasite resistance to drugs (Guintran, Delacollette, & Trigg, 2006). Most of those factors were identified during the major Ethiopian highland epidemics, including the one in 1958, which resulted in an estimated 3 million cases and 150,000 deaths (Fontaine, Najjar, & Prince, 1961) and the epidemic in 2003 (Negash et al., 2005).

In the lowlands below about 1,500 m, malaria transmission is seasonal as determined by rainfall but continuous around irrigation systems and other permanent water sources (Tulu, 1993), reemphasizing the transmission risk in water development schemes. *Plasmodium falciparum* is the main malarial parasite in Ethiopia, followed by *P. vivax*, with sporadic reports of *P. malariae*. *Anopheles arabiensis,* the most important vector in Ethiopia, breeds in both temporary pools and permanent water bodies. *A. pharoensis* and *A. funestus* are also widely distributed. Transmission is largely seasonal and unstable, characterized by wide fluctuations in the monthly number of cases. The main malaria transmission season follows the major *keremt* rains from June through September, and a smaller transmission season follows the small rains *belg* of February to March. Year-round transmission occurs only in the lowlands around permanent water bodies (Adhanom et al., 2006). The creation of reservoirs and irrigation farms in the Rift Valley and other areas created conditions conducive for year-round breeding of *A. arabiensis* and malaria transmission. Labor migrations among irrigation schemes in the Awash

Valley and other endemic localities, poor water management practices, and failure to screen seasonal laborers for malaria further facilitated its spread (Haile Meskal & Kloos, 1989).

Lake Koka, storing irrigation water for the Awash Valley farms, and the recently impounded Gilgel Gibe I Reservoir on a tributary of the upper Omo River have become malaria transmission sites (Lautze et al., 2007; Yewhalaw, Legesse, Gebre-Selassie, & Kloos, 2009). The influx of large numbers of seasonal farm laborers from the highlands and lack of adequate health services and safe domestic water supplies, irrigation inefficiency, and failure to screen seasonal laborers for parasitic diseases resulted in the spread of malaria and schistosomiasis in the Awash Valley downstream of the Koka dams (Haile Meskal & Kloos, 1989; Kloos et al., 1981). Other infectious diseases can also be introduced by dam workers to water development projects. Abayou (2007) reported increasing HIV risk around the Gilgel-Gibe II Hydroelectric Project associated with the presence of laborers from other areas and increasing prostitution around the project site.

The construction of small dams and irrigated plots as part of the Ethiopian government's strategy to eliminate food insecurity prompted a number of epidemiological studies, especially in Tigray Region. Increased malaria risk in several small-scale irrigation systems and microdams in Tigray Region was associated with leaking irrigation canals, seepage at the base of dams, and manmade pools (Adhanom et al., 2006; Behailu, Abdulkadir, Mezgebu, & Yasin, 2004; Yohannes et al., 2005). Malaria incidence rates among children in six other villages in Tigray located near small dams were highest in poor households irrigating land (Ghebreyesus, Hail, et al., 2000).

One of the few studies of the impact of wetland development and vector-borne diseases in Ethiopia associated increasing malaria incidence and diffusion in the Gambela area in the 1980s and 1990s with drainage and cultivation of the wetlands, unexpected rainfall patterns, temperature increase, settlement and land use changes, political insecurity, and poverty (Woube, 1997). Farmer populations from the malaria-free highlands in central and northern Ethiopia settling during the 1983–1984 famine in these wetlands experienced particularly high mortality (Tesfaye & Tsehay, 1987). In Uganda, malaria rates and mosquito densities were higher in communities cultivating wetlands than in villages near natural swamps, a finding that was associated with higher temperatures in the latter (Lindblade, Walker, Onapa, Katungu, & Wilson, 2000) and which may have implications for wetland development in Ethiopia. Similarly, mosquito densities increased sharply in deforested areas that were subsequently farmed (Yasuoka & Levins, 2007), an important malaria risk factor on flood plain wetlands with riparian forests that are cut and developed to irrigate agriculture (Dixon, 2003; Woube, 1997). These two studies further argued for the maintenance and preservation of forests and wetlands.

Relatively few malaria-infected persons in the endemic areas obtain medical treatment, causing much debilitation. One study in a rural village in western Ethiopia reported that malaria reduced household income by 25% and 45% (Bedri Kello, 1995). Around a small town in central Ethiopia, malaria accounted for 26% of illness during a two-year recall period (Abdulahi, Haile Mariam, & Kebede, 2001). In households close to small dams in Tigray Region, lost production time due to malaria and caring for the sick cost these households 150–250 Birr per season (Amacher, Ersado, Grebner, & Hyd, 2004). The extension of community health services since 1991 to provide better coverage to vulnerable segments of the population, especially women, children, migrant workers, and farmers living near water development projects, coupled with early treatment seeking, mosquito net ownership, and house spraying with aerosols have been associated with lower infection rates (Adhanom et al., 2006; Ghebreyesus, Witte, et al., 2000; Deressa, Ali, & Berhane, 2007). Nevertheless, delays in diagnosis and treatment seeking by poorer people and weaknesses of the public health services emphasize the need for health education and upgrading of health services during epidemics (Deressa, 2007; Guthman et al., 2007).

The only study of the health-related costs of establishing small-scale irrigation dams in Tigray, by Ersado (2005), established that agricultural yields and farm profits increased more in villages located near dams than in villages more than 3 km away and that increased incidence of water-related diseases in the former significantly reduced returns from investment in irrigation water. Disease-ridden households were less likely to adopt productivity-enhancing and resource-conserving technologies, earned less from off-farm work, lost more time due to illness, and spent more time caring for household members. Only carefully designed dams in areas with little or no malaria and schistosomiasis could significantly increase agricultural output and food security. Caution must be exercised in interpreting these results because of the failure to consider the pain and suffering of infected people and the increase in salinity and dam sedimentation, which would decrease the benefits and increase the costs of dam investment (Ersado, 2005).

Lymphatic Filariasis

Lymphatic filariasis, another mosquito-borne disease caused by the parasitic filarial worm *Wuchereria bancrofti* in Africa, has increased in frequency in this region in recent decades as a consequence of large-scale irrigation development. In Ethiopia, although reports of infections in humans and mosquitoes date back to the 1930s and 1970s, respectively, little is known about the geographic distribution and prevalence of this disease (Hailu et al., 2006). *Anopheles gambiae* and *Anopheles funestus* were found naturally occurring in the Gambela lowlands, where bancroftian elephantiasis prevalence rates between 19% and 80% were found in rural indigenous Annuak populations (McConnell, Ash, & Dennis, 1976). The widespread occurrences of the vectors suggest that lymphatic filariasis is endemic in a larger area in Ethiopia. Moreover, it may occur in a larger area and overlap with podoconiosis, a nonfilarial form of enlarged lower feet disease caused by barefoot exposure to several chemicals in volcanic clay soils (Hailu et al., 2006; Davey, Tekola, & Newport, 2007). Although lymphatic filariasis is endemic in hot lowland areas in Ethiopia and other tropical countries and podoconiosis is endemic in the Ethiopian and other East African highlands, delimiting their endemic areas is difficult and would require differential diagnosis (Hailu et al., 2006).

Schistosomiasis

Due to the topographic, climatic, geologic, and hydrologic complexity of Ethiopia, the geographic distribution of the two types of endemic schistosomiasis in Ethiopia is discontinuous. *Schistosoma mansoni* tends to be most common at altitudes between 1,200 m and 2,000 m and *Schistosoma haematobium* below 1,000 m in the Awash and Webe Shebele basins and at Kurmuk town at the Sudan border in Benishangul Gumuz Region (Ali, Erko, Woldemichael, & Kloos, 2006). Profound hydrological and socioeconomic changes following the development of water resource projects throughout Africa have resulted in suitable conditions for the spread of intermediate host snails and schistosomiasis (El Khoby et al., 2000; Hunter et al., 1993).

Human infections with the intestinal parasite causing schistosomiasis (*Schistosoma mansoni*), transmitted mostly by the aquatic snail *Biomphalaria pfeifferi* in Ethiopia, were first reported from large irrigation schemes in the Awash Valley in the mid-1960s (Lemma, 1969), a few years after the commencement of commercial irrigation. *S. mansoni* infection rates among farm laborers in the Dubti and Abadir cotton farms and the Metahara sugar estate, as well as among indigenous pastoral nomads, were below 3% and variable (1–60%) in the Wonji sugar estate farm population (Lemma, 1969).

By the mid-1970s, high *S. mansoni* infection rates were found among both farm laborers and indigenous populations in and around all four major irrigation schemes in the upper part of the valley, as well as host snails (Kloos & Lemma, 1977), indicating that intestinal schistosomiasis had become endemic in those farms within a few years of their establishment. Subsequent surveys in Wonji farm revealed a continuous increase of schistosomiasis prevalence, from 5.6% in 1968 (Lemma, 1969) and 11.3% in 1975 (Kloos & Lemma, 1977) among farm laborers to 82% in children 5–14 years in 1989 (Simonsen, Nega, & Furu, 1990). These increases reflect not only the increase in water surfaces and human populations in and around irrigation schemes but also the changing hydrology and ecology of the Awash River and floodplains since the construction of Koka Dam in 1961 (Kloos, 1982).

Endemic *Schistosoma haematobium* is spatially confined to the distribution of the tropical African snail species *Bulinus abyssinicus* in swamps in the Awash and Wabe Shebele basins and apparently to only one locality where *B. africanus* occurs, in a stream near Kurmuk on the Sudan border. High rates of *S. haematobium* infection, transmitted by the swamp-dwelling *Bulinus abyssinicus* snail hosts, were widespread among pastoral Afar in the middle and lower parts of the valley (Kloos & Lemma, 1977). *B. abyssinicus*, absent from irrigation systems in the 1960s and 1970s, was reported in 1985 to have invaded the Dubti and Dit Bahari cotton schemes (Zakaria, 1989). The delay in this snail's invasion reemphasizes the need for systematic monitoring of irrigation schemes.

In Tigray Region, *Schistosoma mansoni* has become more common around the recently constructed microdams. Nearly half of the 41 microdams in Tigray supported *Biomphalaria pfeifferi* intermediate host snails, and, on average, 7.2% of the local villagers were infected with intestinal schistosomiasis, with the highest rates in populations living nearest to dams older than five years (Alemeyehu et al., 1998). Considering the lack of reliable information on the positive and negative impacts of microdams in Tigray, the president and vice president of Mekele University recommended that more research be carried out on system design, farm hydrology, socioeconomic constraints, and system transferability before further up-scaling (Behailu & Haile, 2002).

Other Parasitic Diseases Transmitted by Aquatic Snails

Liver fluke infection, caused by *Fasciola hepatica*, and bovine schistosomiasis (*Schistosoma bovis*) are endemic in Ethiopia. *B. abyssinicus, B. africanus*, and *B. truncatus* snails transmit *S. bovis* to cattle (Ali et al., 2006) and the lowland snail *Lymnaea natalensis* and the highland species *L. truncatula* transmit *F. hepatica*, which infects both humans and livestock (cattle and sheep) (Goll & Scott, 1978; Lo, 1975). Fascioliasis is mainly an animal disease in Ethiopia with few human cases reported in the literature (Bayu et al., 2005). Prevalence rates between 40% and 90% were reported from three localities in the highlands of Gojam and Shewa, with half of all mortality in calves and weight loss and liver condemnation in untreated animals (Heinonen, Rufl, & Kebede, 1992; Lemma, Gebre-ab, & Tedla, 1985). Economic loss from ovine fascioliasis alone was estimated at 48.4 million Birr per year (Ngategize, Bekele, & Tilahu, 1992). Considering that *L. natalensis* snails were found in most administrative regions (Lo, Redda, & Gemeda, 1982) and in small irrigation farms in the Awash Valley (Asrat, 2004), the risk of fascioliasis spreading to other irrigation schemes is great.

Less is known about the prevalence and distribution of bovine schistosomiasis in Ethiopia. The three snail intermediate hosts of *S. bovis* in Ethiopia—*B. abyssinicus*, *B. africanus*, and *B. truncatus*—transmit this parasite to cattle (Lo, 1975) and probably also to sheep. In Sudan and Tanzania, infection rates of around 30% were reported in cattle and sheep, with a mortality rate of 7% in calves in Sudan. *S. bovis* is also endemic in various West, Central, and South African countries, with prevalence rates up to 65% reported from Zambia (Food and Agriculture Organization of the United Nations, 2008). Christensen, Mutani, and Frandsen (1983), who reviewed the epidemiology of bovine schistosomiasis, predicted that water conservation and changing methods of husbandry might increase the problem of this livestock disease in Africa.

Prevention and Control of Vector-Borne Diseases

Environmental and health impacts, which are causally related and have resulted in the failure of many irrigation projects, may be prevented through proper planning and implementation of schemes (Hunter et al., 1993; McCartney et al., 2007). Early detection of epidemics by identifying mosquito breeding sites, locating settlements away from these sites, using impregnated bed nets, and effective health services, can control malaria epidemics in irrigation schemes in the vicinity of reservoirs and in wetlands (Guintran et al., 2006; McCartney et al., 2007).

The continuing development of irrigation schemes and resettlement projects and rapid population growth in the Ethiopian lowlands demand further efforts to reduce the risk of vector-borne diseases. Integrated health programs providing both preventive and curative services, including vector control, bed nets, health education, and treatment of infections, have been able to reduce disease transmission in some projects (Adhanon et al., 2006).

GEOCHEMICAL DISEASES

The highly complex geological structure of Ethiopia, which includes volcanic, ancient crystalline, metamorphic, and various types of sedimentary rocks, and its location on the high-fluoride East African Rift Valley warrant an assessment of potential health impacts even though few data exist on most pathogenic geochemicals. A recent review of groundwater quality studies in Ethiopia concluded that 1) fluoride is the major geochemical health hazard in the Rift Valley, 2) high sodium and chloride levels found in the same waters contribute to rendering them unacceptable for human consumption although by themselves they pose no direct health threat, and 3) uranium and arsenic (both carcinogens) occur in ancient crystalline rocks and in some groundwaters in the southern Rift Valley (British Geological Survey, 2001; Reimann et al. 2003). Mapping the distribution of 70 geochemicals from 138 sites throughout the Ethiopian Rift Valley, Reimann et al. (2003) found that 86% of all samples had concentrations above the drinking water regulations of the European Union and the WHO-proposed maximum allowable concentration (MAC). Redda Teklehaimanot (2003) spatially associated elevated arsenic, as well as chromium and boron levels in the Awash River, with industrial sources in Addis Ababa. Little is known about the health effects of multiple, synergistically acting chemicals, an issue which will need greater attention as water pollution from urban, industrial, and agricultural sources increases in Ethiopia and other African countries, as indicated in chapter 12.

Fluorosis

Fluorosis in humans and domestic animals is caused by the ingestion of high levels of fluoride in water and to a lesser extent in food. Fluoride occurs in all rock types, but the highest concentrations of fluoride in water sources and the highest prevalence and intensity of fluorosis are found around Andean type magmatic belts, tectonic faults, and rift zones. Heavy clay soils, groundwater, and lake water can take up high levels of fluoride from these rocks. The East African Rift Valley contains surface and ground waters with the world's highest fluoride concentrations (World Health Organization, 1984; Tekle-Haimanot et al., 2006).

Fluorosis manifests itself in humans as dental fluorosis and, with higher and prolonged exposure, as skeletal fluorosis. Whereas dental fluorosis is considered mostly as a cosmetic and psychosocial problem, skeletal fluorosis in its severe form is a crippling disease severely impacting the daily lives of people (World Health Organization, 1984). The WHO has set the fluoride guideline value at 1.5 mg/L of water, but the mild form of dental fluorosis begins to occur in hot climates where water consumption is high at levels as low as 0.8 mg/L (Brouwer, Dirks, DeBruin, & Hautvast, 1988). Prolonged ingestion of water above 1.5 mg/L causes marked dental fluorosis, 3–6 mg/L causes skeletal fluorosis, and higher concentrations can cause crippling skeletal fluorosis (Brouwer et al., 1988). Both forms of fluorosis are widespread in the Ethiopian Rift Valley, which constitutes part of the East African Rift Valley. The economic impact of fluorosis is compounded by its crippling effects, which may involve paralysis in people, occurrence of the disease in cattle and other livestock, and high levels of fluoride (around 10 mg/g) in human foods grown in the Ethiopian Rift Valley (Kloos & Tekle-Haimanot, 2006; McDonald & Partners, 1976; Tekle-Haimanot, 2005).

The location of the Ethiopian high-fluoride area in the Rift Valley has put the populations in about a dozen large irrigation schemes at risk of developing fluorosis unless they are provided with safe drinking water. Hard physical work in the Rift Valley confers particularly high risk of fluorosis due to high water consumption by farm laborers and other groups engaged in physically strenuous activities in hot climates (McGill, 1995). The most extensive survey of fluoride distribution in Ethiopia (Tekle-Haimanot et al., 2006) found that 50.0% of 506 deep wells, 27.1% of 162 shallow wells, 50.0% of 20 hot springs, 112.4% of

95 cold springs, and 10 of 12 lakes had fluoride concentrations higher than 1.5 mg/L. Nearly one third of the deep wells, 80.0% of the hot springs, and 66.7% of the lakes had fluoride concentrations above 3 mg/L, reaching 202.4 mg/L in Lake Abijata and 264.0 mg/L in Lake Shala. The upper part of the Awash River has low fluoride levels, but the middle and lower parts of this heavily used river have levels around the 1.5 mg/L threshold due to the inflow of high-fluoride water from hot springs. The plan by the government to divert water from the rising soda lake Beseka to the Awash River, to prevent further flooding of the regional railroad and road, may further elevate fluoride levels in the river and increase fluorosis risk among farming and pastoralist populations. High prevalence rates of crippling fluorosis were reported from Wonji-Shoa irrigation scheme and two small towns in the central part of the rift valley (Tekle-Haimanot et al., 2006). Among 262 retired employees in Wonji-Shoa, 20.0% were diagnosed with skeletal fluorosis on clinical grounds, but 70.3% radiologically, indicating that many cases are asymptomatic (Assefa, Shifera, Melaku, & Haimanot, 2004).

Wonji-Shoa scheme is the only irrigation scheme in the Rift Valley that developed a fluorosis control program. Although this farm has some of the most highly developed social and medical services, the operation has been intermittent due to equipment failures and shortages of chemicals and is considered a failure. Current plans call for the installation of a piped system using treated water from the Awash River, similar to the one already in operation in the nearby town of Adama (Nazaret) (Tekle-Haimanot et al., 2006). Other promising defluoridation technologies include filtration through laterite clay or ground clay chards and immersion of the seeds of the plant *Moringa oleifera* (Kloos & Tekle-Haimanot, 2006). It has recently been suggested that more wells be drilled along large rift faults which act as conduits of low-fluoride water originating in the nearby highlands (Ayenew, 2007). The rapid spread of rainwater harvesting projects in Ethiopia (chapter 5) may provide for an alternative water source cheaper than groundwater or river diversion with its extensive pipe networks and pumps.

Other Geochemical Diseases

Of all 69 other geochemicals potentially causing disease in humans that were tested in Ethiopia, only uranium is likely to carry significant health risks in the Rift Valley, where 47% of the 138 wells tested by Reimann et al. (2003) yielded water with uranium concentrations above the newly suggested WHO MAC of 2 micrograms per liter (μg/L). Only 7% of the samples had arsenic concentrations above the 10 μg/L safety threshold set by the European Union, indicating that this chemical is not a major health problem in the Rift Valley. Nevertheless, studies of the health risks of these and other geochemicals, including possible synergistic effects from exposure to water with multiple chemicals, will be useful to identify high-risk localities and develop alternative water sources.

Conclusion

Water resources development in the form of constructing dam and irrigation systems and drainage of wetlands has direct health implications in Ethiopia in regard to water-related diseases, although their total health implications are far greater. Most health information has been gathered on the vector-borne diseases of malaria and schistosomiasis, but there is evidence from both Ethiopian and other African projects that the transmission of lymphatic filariasis, onchocerciasis, trypanosomiasis, fascioliasis, and dracunculiasis may also intensify in water projects. The specter of climatic change will most likely extend the altitudinal boundary of these and other water-related diseases, and the continuing movement of the Ethiopian population from the overpopulated, disease-free highlands to the endemic lowlands will increase the need for careful monitoring, prevention, and control of these and other infectious diseases. Fluorosis, which also has severely impacted the health of some irrigation farm populations in the Rift Valley, can be prevented by the provision of low-fluoride water supplies. Domestic water supply in the irrigation farms has been greatly improved since the Haile Selassie period (H. Kloos, unpublished data) but its impact on fluorosis, schistosomiasis, and other water-related diseases has not been evaluated. Potential health risks from uranium and possibly other geochemical agents in groundwater in the Rift Valley also require further attention by researchers and planners.

The current rapid pace of dam construction and irrigation development, as part of Ethiopia's goal of achieving water, food, and energy sufficiency, gives urgency to systematic surveillance and monitoring of projects and necessary remedial measures to prevent significant health and ecological damages. Full cost-benefit analyses of large water projects must include comprehensive preproject health and environmental impact studies and subsequent monitoring of disease vectors and infection rates in local populations. This is necessary because of the interrelatedness of and continuous changes in environmental and health conditions throughout the life of projects. In projects that meet impact assessment criteria, intersectoral collaboration focused on integrated disease prevention and control programs around projects, although still without a track record in Ethiopia, hold great potential to be effective, both in terms of potential health benefits and cost effectiveness, if such collaboration indeed translates to on-the-ground activities. The urgency of implementing effective and socioeconomically and ecologically compatible monitoring and intervention programs is indicated by the large number of water projects to be developed with the objective of meeting national power and food needs and the persisting high rates of malaria and other water-related diseases.

References

Abayou, T. K. (2007). *The role of development: Induced projects in the spread of HIV/AIDS, the case of Gilgel-Gibe Hydro-Electric Project in Jimma Zone* (Unpublished master's thesis). Addis Ababa University.

Abdulahi, H., Haile Mariam, D., & Kebede, D. (2001). Burden of disease analysis in rural Ethiopia. *Ethiopian Medical Journal, 39*, 271–281.

Adams, W. (2000). *Downstream impacts of dams*. Working paper prepared for thematic review I. Social impacts of large dams equity and distributional issues. Cape Town, South Africa: World Commission on Dams.

Adhanom, T., Deressa, W., Witten, K. H., Getachew, A., & Seboxa, T. (2006). Malaria. In Y. Berhane, D. Haile Mariam, & H. Kloos (Eds.), *The epidemiology and ecology of health and disease in Ethiopia* (pp. 556–576). Addis Ababa: Shama Books.

Alemayehu, T., Ye-ebicho, Y., Ghebreyesus, T. A., Witten, K. H., Bosman, A., & Teklehaimanot, A. (1998). Malaria, schistosomiasis, and intestinal helminths in relation to microdams in Tigray, northern Ethiopia. *Parassitologia, 40*, 259–267.

Ali, A., Erko, B., Woldemichael, T., & Kloos, H. (2006). Schistosomiasis. In Y. Berhane, D. Haile Mariam, & H. Kloos (Eds.), *The epidemiology and ecology of health and disease in Ethiopia* (pp. 660–673). Addis Ababa: Shama Books.

Amacher, G., Ersado, L., Grebner, D. L., & Hyd, W. (2004). Disease, microdams and natural resources in Tigray, Ethiopia: Impacts on productivity and labour supplies. *The Journal of Development Studies, 4*, 122–145.

Asrat, M. (2004). *The prevalence of ovine fascioliasis in small-scale irrigation schemes along the upper Awash River Basin and effects of strategic anthelminthic treatment in selected upstream areas* (Unpublished master's thesis). Addis Ababa University.

Assefa, G., Shifera, G., Melaku, Z., & Haimanot, R. T. (2004). Clinical and radiological prevalence of skeletal fluorosis among retired employees of Wonji-Shoa sugar estate. *East African Medical Journal, 81*, 638–640.

Ayenew, T. (2007). The distribution and hydrogeological controls of fluoride in the groundwater of central Ethiopian rift and adjacent highlands. *Environmental Geology, 54,* 1313–1324. Retrieved from http://www.springerlink.com/cotent/62m775083714u176/

Bayu, B., Asnake, S., Woretaw, A., Ali, J., Gedefaw, M., Fente, T., … Yitayew, G. (2005). Cases of human fascioliasis in north-west Ethiopia. *Ethiopian Journal of Health Development, 19*, 237–240.

Bedri Kello, A. (1995). The effect of malaria on peasant production: A case study from two Ethiopian villages. *Ethiopian Journal of Development Research, 17*, 1–37.

Behailu, M., Abdulkadir, M., Mezgebu, A., & Yasin, M. (2004). *Preliminary report on community based irrigation schemes. A case study of three small-scale irrigation schemes: Impact assessment* (Unpublished manuscript). Mekelle University, Ethiopia.

Berhane, Y., Haile Mariam, D., & Kloos, H. (Eds.). (2006). *The epidemiology and ecology of health and disease in Ethiopia.* Addis Ababa: Shama Books.

Behailu, M., & Haile, M. (2002, December). Water harvesting in northern Ethiopia: Environmental, health and socioeconomic impacts. Paper presented at the Workshop on Integrated Water and Land Management Research and Capacity Building Priorities in Ethiopia, International Livestock Resarch Institute, Addis Ababa.

Boko, M., Niang, I., Nyong, A., Vogel, C., Githeko, A., & Yanda, P. (2007). Impacts, adaptation and vulnerability. In M. L. Parry, O. F. Canzani, J. P. Palutikof, P. J. van der Linden, & C. E. Hanson (Eds.), *Contribution of Working Group II to the Fourth Assessment Report of the Intergovernmental Panel on Climate Change* (pp. 433–467). Cambridge: Cambridge University Press.

Brewster, D. (1999). Environmental management for vector control: Is it worth a dam if it worsens malaria? *British Medical Journal, 319*, 651–652.

British Geological Survey. (2001). *Groundwater quality: Ethiopia*. London: British Geological Survey.

Brouwer, J. D., Dirks, O. A., DeBruin, A., & Hautvast, J. G. (1988). Unsuitability of WHO guidelines for fluoride concentration in drinking water in Senegal. *Lancet, 1*, 223–225.

Christensen, N. O., Mutani, A., & Frandsen, F. (1983). A review of the biology and transmission ecology of African bovine species of the genus *Schistosoma*. *Parasitology Research*, *69*, 551–570.

Campagna per la riforma della Banco Mondiale. (2008). *The Gilgel Gibe affair: An analysis of the Gilgel Gibe hydroelectric projects in Ethiopia*. Retrieved from http://www.bankwatch.org/documents/gibe-study.pdf

Colson, E. (1971). *The social consequences of resettlement. The impact of the Kariba resettlement upon the Gwembe Tonga* (Kariba Studies No. 4). Published on behalf of the Institute of African Studies, University of Zambia. Manchester: Manchester University Press.

Danish Bilharzia Laboratory. (2007a). *Health impact assessment: HIA e-learning-course development workshop*. Sponsored by DBL in collaboration with WHO. Copenhagen: Author.

Danish Bilharzia Laboratory. (2007b). *HIA Policy Seminar*. Sponsored by DBL in collaboration with WHO, Hanoi, Vietnam. Copenhagen: Author.

Davey, G., Tekola, F., & Newport, M. J. (2007). Podoconiosis: Non-infectious geochemical elephantiasis. *Transactions of the Royal Society of Tropical Medicine and Hygiene, 101*, 1175–1180.

Deressa, W. (2007). Treatment-seeking behavior for febrile illness in an area of seasonal malaria transmission in rural Ethiopia. *Malaria Journal*, *6*, 49.

Deressa, W., Ali, A., & Berhane, Y. (2007). Household and socioeconomic factors associated with childhood febrile illnesses and treatment seeking behaviour in an area of epidemic malaria in rural Ethiopia. *Transactions of the Royal Society of Tropical Medicine and Hygiene, 101*, 939–947.

Dixon, A. B. (2003). *Indigenous management of wetlands: Experiences in Ethiopia*. Aldershot, U.K.: Ashgate Publishing.

El-Khoby, T., Galal, N., Fenwick, A., Barakat, R., El-Hawey, A., & Strickland, G. T. (2000). The epidemiology of schistosomiasis in Egypt: Summary findings in nine governorates. *American Journal of Tropical Medicine and Hygiene, 62*, 88–99.

Erlanger, T. E., Keiser, J., de Castro, M. C., Bos, R., & Utzinger, J. (2005). Effect of water resources development and management on lymphatic filariasis and estimates of population at risk. *American Journal of Tropical Medicine and Hygiene*, *73*, 523–533.

Ersado, L. (2005). *Small-scale irrigation dams, agricultural production, and health: Theory and evidence from Ethiopia*. World Bank Policy Research Paper (No. 3494). Washington, DC: Author.

Fontaine, R. E., Najjar, A. E., & Prince, J. E. (1961). The 1958 malaria epidemic in Ethiopia. *American Journal of Tropical Medicine and Hygiene*, *10*, 795–803.

Food and Agriculture Organization of the United Nations. (2008). *Distribution and impact of helminth disease of livestock in developing countries*. Retrieved from http://www.fao.org/DOCREP/004/T0584E/T0584E03.htm

Ghebreyesus, T. A., Hailu, M., Witten, K. H., Getachew, A., Yohannes, M., & Byass, P. (2000). Household risk factors for malaria among children in the Ethiopian highlands. *Transactions of the Royal Society of Tropical Medicine and Hygiene*, *94*, 17–21.

Ghebreyesus, T. A., Witten, K. H., Getachew, A., Yohannes, A. M., Tesfay, W., & Teklehaimanot, A. (2000). The community-based malaria control program in Tigray, northern Ethiopia: A review of programme set-up, activities, outcomes and impact. *Parassitologia, 42*, 255–290.

Ghebreyesus, T. A., Witten, K. H., Getachew, A., Haile, M., Yohannes, M., & Byass, P. (2002). Schistosome transmission, water resources development and altitude in northern Ethiopia. *Annals of Tropical Medicine and Parasitology, 96*, 489–495.

Goll, P., & Scott, J. M. (1978). The interrelationship of *Lymnaea truncatula* and ovine fascioliasis in the Ethiopian Central Highlands. *British Veterinary Journal, 134*, 551–555.

Guintran, J.-O., Delacollette, C., & Trigg, P. (2006). *Systems for the early detection of malaria epidemics in Africa: An analysis of current practices and future priorities*. Geneva: World Health Organization.

Guthmann, J. P., Bonnet, M., Ahoua, L., Dantoine, F., Balkan, S., & Checchi, F. (2007). Death rates from malaria epidemics, Burundi and Ethiopia. *Emerging Infectious Diseases, 31*, 140–143.

Haile Meskal, F., & Kloos, H. (1989). Vector-borne disease occurrence and spread as affected by labor migrations to irrigation schemes in Ethiopia. In M. W. Service (Ed.), *Demography and vector-borne diseases* (pp. 225–236). Boca Raton: CRC Press.

Hailu, A., Gebre, T., Panicker, K. N., Negash, K., Belete, K., & Medhin, A. (2006). Lymphatic filariasis. In Y. Berhane, D. Haile Mariam, & H. Kloos (Eds.), *The epidemiology and ecology of health and disease in Ethiopia* (pp. 609–614). Addis Ababa: Shama Press.

Heinonen, M., Rufl, G., & Kebede, S. (1992). The effect of antiparasitic treatment against *Fasciola* on cross-bred and zebu cows in Ethiopia—Short communications. Retrieved from http://www.fao.org.docrep/V4440T/v4440T0y.htm

Hira, P. R. (1970). Schistosomiasis at Lake Kariba. I. Prevalence and potential intermediate hosts at Siavonga. *Tropical and Geographical Medicine, 22*, 323–234.

Hubendick, B. (1973). *A plan for a transdisciplinary study of events occurring when a tropical water development scheme is implemented, with special regard to its bearing on schistosomiasis.* Report from the Nobel Workshop. Stockholm: The International Federation of Institutes for Advanced Study.

Hunter, J. M., Rey, L., Chu, K. Y., Adekolu-John, E. O., & Mott, K. E.(1993). *Parasitic diseases in water resources development: The need for intersectoral negotiation.* Geneva: World Health Organization.

Huntingford, C., Hemming, D., Gash, J. H., Gedney, N., & Nuttal, P. A. (2007). Impact of climate change on health: What is required of climate modelers? *Transactions of the Royal Society of Tropical Medicine and Hygiene, 101*, 97–103.

Integrated Regional Information Network. (2008a, January 29). Mozambique: Flood lessons. *IRIN Newsletter.* Office of the Coordination of Humanitarian Affairs, United Nations.

Integrated Regional Information Network. (2008b, April 7). Rich must pay climate change health costs. *IRIN Newsletter.* Office of the Coordination of Humanitarian Affairs, United Nations.

Jobin, W. R. (1999). *Dams and disease: Ecological design and health impacts of large dams, canals and irrigation systems.* London: Francis & Taylor.

Keiser, J., de Castro, M. C., Maltese, M. F., Bos, R., Tanner, M., & Utzinger, J. (2005). Effect of irrigation and large dams on the burden of malaria on a global and regional scale. *American Journal of Tropical Medicine and Hygiene, 72*, 392–406.

Khasnis, A. A., & Nettleman, M. D. (2005). Global warming and infectious disease. *Archives of Medical Research, 36*, 689–698.

Kitron, U., & Spielman, A. (1989). Suppression of malaria through source reduction: Anti-anopheline measures used in Israel, the U.S.A. and Italy. *Review of Infectious Diseases, 11*, 391–405.

Kloos, H. (1985). Schistosomiasis and water resources development in the Awash Valley, Ethiopia. *Social Science and Medicine, 20*, 609–625.

Kloos, H., DeSole, G., & Lemma, A. (1981). Intestinal parasitism in seminomadic pastoralists and subsistence farmers in and around irrigation schemes in the Awash Valley, Ethiopia, with special emphasis on ecological and cultural associations. *Social Science and Medicine, 15B*, 457–469.

Kloos, H., & Lemma, A. (1977). Schistosomiasis in irrigation schemes in the Awash Valley, Ethiopia. *American Journal of Tropical Medicine and Hygiene*, *26*, 899–908.

Kloos, H., & Tekle-Haimanot, R. (2006). Distribution of fluoride and fluorosis in Ethiopia and prospects for control. *Tropical Medicine and International Health*, *4*, 355–364.

Kurttio, P., Auvinen, A., Salonen, L., Saha, H., Pekkanen, J., & Komulainen, H. (2002). Renal effects of uranium in drinking water. *Environmental Health Perspectives*, *110*, 237.

Lautze, J., McCartney, M., Kirshen, P., Olana, D., Jayasinghe, G., & Spielman, A. (2007). Effect of a large dam on malaria risk: The Koka reservoir in Ethiopia. *Tropical Medicine and International Health, 12*, 982–989.

Lemma, A. (1969). Bilharziasis in the Awash Valley. I. An epidemiological study with special reference on its possible future economic and public health importance. *Ethiopian Medical Journal, 7*, 147–176.

Lemma, B., Gebre-ab, F., & Tedla, S. (1985). Studies on fascioliasis in four selected sites in Ethiopia. *Veterinary Parasitology*, *18*, 29–37.

Lindblade, K. A., Walker, E. D., Onapa, A. W., Katungu, J., & Wilson, M. L. (1999). Highland malaria in Uganda: Prospective analysis of an epidemic associated with El Nino. *Transactions of the Royal Society of Tropical Medicine and Hygiene*, *93*, 480–487.

Lindblade, K. A., Walker, E. D., Onapa, A. W., Katungu, J., & Wilson, M. L. (2000). Land use change alters transmission parameters by modifying temperature in a highland area of Uganda. *Tropical Medicine and International Health*, *5*, 263–274.

Lo, C. T. (1975). Studies on *Schistosoma bovis* in Ethiopia. *Annals of Tropical Medicine and Parasitology, 69*, 375–381.

Lo, C. T., Redda, A., & Gemeda, N. (1982). Malacological studies of human schistosomiasis in Ethiopia. In T. Ayele & C. T. Lo (Eds.), *Proceedings of a Symposium on Human Schistosomiasis in Ethiopia* (pp. 40–47). Addis Ababa: Institute of Pathobiology.

Mara, D. D., & Clapham, D. (1997). Water-related carcinomas: Environmental classification. *Journal of Environmental Engineering, 123*, 416–422.

Mara, D. D., & Feachem, R. G. A. (1999). Water- and excreta-related diseases: Unitary environmental classification. *Journal of Environmental Engineering, 125*, 334–339.

McCartney, M. P., Boelee, E., Cofie, O., & Mutero, C. M. (2007). *Minimizing the negative environmental and health impacts of agricultural water resources development in Sub-Saharan Africa*. Colombo, Sri Lanka: International Water Management Institute.

McConnell, E., Ash, W., & Dennis, D. T. (1976). A survey of *Wuchereria bancrofti* in Ethiopia. *Ethiopian Medical Journal, 14*, 31–36.

McDonald and Partners. (1976). Supply of reduced fluoride potable water to Wonji-Shoa and Metahara sugar estates. Report prepared for the Ethiopian Sugar Corporation, Addis Ababa.

McGill, P. E. (1995). Endemic fluorosis. *Bailléire's Clinical Rheumatology*, *9*, 75–81.

Millennium Ecosystem Assessment. (2005). *Ecosystems and human well-being* (Vol. 3, Policy Responses, chapter 12, p. 367). Washington, DC: Island Press.

Negash, K., Kebede, A., Medhin, A., Argaw, D., Babaniyi, O., & Delacollette, C. (2005). Malaria epidemics in the highlands of Ethiopia. *East African Medical Journal, 82*, 186–192.

Ngategize, P. K., Bekele, T., & Tilahun G. (1993). Financial losses caused by ovine fasciolosis in the Ethiopian highlands. *Tropical Animal Health and Production, 25*, 155–161.

Redda Teklehaimanot, R. (2003). *A literature based study into the expected pollution of the Awash River, Ethiopia (TOX-80316): Preliminary toxic risk evaluation of the Awash River, Ethiopia (TOX-80328)* (Unpublished master's thesis). Wageningen University, Wageningen, The Netherlands.

Reimann, C., Bjorvatn, K., Frengstas, B., Melaku, Z., Tekle-Haimanot, R., & Siewers, U. (2003). Drinking water quality in the Ethiopian section of the East African Rift Valley. I. Data and health aspects. *Science of the Total Environment, 311*, 65–80.

Scudder, T. (2005). *The Kariba case study* (Social Science Working Paper No. 1227). Pasadena: California Institute of Technology.

Scudder, T., & Colson, E. (1982). From welfare to development: A conceptual framework for the analysis of dislocated people. In A. Hansen & A. Oliver-Smith (Eds.), *Involuntary migration and resettlement: The problems and responses of dislocated people*. Boulder, CO: Westview Press.

Simonsen, P. E., Nega, A., & Furu, P. (1990). Intestinal schistosomiasis among children in a labour village of Wonji Sugar Estate, Ethiopia. *East African Medical Journal, 67*, 532–538.

Sleigh, A. C., & Jackson, S. (2001). Dams, development, and health: A missed opportunity. *Lancet, 357*, 570–571.

Soils Incorporated (in association with Chalo Environmental and Sustainable Development Consultants). (2000, February). *Kariba Dam case study: Zambia and Zimbabwe*. Draft report prepared for the discussion at the Stakeholder Meeting in Siavonga (Zambia), World Commission on Dams, Cape Town, South Africa.

Steinmann, P., Keiser, J., Bos, R., Tanner, M., & Utzinger, J. (2006). Schistosomiasis and water resources development: Systematic review, meta-analysis, and estimate of people at risk. *Lancet Infectious Diseases, 6*, 411–425.

Tadesse, I., Isoaho, S. A., Green, F. B., & Puhakka, J. A. (2003). Removal of organics and nutrients from tannery effluent by advanced integrated Wasterwater Pond Systems technology. *Water Science and Technology, 48*, 307–314.

Tekle-Haimanot, R. (2005). *Study of fluoride and fluorosis in Ethiopia with recommendations on appropriate defluoridation technologies: A UNICEF sponsored consultancy* (Unpublished manuscript). Addis Ababa University.

Tekle-Haimanot, R., Melaku, Z., Kloos, H., Reimann, C., Wondwossen, F., & Bjorvatn, A. (2006). The geographic distribution of fluoride in surface and groundwater in Ethiopia with an emphasis on the Rift Valley. *Science of the Total Environment, 367*, 182–190.

Tesfaye, B., & Tsehay, A. (1987). *Report on the Gambella settler communities*. Report prepared for the National Research Institute of Health, Addis Ababa.

Tulu, A. N. (1993). Malaria. In H. Kloos & Z. A. Zein (Eds.), *The ecology of health and disease in Ethiopia* (pp. 341–352). Boulder and Oxford: Westview Press.

World Health Organization. (1984). *Fluorine and fluorides*. Environmental Health Criteria No. 36. Geneva: Author.

Woube, M. (1997). Geographical distribution and dramatic increases in incidence of malaria: Consequences of the resettlement scheme in Gambela, SW Ethiopia. *Indian Journal of Malariology, 34*, 140–163.

Yasuoka, J., & Levins, R. (2007). Impact of deforestation ad agricultural development on anopheline ecology and malaria epidemiology. *American Journal of Tropical Medicine and Hygiene, 76*, 450–460.

Yewhalaw, D., Legesse, W., Gebre-Selassie, S., & Kloos, H. (2009). Malaria and water resources development in Ethiopia: The case of Gilgel-Gibe Hydroelectric Dam. *Malaria Journal*, *8*, 21. doi: 10.1186/1475-2875-8-21

Yohannes, M., Haile, M., Ghebreyesus, T. A., Witten, K. H., Getachew, A., & Lindsay, S. W. (2005). Can source reduction of mosquito larval habitat reduce malaria transmission in Tigray, Ethiopia? *Tropical Medicine and International Health*, *10*, 1274–1285.

Zakaria, K. (1989). New foci of *Bulinus abyssinicus* in lower Awash. *Ethiopian Medical Journal*, *27*, 43–44.

Zimmerman, R. H. (2001). Wetlands and infectious diseases. *Cadernas de Saúde Pública*, *17* (Suppl.), 127–131.

Chapter 14

Conclusion

Helmut Kloos and Worku Legesse

This book reveals that a wide range of water resources uses, management practices, and, increasingly, conservation methods are being employed in Ethiopia and the wider Nile Basin that must be considered by planners and managers in community, national, and transboundary water resources management. The study identifies many conditions, processes, and activities that affect water availability and quality, societies, and ecosystems. It clearly shows that deforestation and prevailing land use practices in rain-fed agriculture may exacerbate soil aridity and flood events, that irrigated agriculture and hydropower development, while increasing water availability locally, are associated with floodplain desiccation, socioeconomic impacts, and the spread of water-related diseases, that industrial and agricultural development are starting to heavily pollute freshwater resources, and that all water development and management activities are hampered by general underdevelopment in Ethiopia. Although some of these relationships have been reported from Ethiopia and other Nile Basin countries, this study provides a comprehensive assessment of the main drivers and impacts of a wide range of water problems across all sectors of the economy and in the environment that planners and managers need to consider in the quest for effective and sustainable water resources management strategies.

While most elements of an integrated water resources management approach have been specified in Ethiopia's water policy and the pending Nile Basin Initiative, a combination of weak institutional capacity and infrastructure, widespread poverty, and misguided government policies have prevented efficient and equitable water management and have failed to prevent a number of negative environmental and health impacts. Further water resources development and management efforts will require improved socioeconomic conditions, increased access to land, water, markets, farm credit, and other support structures and a policy environment that not only promotes sustainable land and water use practices but also prevents water pollution and water-related diseases. Consideration of the country's socioeconomic, cultural, and environmental assets and vulnerabilities, institutional capacity, and development of mutually agreeable partnerships with local communities are essential in this regard. The increasingly serious negative environmental, socioeconomic, and health impacts of water development must be considered as part and parcel of assessments of water projects prior to their implementation and throughout the period of their operation. Particularly, hydropower and irrigation development in the Awash and Omo valley, designed to significantly boost national food and energy needs, remains suboptimal due to inadequate integration of the needs of local populations and ecologies. These two projects are examples of unilateral water management of transboundary water resources that urgently demand bilateral agreements between Ethiopia, Kenya, and Djibouti in view of ongoing negotiations on the Nile Basin Initiative.

Agriculture will continue to be in the foreseeable future the sector using most water in the Nile Basin and may generate greater water savings than any other sector if properly managed. It is uncertain, however, how agricultural systems and policy makers in this region will respond to increasing demands, weak institutional capacity, and climate change to support higher water productivity (Cook, Harrington, & Huber-Lee, 2008). The responses of the various Nile riparian countries to water scarcity are still largely confined to dam construction to store more water, with mostly subsistence farmers conserving water, but fewer of them raising its productivity due to lack of adaptive capacity. The evolution of water development and management in Egypt suggests that difficult choices are required to achieve greater efficiency and productivity of water that may not, however, be fully applicable to the other riparians because of their different economies and social structures. Additional water savings, as well as environmental benefits, may be obtained in the Nile Basin through the application of alternative water harvesting/storage and energy technologies. Whether alternative water storage and energy technologies and traditional soil and water management approaches can be effectively developed will largely depend on political will and the provision of support structures.

Efforts by the Ethiopian government to address these various issues have been gradually increased by the last three administrations, and the current regional water development plans provide the necessary road map to achieving water, energy, and food security. The recent pursuit of an approach more amenable to meeting the needs of local populations and ecosystems has been marked with some progress. Increased water supply coverage in rural areas, and to a lesser extent small-scale irrigation and rainwater harvesting, are contributing to meeting the water needs of the broad population and the Millennium Development Goals. However, achievements in most other sectors are uneven and further progress is in danger of being compromised by economic, political, and demographic developments and international interests running counter to good stewardship and conservation practices traditionally pursued by Ethiopian farmers. The push at all costs by the current government to rapidly develop the country's blue water resources toward meeting water, energy, and food security targets carries with it a range of inherent risks, from inefficiency and nonsustainability of projects to the perpetuation of water and energy scarcity and negative economic and health impacts. Inadequate participation of the population and the private sector related to the lack of incentives, support services, and a tradition of heavy-handed, top-down government management have rendered the implementation and upscaling of programs difficult.

These findings are in sharp contrast to the periodic reports by the international press and many relief organizations attributing Ethiopia's water scarcity primarily to climatic drought and also dispel the notion that deforestation and soil erosion are recent phenomena. Many of these findings constitute misinformation that was used by foreign aid and development organizations to justify their programs and by the Derg socialist government to carry out its political agenda. The rural population became the scapegoat for environmental problems, particularly deforestation and soil erosion in the highlands. But the ineffectiveness and outright failure of the massive, top-down conservation and related programs of the Derg regime, some of which even contributed to the water crisis, points out their inherent flaws. The real, underlying causes of the country's water problems, mostly related to poverty and lack of property rights, rural infrastructure, and technical support, were not addressed until a few years ago. This scenario unfolded many times during the last century in Africa and other developing regions where farmers were considered mismanagers of water and soil resources (Pretty & Shah, 1997; Ives, 2006). The relative success and sustainability of indigenous soil and water conservation systems, in spite of their shortcomings and setbacks after Ethiopia's move to socialism, belies the widespread attitude among Ethiopian administrators toward rural people, which Clapham (1990) succinctly described as: "Government in Ethiopia is a matter of experts who know what to do; the ignorant peasant, by contrast, is there to be organized, conscripted, taxed, or in one word, governed" (p. 39). Much information presented in this book clearly shows that Ethiopian peasants are not the cause of the water problems or passive bystanders, but the solution and active innovators if given the opportunity to pursue the production and conservation activities that are best adapted to their local environments, socioeconomic situation, and cultural traditions. In the absence of an enabling environment, many of them continue to exploit water and other natural resources in a survival mode.

The various hastily planned government programs implemented on a large scale during the socialist era, some of which have persisted until today, largely failed to consider environmental characteristics of project areas and community needs and expectations. Attempts to apply centrally planned conservation and development programs in a "one-size-fits-all" blanket approach aimed at quantitative achievements rather than quality and sustainability of interventions have a long history in Ethiopia but are highly misguided (Kruseman, Ruben, & Tesfay, 2006; Zeleke, Kassie, Pender, & Yesuf, 2006) in this highly diverse biophysical, economic, and cultural country. The use of heavy-handed, top-down approaches continues to be used by the current administration in several areas in spite of significant political, administrative, and financial decentralization. This includes the controversial hydroelectric program, carried out to achieve national energy security but also with transboundary water resources development objectives. The construction of increasingly disfavored large dams rather than the development of more environmentally friendly alternative energy sources that are readily accessible in rural areas and failure to comprehensively assess dam impacts at pre- and post-inundation phases may negate intended benefits. The negative impacts of Koka Dam and large irrigation schemes in the Awash Valley and potential threats associated with the development of new hydropower projects on the Awash, Omo, and Blue Nile river basins illustrate the need for transparency and community participation in the planning and implementation processes. These requirements must become part and parcel of plans to harness Ethiopia's large water resources to ensure the social equity of projects. The need to integrate the needs and resources of all water users at the community, regional and transboundary levels is becoming increasingly urgent with population and economic growth, increasing competition between different sectors of the economy, as well as between urban and rural areas for water resources, and the specter of climate change.

Our findings from Ethiopia and those by social, environmental, and biological scientists in other developing countries attest to the effectiveness and sophistication of many indigenous natural resource management systems that demand greater attention. Rejection of government-prescribed conservation and rehabilitation measures during the 1980s was based not only on political opposition to the socialist Derg regime but more fundamentally on farmers' knowledge of appropriate farming techniques in their particular environments. A backlash against top-down approaches to environmental management using exclusively professional and scientific "expert" knowledge has been described throughout the developing world (Smith, 2008). Nevertheless, it is increasingly recognized that the bottom-up approach, as well, has shortcomings, including widespread lack of community coherence and cohesion, tokenism, and the lack of resources in supporting and scaling up initiatives and programs, making it necessary that a certain degree of state authority is integrated in bottom-up approaches (Migdal, 2001; Smith, 2008). The development of more participatory, egalitarian approaches and local empowerment as a basis for a creative two-way dialogue between the people and the government are destined to be slow in Ethiopia, a country where policy making has traditionally been patterned by centralized, hierarchical systems of control.

Water pollution from industrial and agricultural sources, constituting one of the most recent threats to water availability, human health, and ecosystem function in Ethiopia, is increasingly affecting cities, smaller towns, and commercial farms and rural areas. Major factors in this trend are industrial and agricultural practices in a weak regulatory environment and low awareness of the population of the health implications of chemically polluted water. Development of technologically and environmentally sound remediation methods tested in Ethiopia requires political will through water policy provisions and the development of technical support. Judging from recent attempts to secure the support of policy makers for the defluoridation of drinking water, the most intensively researched water treatment method in Ethiopia, and to operationalize the production and application of an effective community-based method (Teklehaimanot, 2005), scaling up of remedial methods for use at the regional or national levels may be difficult.

The risk of water-related infectious and geochemical diseases has increased in recent years in Ethiopia with the movement and resettlement of many people from the overpopulated but relatively salubrious highlands to the disease-ridden lowlands. This problem has been exacerbated by the stepped-up development of water resources. The planned construction of additional dams and irrigation systems in the Rift Valley is further increasing exposure to both disease groups, and the massive small-scale irrigation and microdam construction program may extend the threat of vector-borne diseases in much of the country and compromise the benefits of these projects if preventive measures are not implemented. Water-related diseases may be prevented and successfully treated, although persistent deficiencies in water resources development and health services, as well as the spread of HIV/AIDS (Berhane, Haile Mariam, & Kloos, 2006) have curtailed prospects.

Progress in the management of Ethiopia's water resources is evident, although haltingly, in the areas of small-scale irrigation, water harvesting, domestic water supplies, and alternative energy development. These programs may significantly contribute to overcoming water and food insecurity if designed to address not only supply and technical aspects but also user demands emanating from local socioeconomic, cultural, and ecosystem needs. Small-scale irrigation and rainwater harvesting are now receiving the attention they deserve, albeit inadequately, as locally manageable systems that may contribute significantly to meet food and water needs. However, the potential of small-scale irrigation may be reached only with increased tenure security, the formation of viable and egalitarian water associations drawing on indigenous knowledge, technological inputs, and better access to markets (including extension of the road network). Modern rainwater harvesting systems can contribute as a supplementary water source to meeting water needs in rural areas where surface water supplies are scarce or absent. Development of cost-efficient variants of this newly introduced technology, technical support from government and NGO sources, and local production of rainwater harvesting equipment and spare parts may make these systems affordable and sustainable. Development of water harvesting is all the more important as the ambitious goal of the government's Universal Access Program to achieve 98% coverage with safe water supplies in rural areas by 2012 is unlikely to be reached.

This book contributes to a better understanding of the occurrence of the 2006 floods in Ethiopia by elucidating the relationship between rainfall and river flow in different-sized catchments and the role of environmental degradation and poverty in their devastating impacts, the findings of which may be applicable particularly to upper Nile countries. Documentation of the spatio-temporal relationship between rainfall and flooding events in the Omo can help to encourage further flood modeling efforts and monitoring toward flood prevention and efficient operation of the Gilgel Gibe and other dams. The micro study of the Dire Dawa floods suggests that while floods may not be prevented, their impacts may be mitigated using an integrated approach, including the construction of physical structures on fields and in urban areas, urban planning, and economic development, reducing the vulnerability of the poor. Information on the incidence and severity of floods and silt content of flood waters is of increasing relevance also to hydroelectric and irrigation development.

Although the prospects of increasing water security in the various sectors may be further increased under the revised policies, ultimate outcomes will depend to a large degree on the improvement of underlying socioeconomic conditions through management approaches that give a voice and decision-making power to local populations in the planning and implementation of projects and in improving their livelihoods. Considering the long history of centralized planning approaches in Ethiopia, its still fledgling democracy, and the failure of various state-sponsored projects in the past, the current government's initiatives toward greater community participation in soil and water conservation, decentralization, and community development will take time. The emergence of Ethiopia as a stronger power in hydropolitics in the Nile Basin and its determination to claim a larger share of the Nile's water and develop other river basins as well is putting to the test the government's recently formulated people-oriented approach versus the traditional top-down approach.

References

Berhane, Y., Haile Mariam, D., & Kloos, H. (Eds.). (2006). *The epidemiology and ecology of health and disease in Ethiopia.* Addis Ababa: Shama Books.

Clapham, C. (1990). State, society and political institutions in revolutionary Ethiopia. *IDS Bulletin, 21*(No. 4), 35–45.

Cook, S., Harrington, L., & Huber-Lee, A. (2008). Water and food in the river basins of Africa, Asia, and Latin America: A comparative analysis. In R. S. Bayout & E. Humphreys (Eds.), *Proceedings of the CGIAR Challenge Program on Water and Food and 2nd International Forum on Water and Food* (pp. 56–60). Addis Ababa: CGIAR.

Ives, J. D. (2006). *Himalayan perceptions: Environmental change and the well-being of mountain peoples* (2nd ed.). Lalitpur, Nepal: Himalayan Association for the Advancement of Science.

Kruseman, G., Ruben, R., & Tesfay, G. (2006). Village stratification for policy analysis: Multiple development domains in the Ethiopian Highlands of Tigray. In J. Pender, F. Place, & S. Ehui (Eds.), *Strategies for sustainable land management in the East African Highlands* (pp. 81–106). Washington, DC: International Food Policy Research Institute.

Midgal, J. S. (2001). *State in society: Studying how states and societies transform and constitute one another.* New York: Cambridge University Press.

Pretty, J. N., & Shah, P. (1997). Making soil and water conservation sustainable: From coercion and control to partnership and participation. *Land Degradation and Development, 8*, 39–58.

Smith, J. L. (2008). A critical appreciation of the "bottom-up" approach to sustainable water management: Embracing complexity rather than desirability. *Local Environment, 13*, 353–366.

Teklehaimanot, R. (2005). *Study of fluoride and fluorosis in Ethiopia with recommendations on appropriate defluoridation technologies: A UNICEF sponsored consultancy*. Paper prepared for UNICEF, Addis Ababa Office.

Zeleke, G., Kassie, M., Pender, J., & Yesuf, M. (2006). *Stakeholder analysis for sustainable land management (SLM) in Ethiopia: Assessment of opportunities, strategic constraints, information needs, and knowledge gaps*. Report prepared for the Environmental Economic Policy Forum for Ethiopia (EEPFE), Addis Ababa.

Further Reading

Awulachew, S. B. (2000). *Investigation of water resources aimed at multi-objective development with respect to limited data situation: The case of the Blue Nile and Chamo basins, Ethiopia* (Unpublished doctoral dissertation). Technische Universität, Dresden, Germany.

Baxter, P. T. W. (1991). *When the grass is gone* (Seminar proceedings No. 25). Uppsala: The Scandinavian Institute of African Studies.

Bekele, M. (2003). *Forest property rights, the role of the state, and institutional exigency: The Ethiopian experience* (Unpublished doctoral dissertation). Swedish University of Agricultural Sciences, Uppsala.

Berhane, Y., Haile Mariam, D, & Kloos, H. (Eds.). (2006). *The epidemiology and ecology of health and disease in Ethiopia.* Addis Ababa: Shama Books.

Bewket, W., & Sterk, G. (1999). Dynamics in land cover and its effects on stream flow in the Chemoga watershed, Blue Nile Basin, Ethiopia. *Hydrological Processes*, *19*, 445–459.

Devi, R., Tesfahune, E., Legesse, W., Deboch, B., & Beyene, A. (2005). Assessment of siltation and nutrient enrichment of Gilgel Gibe Dam, southwest Ethiopia. *Bioresource Technology, 99*, 975–979.

Disk, A. I. J. M., &, Keenan R. J. (2007). Planted forests and water in perspective. *Forest Ecology and Management, 252*, 1–9.

Dumond, H. J. (2009). *The Nile: Origin, environments, limnology and human use*. Dortrecht: Springer Netherlands.

Earth Policy Institute. (2008). *World geothermal power nearing eruption*. Retrieved from http://www.earthpolicy.org/Updates/2008/Update74_data.htm

Elia, E. (2003, October). *Environmental role of agriculture in Ethiopia.* Paper presented at the Roles of Agriculture Project International Conference sponsored by the Agricultural Economics Division, FAO, Rome.

El Swaify, S., & Hurni, H. (1996). Transboundary effects of soil erosion and conservation in the Nile basin. *Land Husbandry*, *1*, 7–21.

Environmental Resources Management Ltd. & Government of Ethiopia, Ministry of Water Resources. (2007). *Ethiopia Irrigation and Drainage Project: Environmental and social management framework*. Washington, DC, and Addis Ababa, Author.

Ethiopian Agricultural Research Organization. (2002). *Concept paper on agricultural water management* (Unpublished manuscript). Addis Ababa.

Falkenmark, M. (2007). *Good ecosystem governance: Balancing ecosystems and social needs*. Heidelberg: Springer Verlag.

Gebre Michael, Y. (1998). Indigenous soil and water conservation practices in Ethiopia: New avenues for sustainable land use. In H. P. Blume, H. Eger, H. Fleischauer, A. Hebel, C. Re, & K.G. Steiner (Eds.), Towards sustainable land use furthering cooperation between people and institutions (Vol. 2). *Advances in Geoecology, 31*, 1359–1366.

Haile, M., Gebreyesus, T. A., Witten, K., Yohannes, M., Byass, P., & Lindsay, S. (2003). Water harvesting in micro dams: Its impact on the socio-economic condition of the community and the salinity of the irrigated fields in Tigray. In M. Haile & S. N. Merga (Eds.), *Proceedings of the Workshop on the Experiences of Water Harvesting in the Drylands of Ethiopia: Principles and practices* (pp. 77–88). DCG Report No. 19. Ås, Norway: Dryland Coordination Group.

Hanjra, M. A., Ferede, T., & Gutta, D. G. (2009). Pathways to breaking the poverty trap in Ethiopia: Investments in agricultural water, education, and markets. *Agricultural Water Management*, doi. 10.1016/j.agwat.2009.06.008

International Water Management Institute. (2007). *IWRM challenges in developing countries: Lessons from India and elsewhere.* Water Policy Briefing (No. 24). Colombo, Sri Lanka: Author.

Kloos, H. (1982). Development, drought and famine in the Awash Valley of Ethiopia. *African Studies Review, 25*, 21–48.

Liu, B. M., Abebe, Y., McHugh, O. V., Collick, A. S., Gebrekidan, B., & Steenhuis, T. S. (2007). Overcoming limited information through participatory watershed management: Case study in Amhara, Ethiopia. *Physics and Chemistry of the Earth, 33*, 13–21.

MacDonald, A. M., Calcow, R. C., Nicol, A., Hope, B., & Robin, N. S. (2001). *Ethiopia: Water security and drought* (BGS Technical Report WC/01/02). Nottingham, U.K.: British Geological Society.

Mengistu, S., & de Stoop, C. (Eds.). (2007). *Environment for survival: Taking stock of Ethiopia's environment. Proceedings of the First Green Forum Conference.* Addis Ababa: Green Forum.

Ministry of Finance and Economic Development. (2006). *Ethiopia: Building on progress, a plan for accelerated and sustained development to end poverty (PASDEP) 2005/06-2009/10*. Addis Ababa: Author.

Ministry of Health and Ministry of Education on the integrated implementation modality of water supply, sanitation and hygiene education (WASH) programs in Ethiopia. (2005). Addis Ababa: Author.

Ministry of Mines and Energy. (2007). *The Ethiopian energy sector status and strategic plan.* Addis Ababa: Author.

Namara, D. C., Awulachew, S. B., Namara, R. E., Babel, M. S., & Gupta, A. D. (2007). Indigenous systems of conflict resolution in Oromia, Ethiopia. In B. C. P. Koppen, M. Giordano, & J. Butterworth (Eds.), *Community-based water law and water resources management reform in developing countries* (pp. 145–157). Wallingford, CT: CABI.

Pankhurst, A., & Piguet, F. (Eds.) (2009). *Moving people in Ethiopia: Development, displacement & the state*. London: James Currey.

Rahmato, D. (1999). *Water resources development in Ethiopia: Issues of sustainability and participation*. Addis Ababa, Forum for Social Studies.

Sadoff, C., & Revels, C. (2003, March). *Ethiopia country water resources assistance strategy*. Paper presented at the World Bank Symposium Water and Development, Water Week, Washington, DC.

Snowy Mountains Engineering Corporation. (2007). *Hydrological study of the Tana-Beles sub-basin*. Report prepared for the Ministry of Water Resources, Addis Ababa.

Swain, A. (2002). The Nile River Basin Initiative: Too many cooks, too little broth. *SAIS Review, 22*, 293–308.

Teklehaimanot, R., Melaku, Z., Kloos, H., Reimann, C., Fantaye, W., Zerihun, L., & Bjorvatn, K. (2006). The geographic distribution of fluoride in surface and groundwater in Ethiopia with an emphasis on the Rift Valley. *Science of the Total Environment, 367*, 182–190.

Tessera, A. (2004). The imperative need for watershed management in the Eastern Nile basin. In M. Feseha & T. Tafesse (Eds.), *Proceedings of the Conference on Management of Water Resources in Ethiopia* (pp. 67–155). Addis Ababa: Addis Ababa University, Institute of Development Research.

United Nations Development Program (2007). *Climate change and human development in Africa: Assessing the risks and vulnerability of climate change in Kenya, Malawi and Ethiopia*. New York: Auhor. Retrieved from hdr.undp.org/en/reports/global/hdr2007-2008/papers/igad.pdf

United Nations Environmental Programme (2006). *Sustainable consumption and production activities in Africa: Second regional status report*. Nairobi: Author.

United Nations Environmental Programme. (2007). *Sudan post-conflict environmental assessment*. Nairobi: Author.

UN-Water. (2008). Status report on integrated water resources management and water efficiency plans. New York: Author.

Wolde Mariam, M. (1984). *Rural vulnerability to famine in Ethiopia 1958–1977*. New Delhi: Vikas Publishing House in collaboration with Addis Ababa University.

World Bank. (2001). Ethiopia woreda studies. Vol. 1.

List of Contributors

Aynalem Adugna is a lecturer at Sonoma State University, California. He holds a PhD in medical demography from the University of London, an MA from the University of Durham, UK, and a BA in Geography from Addis Ababa University. His interests are demography, medical geography, and spatial analysis. He published a chapter in the book *The Epidemiology and Ecology of Health and Disease in Ethiopia*, as well as articles on the Ethiopian population in the journals *Northeast African Studies*, the *Geographical Journal*, and the *International Journal of Health Geographics*.

Yirga Alemu received a MSc in Irrigation Engineering from Arba Minch University in Arba Minch and a BSc in Agricultural Engineering from Alemaya University in Ethiopia. Since then, he has conducted research on soil and water management in the northwestern Ethiopian highlands for the Amhara Regional Agricultural Research Center and has been responsible for the establishment of the Gonder Agricultural Research Center, where he currently serves as a Center Manager. The results of his research on alternate furrow irrigation, water productivity, and water conservation structures have been presented at several workshops and conferences and also included in reports by the Amhara Region Agricultural Research Institute.

Zebene Asfaw is an assistant professor at Wondo Genet College of Forestry and Natural Resources, Hawassa University in Shashement, Ethiopia. He received a PhD in Agroforestry from the Swedish University of Agricultural Sciences, an MPhil in Tropical Forestry from the University of North Wales, UK, and a BSc in Plant Sciences from Addis Ababa University. He has taught courses in silviculture, agroforestry and soil biology for undergraduate and graduate programs and conducted research on forest structure and natural resources management, including gender issues. He has published articles in the journal *Acta Universitatis Agriculturae*, research reports, as well as papers in conference and workshop proceedings.

Badege Bishaw is the director of International Programs at the College of Forestry at Oregon State University and has over 25 years of teaching and research experience in forestry and natural resources. He received his PhD in Forestry from Oregon State University, an MSc in Tropical Forestry from the University of Dresden, Germany, and a BSc in Plant Sciences from Addis

Ababa University. Dr Bishaw has published on forestry education, training and professional development in Africa, and deforestation and land degradation in the Ethiopian Highlands in *Journal of Northeast African Studies*, *Mountain Research and Development*, *International Forest Review*, and other journals.

Yohannes Gebre Michael is an associate professor in the Department of Geography and Environmental Studies at Addis Ababa University. He received a PhD in natural resources management from the Institute of Geography at the University of Berne in Switzerland, an MA in geography from Addis Ababa University, Ethiopia, an MSc in development studies from the University of East Anglia, UK, and a BA in Geography from Addis Ababa University. He published several monographs on water and soil conservation, including *The use, maintenance and development of soil and water conservation measures by small-scale farming households in different agro-climatic zones of northern Shewa and southern Wello, Ethiopia,* and reports on flooding and pastoral nomadism in Ethiopia for Earthscan Publications.

Mark Giordano is a water resources specialist holding PhD and MA degrees in applied economics from the University of Minnesota and a BA in Economics from Whitman College. He is currently principal investigator and a research director at the International Water Management Institute in Colombo, Sri Lanka. His has wide experience in water management, economic development, and transboundary resources in the Nile Basin and other developing regions. He has published a number of articles on these subjects, the more recent ones in the journals *Climate Change*, *The Geographical Journal*, and *Annual Review of Environment and Resources*.

Mark Harvey has extensive work experience as engineering expert in several African and Asian countries and currently works as water & sanitation policy adviser for the Ministry of Water in Ethiopia. He received an MSc from University of Technology, UK, and BSc from the University of Leeds, UK.

Helmut Kloos has been a research associate at the University of California, San Francisco, since retiring as an associate professor of geography at Addis Ababa University, Ethiopia, where he was also a research associate at the Institute of Pathobiology. He holds a PhD in geography with specialization in medical geography from the University of California at Davis, and master's and bachelor's degrees from California State University, Fresno. He has carried out research on water resources development and management, water use behavior, population distribution and resettlement, and the transmission and spread of water-related diseases in relation to socioeconomic, behavioral and environmental factors in Ethiopia, Egypt, Kenya, and Brazil. He is coeditor of the books *The Ecology of Health and Disease in Ethiopia* and *Schistosomiasis in Ethiopia* and has published articles in *The Geographical Journal*, *African Studies Review*, *International Migration Review*, *Ethiopian Journal of Development Research, Disasters*, *Environmental Monitoring and Assessment*, *Northeast African Studies*, *Ethiopian Journal of Agricultural Science*, and *Waterlines*. He has also published in health journals, including *Social Science and Medicine, Ethiopian Journal of Health Development*, *Geospatial Health*, *American Journal of Tropical Medicine and Hygiene*, and *Malaria Journal*.

Jonathan Lautze has is an interdisciplinary background in water management, the environment, and development. He is a water resources specialist at the International Water Management Institute in Colombo, Sri Lanka. He holds a PhD in water resources engineering, with specialization in environmental policy and development economics, and an MA in law and diplomacy; both degrees are from Tufts University. He has lived and worked in Benin, Ethiopia, Ghana, South Africa, and Sri Lanka. His research has been published in the *Journal of Environment and Development*, *Water Policy*, *International Journal of River Basin Development*, *Water International*, and other journals.

Worku Legesse is an associate professor of applied ecology at Jimma University in Ethiopia. He holds a PhD from University College Cork in Ireland, an MSc from Gent University in Belgium, and a BSc from Asmara University in Eritrea. Dr Worku has written a book chapter on water supply and sanitation in Ethiopia. He also published articles in the *Malaria Journal*, *Environmental Monitoring and Assessment*, *Ecological Indicators*, and *Bioresource Technology*. He also initiated a partnership between Gent University and Jimma University, Ethiopia, to study the impact of Gilgel Gibe I Hydroelectric Dam.

Stuart McFeeters is an assistant professor at California State University Fresno, where he teaches GIS courses. He holds a PhD in Geography with specialization in remote sensing and GIS from the University of Nebraska-Lincoln, as well as a master's degree in geography from California State University Fresno. He has done field research on remote sensing and GIS analytical techniques for environmental applications. He published articles in the journals *California Agriculture* and the *Journal of the American Water Resources Association.*

Sorssa Natea Merga is an agriculture and natural resources specialist for the Ethiopian government's Productive Safety Net Program. Prior to this appointment, he conducted research on irrigation and rainwater harvesting issues for the Ethiopian office of the World Bank. He received a BSc from Addis

Ababa University, Alemaya Agricultural College, Ethiopia, and a MSc from Utah State, University, USA. He has extensive field experience in rainwater harvesting and irrigation management in Ethiopia and has developed several technical guidelines on small-scale irrigation, rainwater harvesting and rural water supply and sanitation for the World Bank.

Semu Moges is a hydrologist in the engineering faculty at Addis Ababa University. Prior to this appointment, he taught water resource management and hydrology courses at the Arba Minch Institute of Water Technology, both at graduate and undergraduate levels. He received his PhD and MSc degrees from the University of Dar es Salaam, Tanzania, and a BSc from Arba Minch Water Technology Institute, Ethiopia. Dr. Moges has initiated and lead research on various water resources and hydrological issues. He has also published papers on remote sensing, estimation of evapotranspiration, and flood monitoring in the Blue Nile Basin

Alula Pankhurst holds an MA and a PhD in social anthropology from the University of Manchester, UK, and a BA in Oriental studies from the University of Oxford. Dr. Pankhurst is affiliated with the University of Wageningen Disaster Studies Program and is the Ethiopia Country Director of the University of Oxford Young Lives Research Project. Prior to this appointment, he was an associate professor in the Department of Sociology and Anthropology, Addis Ababa University. His research interests are in the areas of resource management, migration and displacement, social institutions, and HIV/AIDS. He coedited the books *Natural resource management in Ethiopia* and *Moving people: displacement, development and the State in Ethiopia* and has published articles in journals such as *Disasters*, *World Development*, and *Northeast African Studies*.

Cherkos Tefera is an expert in irrigation engineering with more than 17 years of professional experience in planning, design, implementation, supervision and evaluation of a variety of civil engineering works in Ethiopia. He received his MSc from Addis Ababa University and a BSc from Arba Minch Water Technology Institute, Arba Minch, Ethiopia. He has published a number of technical reports and manuals, including *General aspects of domestic rainwater harvesting and multiple water uses*.

David Turton has a PhD in anthropology and a BSc in sociology from the London School of Economics and a Lic. Phil. from Gregorian University, Rome. He is a Senior Research Fellow in the African Studies Center at the University of Oxford. He was formerly a Reader in forced migration, director of the Refugee Studies Center, and Fellow of Green-Templeton College, University of Oxford. Before moving to Oxford, he taught in the Department of Social Anthroplogy at the University of Manchester. He has edited eight books, including *Ethnic federalism: The Ethiopian experience in comparative perspective*, and has published many articles in journals such as *Anthropology Today*, *Oxford Development Studies*, *Cambridge Anthropology*, *Disasters*, and *African Affairs*.

Index

Made in the USA
Monee, IL
12 November 2026

bbb203ba-61d7-4a82-aafe-5a325a760e33R01